Enlightenment Geography

Studies in Modern History

General Editor: **J. C. D. Clark**, Joyce and Elizabeth Hall Distinguished Professor of British History, University of Kansas

Titles include:

Bernard Cottret (*editor*)
BOLINGBROKE'S POLITICAL WRITINGS
The Conservative Enlightenment

Philip Hicks
NEOCLASSICAL HISTORY AND ENGLISH CULTURE
From Clarendon to Hume

William M. Kuhn
DEMOCRATIC ROYALISM
The Transformation of the British Monarchy, 1861–1914

Kim Lawes
PATERNALISM AND POLITICS
The Revival of Paternalism in Early Nineteenth-Century Britain

Nancy D. LoPatin
POLITICAL UNIONS, POPULAR POLITICS AND THE GREAT REFORM ACT OF 1832

James Muldoon
EMPIRE AND ORDER
The Concept of Empire, 800–1800

W. D. Rubinstein and Hilary Rubinstein
PHILOSEMITISM
Admiration and Support for Jews in the English-Speaking World, 1840–1939

Lynne Taylor
BETWEEN RESISTANCE AND COLLABORATION
Popular Protest in Northern France, 1940–45

Studies in Modern History
Series Standing Order ISBN 0–333–79328–5
(*outside North America only*)

You can receive future titles in this series as they are published by placing a standing order. Please contact your bookseller or, in case of difficulty, write to us at the address below with your name and address, the title of the series and the ISBN quoted above.

Customer Services Department, Macmillan Distribution Ltd, Houndmills, Basingstoke, Hampshire RG21 6XS, England

Enlightenment Geography

The Political Languages of British Geography, 1650–1850

Robert J. Mayhew
Lecturer in Human Geography
Institute of Geography and Earth Sciences
University of Wales
Aberystwyth

First published in Great Britain 2000 by
MACMILLAN PRESS LTD
Houndmills, Basingstoke, Hampshire RG21 6XS and London
Companies and representatives throughout the world

A catalogue record for this book is available from the British Library.

ISBN 0–333–79186–X

First published in the United States of America 2000 by
ST. MARTIN'S PRESS, LLC,
Scholarly and Reference Division,
175 Fifth Avenue, New York, N.Y. 10010

ISBN 0–312–23475–9

Library of Congress Cataloging-in-Publication Data
Mayhew, Robert J. (Robert John), 1971–
Enlightenment geography : the political languages of British geography, 1650–1850 / Robert J. Mayhew.
p. cm.
Includes bibliographical references (p.).
ISBN 0–312–23475–9 (cloth)
1. Great Britain—Historical geography. 2. Great Britain—Politics and government. 3. Geography—Great Britain—History. 4. Great Britain—Intellectual life. 5. Political geography—History. 6. Enlightenment—Great Britain. I. Title.

DA600 .M295 2000
911'.41—dc21

00–027247

10 9 8 7 6 5 4 3 2 1
09 08 07 06 05 04 03 02 01 00

Printed and bound in Great Britain by
Antony Rowe Ltd, Chippenham, Wiltshire

Contents

List of Illustrations

All illustrations are reproduced by permission of the Syndics of Cambridge University Library.

Acknowledgements

My study begins with Peter Heylyn, a lecturer in geography at Hertford College, Oxford from the 1610s. My thanks are owed to the same institution, which fostered my humanist approach to geography and eventually appointed me to a college lectureship some 375 years after Heylyn. As Heylyn sought patronage from William Laud at St John's, so I am also delighted to acknowledge my debt to that college, which supported my doctoral research through its generous award of a North Senior Scholarship. The award of a British Academy post-doctoral fellowship has allowed this project to reach fruition in Cambridge, where Corpus Christi College has given me a third congenial home, of which I can say with Richard Gough: 'while I erect this monument to the votaries of antiquity, can I forget to acknowledge that my passion for these pursuits was fostered within those venerable walls?' Finally, as John Ogilby's first route-map passes through Oxford on its way to Aberystwyth, so I have traversed that trajectory and would like to express my thanks to the geographers at Aberystwyth for appointing me to a lectureship.

Philip Howell and Charles Withers read and commented on an earlier and shorter version of this book with characteristic rigour and generosity. David Livingstone made important points which have much improved the continuity and focus of the final draft; this is but one of many scholarly debts I owe him. Paul Langford and Howard Erskine-Hill have encouraged me to develop my approach to Samuel Johnson. Boyd Hilton made useful comments on Chapter 11. Quentin Skinner's encouraging comments on my project proved a fillip. I owe a great debt to Jack Langton, who has always exemplified in his work and his criticism the interweaving of geography and humanism which I aspire to achieve. My thanks must also go to Jonathan Clark, whose writings have greatly influenced my own, and who has consistently supported my attempts to recover geography's humanist history.

My final thanks must go to Yvonne: if Enlightenment and humanism are the key terms in this study, they take on a living meaning in and through her.

1
Effacing and Recovering the History of Geography

This book is an enquiry into the history of geography. More specifically, I seek to analyse the political languages which were encoded in British geographical discourse, c.1650–1850. In order to engage in such an enquiry, we need to start with a clear understanding of what distinguishes history as a mode of scholarship from other modes, and therefore what an enquiry into the history of geography can and cannot be about. This may appear self-evident, but, as we shall see, most enquiries which have nominally been conducted under the rubric of the history of geography have not, in fact, been historical in their ambition. On the contrary, they have effaced geography's history.

Oakeshott and the historical mode of knowledge

A lucid account of what distinguishes historical enquiry comes in the work of Michael Oakeshott. For Oakeshott, all experience is of a perceptual variety, making it inseparable from mental processes, and therefore not 'objective' if this is taken to mean independent of human subjectivity. There are a number of modes of experience, that is, of self-contained and logically coherent ways of understanding the world, which by their nature actually construct that world: 'what is achieved in experience is a coherent world of ideas. This world is itself the arbiter of fact, for to be a fact means to have a necessary place within it'.[1]

Because each mode of experience is coherent and creates its own world, there is no way in which different modes can adjudicate one another's validity. Such a project of criticism between modes is described by Oakeshott as a confusion of categories, *ignoratio elenchi*: 'it is impossible to pass in argument from any one of these worlds to any other ... [T]he result of all such attempts is the most subtle and insidious of all forms of error – irrelevance'.[2]

For Oakeshott, history is one of the coherent, constructed modes of experiencing the world which humans have developed. It is a sophisticated mode of experience, only being developed when the basic needs of survival are no longer pressing.[3] The kernel of the historical mode of experience is the attempt to understand: 'the past for its own sake. History is the past for the sake of the past. What the historian is interested in is a dead past; a past unlike the present. The *differentia* of history lies in its very disparity from what is contemporary'.[4]

Such an ambition places history in a paradoxical position: all evidence is only available to be incorporated into a perceptually grounded mode of experience by virtue of its presentness, where history is exclusively interested in the past. The resolution is to see that history is a constructed activity in the present, wherein present evidence is viewed purely in terms of its survival from a past: '"the past" is a consequence of understanding the present in a particular manner'.[5] As such, history as a coherent mode of experience views all things purely in terms of what they tell us about the past.

Of course, not all material from a past society survives into the present: two consequences for the nature of history as an activity follow from this. First, history does not seek 'the truth' *per se*:

> the past in history varies with the present, rests upon the present, is the present. 'What really happened' (a fixed and finished course of events, immune from change) as the end in history must, if history is to be rescued from nonentity, be replaced by 'what the evidence obliges us to believe'.[6]

This reinforces the point that history is a humanly constructed activity, and that no such activity can be 'objective' in the sense of independent of human judgements as these latter constitute rather than reflect the world. Secondly, as history wishes to give us a picture of an entire society, and its materials are the incomplete present survivals from such a society, the ambition of history as an activity is essentially inferential. In other words, history tries to discern what limited evidence obliges us to believe about the whole society which produced that evidence: 'the past of history is, then, always an inference; it is the product of judgment [*sic*] and consequently belongs to the historian's present world of experience.'[7] Oakeshott elaborates on the nature of the inferential procedure in which history is engaged. The historian initially tries to construct an 'historical situation': 'a composition of notionally contemporaneous, mutually related, historical occurrences'.[8] But this is 'an unstable level of historical understanding', as, in virtue of its pastness, it implicitly recognises that change has destroyed that situation, but it does not analyse that change.[9] A more adequate historical analysis, then, moves beyond the historical situation to explain it 'in terms of the mediation of its emergence' and its decay. The result of such an analysis is an understanding of the 'historical event': 'a past constituted not in terms of its situational immobility but of time and change'.[10] The historian's aim, then, is an inferential movement from limited evidence in the present, to a constructed description of a past society's engagement in a specific sphere and the manner in which that engagement emerged and declined.

Three things need to be said about history's inferential procedure, as depicted by Oakeshott. First, to pick out an historical event or situation requires some principle of individuation, which constructs a meaningful historical entity on which the inferential procedure can operate. This principle is one largely assumed by history, but involves sufficient continuity to distinguish an entity, and sufficient discontinuity at the beginning and end to mark it off from predecessors and successors:

> the identity in terms of which an assembled passage of historical events, recognised as differences, may be understood as a passage of change is nothing other than its inherent continuity; this continuity to be distinguished from some changeless item in the situation, from an enduring purpose or end to be realised and from the normalities or the 'law' of a process of change.[11]

In other words, constructing an historical entity does not necessitate the discovery of a common denominator, but a more sophisticated conception of individuation as a passage of related but distinct occurrences. It is such a 'passage of differences' which constitutes the entity/individual which historical enquiry analyses.[12] Secondly, the constructs 'historical situation' and 'historical event' are not the self-understandings of the actors involved,[13] although all the evidence a historian has will be couched in that form. History, then, begins with self-understandings, and needs to comprehend those initially, but seeks to move beyond them to a characterisation of society which the actors could not have made. This is because actors had practical purposes in their utterances and understandings, where the historical mode, by its nature, is not engaged in such a project:

> The idiom of happening is always that of practice, and the record of happening is usually in the idiom of practice: and 'practice' and 'history' are two logically distinct universes of discourse. The task of 'the historian' is, thus, to create by a process of translation: to understand past conduct and happening in a manner in which they were never understood at the time; to translate action and event from their practical idiom into an historical idiom.[14]

Thirdly, the outcome of the inferential movement of historical reasoning is a generalisation about an historical individual that is very tentative: 'if when an historian has managed to assemble a continuity of change, he identifies it by giving it a name (usually not of his own invention and reluctantly appropriated) ... we must understand him to be begging us not to place too much weight upon these identifications'.[15]

In other words, the end point of an historian's quest to delimit and describe an historical event, as encapsulated in such terms as a 'scientific revolution' or an 'Enlightenment attitude', should not in turn be taken as an historical actor; it is simply a covering description. Such a description is the termination of an inferential procedure in history, not the starting point for another round of inference, because at this point abstraction departs from the historian's aim to describe a society, entering another, scientific, mode of experience, wherein law-like generalisations are sought rather than individuals described.

The perils of over-generalisation lead to an important point: Oakeshott was keen to depict what history is because he saw it as a mode of knowledge construction particularly liable to collapse into other modes and because he saw other modes of experience perennially and mistakenly being described as history. History, understanding the present in terms of what it tells us about the past, is 'perhaps the most sophisticated of all presents, difficult to achieve and difficult to sustain. ... Thus, historical understanding is especially prone to relapse into some other engagement'.[16] It is for this reason that Oakeshott distinguishes 'history' from other 'pasts', such that he can comment that 'the historical past is not the only past' nor even, in fact, the most common past.[17] Far more common is 'the practical past', whose interests are diametrically opposite to those of an historical enquiry into the past for its own sake:

> Wherever the past is merely that which preceded the present, that from which the present has grown, wherever the significance of the past lies in the fact that it has been influential in deciding the present and future fortunes of man, wherever the present is sought in the past ... the past involved is a practical, and not an historical past.[18]

Such a practical past is normally the servant of a political or religious position, wherein the 'predominant interest is not in "history" but ... retrospective politics'.[19]

The practical approach to the past ramifies into a whole series of historiographical presuppositions which are logically incompatible with the historical attitude to the past, but which are also rife in enquiries which are purportedly historical. First, such an approach assumes that history is or can be 'relevant'. But where the past is merely taken as a storehouse of wisdom and examples for present consumption, it is not being understood for its own sake. As such, the past's purported relevance has nothing to do with the practice of historical enquiry. Furthermore, history cannot therefore be dismissed as 'irrelevant', this judgement coming from within the practical mode of experience and, as such, itself modally irrelevant to the judgement of the historical mode (it is an example of *ignoratio elenchi*).[20] Secondly, the practical apprehension of the past constructs a 'mainstream', constituted by the events which lead to the present situation. It is in relation to this mainstream that the practical past can discern 'turning points' and 'key moments' wherein the past became recognisably more like the present, while dismissing other moments as 'irrelevant' or 'uninteresting'. Furthermore, and finally, it is the practical attitude which allows for the ascription of moral praise and blame with respect to the past: 'the categories of "right" and "wrong", "good" and "bad", "justice" and "injustice" etc. relate to the organization and understanding of the world in respect of its relationship to ourselves, in respect (that is) of its habitableness [in the present]'.[21] The historiographical contrast with the presuppositions of the historical mode is stark:

> Everything that the evidence reveals or points to is recognized to have its place; nothing is excluded, nothing is regarded as 'non-contributory'. The place of an event is not determined by its relation to subsequent events. ... In 'history' no man dies too soon or by 'accident'; there are no successes and no failures ... Nothing is

approved, there being no desired condition of things in relation to which approval can operate; and nothing is denounced.[22]

In other words, it is only when there is some approved situation which is being used in implicit comparison with the past that the past can be praised or blamed, and that past events can be found to be important or otherwise. Such a comparative benchmark is provided by the practical mode in the form of the present situation. The historical mode, studying as it does the past for the sake of the past, is only concerned with one entity – past society – and thus provides no second and comparative entity with respect to which moral or evaluative judgements can be made.

For Oakeshott, then, the historical mode of experience is a delicate and late blossom in the history of civilisation, prone to being overwhelmed by older and hardier growths on the tree of knowledge. History is the inferential enquiry which moves from limited evidence surviving in the present, which history is interested in exclusively by virtue of its vestigial nature, to a characterisation of the society which produced that evidence. That characterisation is a scholarly one, not one which could have been recognised at the time, because the evidence involved was overwhelmingly created in practical engagements divorced from the modal ambition of historical enquiry. Further, that characterisation is not one of any relevance to the present, as that is to view past-practical creations in the light of present-practical desires, which is once more a modally separate ambition from that of historical enquiry.

Effacing the history of geography: a study in practical pasts

Considerable amounts of scholarly attention have been paid to geography's past, notably in the period 1650–1850, but, as we shall now see, very little of that attention has been concerned with the historical past as opposed to some version of the practical past. The result has been a historiography which has tended to efface the history of geography, by eliding and diminishing the *differentia* of geography's history.

Essentialism and 'traditional' histories of geography

A number of the 'founding fathers' of modern geography as an academic discipline made comments *en passant* about geography in the seventeenth and eighteenth centuries. While no-one would claim that these individuals were centrally defined by their activities as historians of geography, their basic outlook on geography's past was both non-historical in an Oakeshottian sense and exerted a powerful influence over the ways in which purported histories of geography were written, an influence they maintained until surprisingly recently.

Perhaps the most influential view of geography in this period was that established by Richard Hartshorne in his *Nature of Geography* (1939). For Hartshorne, geography lacked a coherent definition in the seventeenth and eighteenth centuries, being in a 'pre-critical' stage of development, where many were moving towards an adequate definition of the subject, but it had yet to be codified. The transition to maturity was defined for Hartshorne by a concern to make geography an independent discipline:

> [Geography's] establishment as a modern science was essentially the work of the century from 1750 to 1850 ... we are not here concerned with the majority of writers of geographical works during the two or three centuries preceding 1800. Our interest is with that smaller, but still considerable number of students of the second half of the eighteenth century who consciously strove to convert a more or less miscellaneous and useful study into an independent science.[23]

That Hartshorne viewed geography as lacking any definition until the late eighteenth century is clear from his argument that it acted as the 'mother of the sciences' in this period, and then faced a crisis, having to find a space among the progeny it had spawned. Geography would find its new role as a synthesising and regionalist discipline, which formed the foundation of Hartshorne's own conception of the 'independent science' of geography, and which the select few had anticipated in the centuries before 1800.

For all their differences over the nature of geography, Hartshorne and Carl Sauer were in essential agreement about the status of geography prior to the late nineteenth century. Sauer in his influential essay, 'The Morphology of Landscape' (1923), also wrote of previous geographical endeavours as 'pre-critical': 'During the period of discoveries a bona fide but uncritical geography attained its greatest development ... attempts at systematic ordering were numerous, but often grotesque rather than successful.'

Again, for Sauer as for Hartshorne, the positive contribution of geography in that period was proleptic: 'without such a preliminary synthesis of the facts of geography the work of the next period would have been impossible'.[24] The 'work' in question was 'the establishment of a critical system which embraces the phenomenology of landscape', a task Sauer saw as originating with Thales and which his own work sought to realise.[25]

The historiographical structure developed by Sauer and Hartshorne, postulating what geography is as a science 'in essence', judging geography's history in the light of this essence, and placing the seventeenth and eighteenth centuries in a pre-critical era with certain anticipations of geography's true task, has been highly influential among geography's historians. One example is David Stoddart's *On Geography* (1986). Stoddart finds 'three prerequisites for the development of modern geography', all of them emergent in the later eighteenth century. All material before this era is dismissed as it made no contribution to geography as a scientific endeavour: 'Our standard histories speak of the work of Strabo and Eratosthenes, Varenius, Hakluyt, Purchas. But all these figures seem to us remote. Their contributions have meaning in the context of their own time, not of ours. Their significance to us is as precursors, its study largely antiquarian.'[26]

For Stoddart, a scientific form of geography originated in '1769 – when Cook first entered the Pacific, and in that brief decade before his death in Hawaii'. That form was subsequently defined by Forster, Humboldt and Darwin, and centred around quantification, comparative method and social concern, all unified by an ecological understanding. This project, notably akin to Stoddart's own practice, was finally codified by the institutional creation of geography as a discipline in the late nineteenth century.[27]

Perhaps the most obvious example of this essentialist historiographical format, however, is the only recent book-length study of geography in the seventeenth and eighteenth centuries in the English language, namely, Margarita Bowen's *Empiricism and Geographical Thought* (1981). Bowen's basic premise is that 'as far as geography was concerned, the initial encounter with scientific empiricism was close to disastrous'.[28] The antidote to this came in a true understanding of the nature of geography's task, and this was embodied in Von Humboldt: 'as physics had its Newton in the seventeenth century and chemistry its Lavoisier in the eighteenth, so, early in the nineteenth century, geography attracted the attention of one of the intellectual giants of the time, the German naturalist Alexander von Humboldt'.[29]

Von Humboldt is seen as jettisoning naïve empiricism in favour of a method of 'greater scientific distinction', involving the integrated study of physical and human phenomena coupled to an awareness of the geographer's social responsibilities. This vision, in turn, is seen as precisely what late twentieth-century geography needs to recapture to achieve scientific status: 'the approach of von Humboldt offers a beginning, and the task for geography, as for all science, is to make the most effective use of past traditions in responding to the issues of today'.[30]

The rest of the book, a study of geography c.1600–1830, operates within the framework this establishes: empiricism as problematic, von Humboldt as the standard to which all should aspire, and instructive parallels between the situation of geography in the 'age of empiricism' and the dilemmas of geography in the present day. On this analysis, as on Hartshorne's, the majority of those who wrote geography books in the period contributed nothing to the subject as they lacked an adequate conception of geography. After Varenius, for example, 'as the subject in general lost its theoretical basis, then, geography suffered a decline as a science, virtually until the time of Kant'.[31] Thus Bowen dismisses over a century of geography books in a sentence. The 'low ebbs' and 'turning points' in geography's history are mapped as deviations from and advances towards the Humboldtian approach: thus Nathaniel Carpenter's geography is praised as 'one of the most intelligent and mature attempts ... to clarify the theoretical position of geography', since he 'indicat[ed] at least the precedent for some of Humboldt's ideas'.[32] Carpenter is the exception to the rule, however: Skinner once quipped that 'in Cassirer's analysis ... the whole Enlightenment was striving to make Kant possible', and similarly for Bowen, geography in the seventeenth and eighteenth centuries was a series of hopeless failures to realise Alexander von Humboldt. These failures are assessed by Bowen just as Carpenter's success is; by a proleptic standard. Thus she comments that Laurence Echard encountered problems in organising evidence in his early eighteenth-century geography books, adding that: 'the parallel with the mid twentieth century is remarkably close, for, then also, geographers were too overwhelmed with the volume of research and the task of quantification to participate actively in re-examining the philosophy of science'.[33]

Finally, Bowen tries to characterise geography in the period in political terms. For her, geography was constantly held back from its scientific task by 'political pressures ... when the power of the ruling class and the established church seems to have put a damp on the quill of many writers'. As such, geography was a 'conservative' field, which by the mid eighteenth century 'might be called a bourgeois geography, appealing to the general curiosity and current imperialist interests of the well-to-do classes'.[34]

If Bowen's work carried Hartshorne's essentialist presuppositions into a specialist work, they have also received a far wider audience through standard textbook histories of geography, perhaps most notably Martin and James's *All Possible Worlds*. For Martin and James, as for Hartshorne, geography was the 'mother of the sciences' in the eighteenth century, and then needed an independent role, which it discovered as a 'bridge' subject between the arts and sciences. Their story, then, is in the main of the process of carving out an institutional space for geography as an independent discipline in the century up to World War II.[35] But Martin and James are notably implicated in the drive to achieve academic respectability for geography which they chronicle. Their essential definition of geography is as both scientific and utilitarian: 'the history of geographical ideas is the record of human efforts to gain more and more logical and useful knowledge of the human habitat'.[35]

In the light of this definition, Martin and James assess earlier geographical writings by the extent to which they gave the discipline scientific respectability. Thus, for the period 1600–1800, they give high praise to Varenius, going on to opine that he was 'far in advance of his time' because he 'insisted that special geography only becomes intelligible when the specific features are explained in terms of abstract concepts or universal laws'.[37] Similarly, Büsching is praised as 'ahead of his time' for introducing quantitative and statistical analyses which led towards the modern, institutional and independent discipline of geography.[38]

As my concern is with the politics of geography, it is worth noting at this point that introductions to political geography adopt the same historiographical line. Taking their approach from Mackinder, the 'founding father' of political geography in the late-nineteenth century, they assume that geography was not tied to politics prior to that period, or, if it was, not in any way of interest to a more rigorous political geography, which emerged as one facet of the drive to institutional respectability which Martin and James chronicle. In other words, following the historiographical lead given by Mackinder, all engagements connecting politics and geography prior to that are seen as pre-critical and unscientific by most textbook treatments of political geography. A polemical position adopted by one of geography's founding fathers has become the basic historical framework within which political geographers are still educated a century later.[39]

Abstracting from the specifics of the various approaches detailed above, we can discern certain recurrent patterns which render them all non-historical engagements with geography's past from an Oakeshottian perspective. In short, we can show that they all operate in the practical, not the historical mode of experience. The key weaknesses in all the works canvassed so far are twofold and related: first, a presentist conception of the role of history; and secondly, an impoverished notion of historical individuation.

All the writers discussed establish a timeless essence which defines geography, that essence relating to twentieth-century debates within geography, and varying with the predilections of the writer in question, being an ecological approach for Stoddart, a regionalist one for Hartshorne and so on. Being timeless, this essence is conflated with what geography has always been, and, switching from indicative to imperative, with what geography ought to be. Clearly, such a procedure lies within the ambit of Oakeshott's practical approach to the past, evidence being looked at in relation to a present-day concern with defining geography. This is most apparent, because most explicit, in Bowen's work, but it is the principle behind all these deployments of the past. Consonant with this practical approach to geography's past, the writers I have discussed can deploy the historiographical methods which Oakeshott depicts as practical: the histories written are often seen as 'relevant' to understanding geography's current predicament (notably in Bowen's case); having defined an 'essence' of geography, material can be arranged around this 'mainstream', such that writers like Varenius have praise heaped upon them for being 'ahead of their time', that judgement being made relative to a subsequently-defined conception of geography; and moral blame can be apportioned on similar grounds to those who retarded the development of the discipline towards its essence. Looking for twentieth-century geographical practices in the past, modern commentators have been led to focus on exceptional geography books at the expence of the majority. This is particularly noticeable in the disproportionate attention lavished upon Varenius. The quality of Varenius's work is not in question, but his belief in geography going back to the first principles of natural philosophy was highly unusual. Moreover, it is not clear that his influence in Britain was as great as has been suggested. He has been called '*the* European scholar who profoundly influenced the content and scope of geography for more than a century',[40] but his *Geographia Generalis* only went into four English editions, while Gordon's *Geography Anatomiz'd* had 23 editions and Guthrie's *Geographical Grammar* had 46 editions. Clearly, eighteenth-century readers valued geography books for different reasons from their twentieth-century commentators.

The weakness of the concept of historical individuation implicit in these histories of geography becomes apparent if we consider their central aim: in each case, the drive is to chronicle geography's emergence as a scientifically 'respectable' subject, thereby justifying its status as an independent discipline in the modern division of knowledge. Yet, if geography is defined in terms of disciplinary independence, geography before the late nineteenth century is a contradiction in terms, as it was only at this time that the subject was organised into a separate school and university discipline. As such, essentialist historiography teleologically locates geography's historical identity in an institutional framework, this nineteenth-century creation mistakenly being taken as a timeless object, rather than locating that identity in a 'passage of differences', wherein what geography 'is' has shifted over time, geography's identity resting in its continuity, the Oakeshottian conception of an historical individual. Essentialist histories of geography, then, have too simplistic a conception of historical identity, wherein what is presently definitive of geography is seen as always having held that role. The possibility that what geography is has changed over time is not considered, the result being the occlusion of geography's historical identity as a passage of differences. This explains the focus on the exceptional in seventeenth and eighteenth century geography at the expense of the majority of geographical writings. By their refusal to recognise that the scope and purpose of geography changes over time rather than residing in an essence, historians of geography were forced to look for geographers who would fit an anachronistic conception of geography as an independent discipline, and were therefore obliged to look to writers who did not adhere to the dominant contemporary definition and generic conventions of geography. Further, just as geography's historical identity is impoverished by this procedure, so is the historical identity of the political sphere with which geography interacts. To the extent that histories of geography have dealt with its political positioning in the period 1650–1850, they have done so in terms derived from a lexicon largely defined by later configurations of political language, using terms such as 'conservative', 'bourgeois', and 'ruling class' with little or no recognition of the historical situatedness of political languages. In other words, the historical identity of politics as a passage of differences has been effaced. Oakeshott argued that most history, being written in the practical mode, was simply retrospective politics. Similarly, we can say that most histories of geography have been retrospective impositions of contemporary definitions, and most asides about the 'politics of geography' have been retrospective politics grafted onto retrospective geography.

Because of the manner in which histories of geography have been constructed, a project which originated with the founders of an institutional discipline of geography as they created a *raison d'être* for themselves, geography's past has simply been dragooned to accord with subsequent definitions. As such, 'histories' of geography, or more properly deployments of geography's 'past', have, paradoxically, effaced geography's history.

New pasts for geography

In one sense, recent histories of geography have moved away from the 'essentialist' approach sketched in the previous section, which reached its apogee in Bowen's work. Historians no longer tend to give transcendental definitions of what geography is and ought to be, using this as an interpretative pivot around which to centre analyses of geography's past practitioners. In another sense, however, recent writings in the history of geography bespeak an important continuity with Bowen's project specifically and with the essentialist tradition more generally. For, just as Bowen sought to reread geography's past through an intellectual current sceptical of Enlightenment empiricism as part of an attempt to find a scientifically reputable approach to geography, so current writings take their inspiration from critical theory within the social sciences which has questioned Enlightenment rationality, and interrogate the construction of geographical knowledge as a product of that rationality.

We can begin with those who might be called the 'new founding fathers' of geography, several of whom have made passing comments about geographical knowledge in the period 1600–1850. David Harvey's historical materialism has proved an influential model. In response to Hartshorne's idealist trajectory, which saw geography as a subject with a correct and permanent definition created by its intellectual function with respect to other aspects of knowledge, historians of geography inspired by Marxism and led by Harvey took a rather different tack in the 1970s and 1980s. Geography was a subject in a division of knowledge created in a socio-economic context, and its definition was not a transcendental one, but one which evolved with modes of production and resultant social formations. This approach was also concerned to show geography as a practical subject, responding to the needs of government and capital rather than being set off in a rarefied realm of the intellect.[41] But this work relating geography to its socio-economic context has retained the Hartshornian temporal framework, focussing on the actions of institutions such as the Royal Geographical Society and on the role of geography in colonialism, both of which necessitated emphasis on the nineteenth century.[42] Once more, geography in the eighteenth century was proleptically interesting for its anticipation of nineteenth-century preoccupations: thus Cook's voyages see the interaction of scientific, geographical and imperialist motifs,[43] which makes them fit Marxisant historiographical presuppositions. Similarly, the fact that geography was seen as vital to commerce in the eighteenth century has been noted, as it helps to bring geography back from the intellectual to the material realm.

In his more recent work, the rubric of Harvey's historical materialism has altered somewhat, at least in its treatment of seventeenth and eighteenth century geography. In his *Condition of Postmodernity* (1989), Harvey discusses the 'Enlightenment project', describing it as: 'the first great surge of modernist thinking, [which] took the domination of nature as a necessary condition of human emancipation. Since space is a "fact" of nature, this meant that the conquest and rational ordering of space became an integral part of the modernizing project'.[44]

As such, Enlightenment knowledge, as anatomised by the critical social theory of Adorno and Foucault, had its geographical incarnation in a new conception of space as something to be reified and allocated within a capitalist mode of production. Such a view has also been developed in Lefebvre's influential work, *The Production of Space*, which sees an 'abstract space' being created between the sixteenth and nineteenth centuries, and, with few significant variations, in Cosgrove's work on the concept of landscape.[45] More recently still, Gregory has added some *caveats* from the postcolonial perspective, arguing that this mode of understanding space, this mode of geographical knowledge, was also intrinsically European, and was forged in the context of colonial encounters and resultant oppression.[46]

None of these figures, of course, would claim to be writing a history of geography *per se*, any more than Hartshorne or Sauer would have: their project, true to the subtitle of Hartshorne's *Nature of Geography*, is a 'critical survey of current geographical thought in the light of the past'. Yet, as with the earlier generation of founding fathers, the historiographical presuppositions behind Harvey's or Gregory's incursions into geography's past have been influential with practitioners of the history of geography. This can be seen, for example, in the work of Charles Withers. Withers has analysed geographical knowledge in the seventeenth and eighteenth centuries as a project of classifying and ordering, consonant with Harvey's 'Enlightenment project': 'geography, in the eighteenth century especially, ... [was] encyclopaedic – part of attempts to bring the world and knowledge of the world to order'.[47]

For Withers, drawing on Foucault, geographical knowledge was implicated in the construction of a disciplinary regime in the eighteenth century, a project which also assisted in the construction of national identities, simultaneously controlling and constructing both peoples and their spaces.[48] Furthermore, and in line with Gregory's reflections, Withers links the mode of geographical knowledge developed in the period with gendered and Eurocentric disciplining:

> If ... the Enlightenment's selfconsciousness was to some extent a geographical consciousness based on the distinctiveness of the part of the world that came to be known as 'Europe', then those ideas of rationality and modernity inscribed by 'the Enlightenment project' must also be seen as Eurocentric ... [and] Founded upon an imperializing masculinist gaze.[49]

Most of Withers's work, as will be apparent, has been on modes of geographical knowledge, rather than on geography books and geography in the period *per se*. These latter are, however, incorporated in the analysis:

> These compendiums have a much more purposive (if varying sense) of utility than has been implied by the hitherto unexamined use of terms like 'dictionary', 'gazetteer' or geographical 'grammar'. In title, they sought to order the world of geographical learning. In content and method, the systematized descriptive listing sought a global view; they tried to provide new interrogative maps of knowledge at a time when the world was being brought to order via discovery and global information was being given order via classification.[50]

The picture, in fact, is not dissimilar to that provided by Hartshorne or Bowen, in that the interest which geography books in the eighteenth century possess to the present day is not in their content as such, but in the organisational effort which they embody. The historiographical shift Withers represents is away from seeing classification as the first stage in a scientific procedure, as in say Sauer, towards seeing ordering as representative of a mode of knowledge, Enlightenment knowledge, tied up with the projects of capitalism, colonialism, and patriarchy.

Withers has in the main used the Foucauldian rubric of ordering and disciplining in his work, but this is by no means the only tool which critical theory has given to recent historians of geography. Closely related is the concept of 'modernity', in which, however it is placed chronologically, the seventeenth and eighteenth centuries play an important role. It is in this context that Miles Ogborn has examined 'historical geographies of London in the period 1680–1780 ... through the lens of modernity'.[51] Ogborn uses the numerous concepts in the orbit of 'modernity' – commodification, rationalisation, Habermas's 'public sphere' and so on – as frameworks for his analyses, studying spaces which he accepts 'might be seen as anachronisms: spaces that were self-consciously novel, spaces that didn't fit, spaces that sought to constitute a different future'.[52] Ogborn, then, like Withers, is interested in questions of geographical knowledge rather than geography *per se* (he discusses no geography books), and deploys the rubrics of a social-theoretical critique of Enlightenment to understand the spaces which were historically constructed, shifting the focus from 'Enlightenment' to 'modernity'.

As the upshot of essentialist histories of geography could be seen in Martin and James's summary textbook, so the result of the new approaches can be seen in David Livingstone's work, *The Geographical Tradition* (1992). Livingstone's chapter on the period 1660–1800 is notable for pivoting around Robert Boyle, who was, as he admits, 'not really a geographer at all'.[53] He sees Boyle's work, however, as representing many of the key strands of geographical knowledge in the period: scientific and classifying in his instructions for travellers, Boyle also tied natural and theological knowledge together as a unity. Livingstone's focus, then, is on geographical knowledge rather than geography books, the latter only being discussed briefly as presenting 'a bird's eye view of the world ... cataloguing the whole gamut of human activities' and by this fact keeping 'the science of geography closely tied to commercial and imperial enterprises'.[54] Livingstone's account of geography in the 'long' eighteenth century, then, is of the progress of a rational classifying enterprise, closely tied to the construction of modern science as embodied in the Royal Society, and amenable to political critique through the rubrics of critical social theory for its implication in the creation of new forms of power by surveillance.

Abstracting from these new histories of geography, several points can be made about their historicity. Before making these points, I must emphasise that there can be no doubt that Ogborn, Withers *et al.* are engaged in a far more thoroughgoing historical practice than any of their essentialist predecessors. This does not mean, however, that they are practising the history of geography in Oakeshottian terms for several reasons. First, all this work demonstrates the same problems of historical individuation as does its predecessors', failing to view geography's identity as a passage of differences. Instead, the tendency has been to take key arenas of debate in contemporary human geography, and then analyse the ways in which these debates can be illuminated through the past. This is notably true of the arenas of landscape and of space, both of which are key to twentieth-century geography, but which (as we shall see in Chapter 2) have no bearing on the ways in which geography was defined in previous centuries. Such an approach is either itself essentialist, in which case it simply takes the present definition of realms of geographical interest to be of trans-historical validity, or, if it eschews essentialism, it works on the old presupposition that geography prior to the nineteenth century had no definition, this latter being Withers's contention: 'Early modern geography ... drew on a complex array of data and should not be seen as possessing some clearly identifiable and immutable core.'[55]

Withers here elides two separate points: it is true that geography does not, contrary to essentialist presuppositions, possess an immutable core, but this is not to say it lacked an identifiable definition at any point in the past. As we shall see in Chapter 2, geography was precisely, albeit differently, defined in the period 1650–1850, and this definition left geography categorically distinct from the spheres of geographical knowledge with which Withers, Livingstone and Ogborn have been concerned. As such, these writers may be practising a form of history, but it is not a history of geography because, by means of a defective view of historical individuation, they are unable to discern precisely how geography was defined at the time or simply choose to ignore this. In either case, geography's history remains effaced.

But there are more serious problems with recent writings on geography's past when they are measured against an Oakeshottian conception of history as a mode of experience. As stated previously, Oakeshott saw any attempt to mix modes of reasoning as leading simply to *ignoratio elenchi*, a meaningless hybrid of different ways of constructing the world. Recent work on geography's past seems to fall under this stricture, attempting to draw on the categories of critical theory in an historical engagement, where it is clear, for example, that Adorno's conception of 'Enlightenment', encompassing as it does a time span from Homer to Hollywood, was not an historical one, but a philosophical one defined in terms of instrumental reason.[56] To simply take philosophical constructs and apply them in an historical analysis can only lead to confusion. That geographical writing has taken this route reflects a desire akin to that which drove earlier, essentialist histories: for just as an earlier generation constructed histories of geography to prove the scientific respectability of the discipline, so contemporary historians of geography seek to prove the discipline's social-scientific respectability. The result is that social-scientific categories are prioritised over historical evidence, that evidence being used to illuminate modern categories of critical reasoning, perhaps most notably in Ogborn's work. The categories in question, 'modernism', 'Enlightenment' and so on are clearly those which would be the final generalisations in an historical analysis, and would as such be highly tentative covering descriptions, where in recent histories of geography they are operative entities to be invoked as causal actors. It is clear that the ways in which categories and individuals are invoked in this writing adheres to some mode of experience other than the historical. This mode of reasoning might be a valuable contribution to social scientific debate, but it is clearly not, for all its deployment of evidence from the past, an historical enquiry. This adherence to the categories of critical theory also explains the political vocabulary within which geography's past has been framed: the language has been, as we have seen, of 'imperialism', of 'masculinity' and of the 'ordering impulse to domination'. Again, these labels might be perfectly valid within a critical theoretical engagement, but they are clearly transhistorical and insensitive to the development of political languages seen as a passage of differences through time. The engagement with the politics of geography's history, in other words, is like its essentialist forebears to the extent that it describes a double effacement of the history of geography and of politics through an inadequate notion of historical identity.

Recovering the history of geography

The upshot of the foregoing historiographical survey is that we are still a long way from writing what Oakeshott might call a history of geography. Oakeshott's conception was, of course, and as he was the first to recognise, an ideal typical one,[57] but to date there has not even been a full awareness of what the task of an historian of geography might be, let alone an attempt to realise it.

Oakeshott was also aware that his work on the tacit presuppositions of any rigorously historical form of enquiry did not amount to a demonstration of how that enquiry might actually function: 'the conditions which may constitute it a mode of understanding are not formulae for conducting an historical enquiry'.[58]

How then, do we enact Oakeshottian desiderata in a history of geography? Two starting points can be gleaned from Oakeshott. First, that history is what the evidence obliges us to believe; and, secondly, that there are separate universes of discourse which can be taken as discrete individuals for historical analysis.[59] In the case of geography's history, the evidence we are obliged to believe comes in the form of geography books: the period 1650–1850 saw an uninterrupted sequence of books called 'geographies'. The *Eighteenth Century Short Title Catalogue* lists some 657 works with geography in the title. Many of these are reprints or books which simply have the word in the title without being about geography, but Sitwell shows that each decade between 1650 and 1850 saw between two and fourteen special geographies published.[60] As has been said in another context, 'put bleakly, what is very largely left of early modern society is words'.[61] In the case of the history of geography, these words are largely to be found in the geography books which previous studies of the history of geography have dismissed as pre-critical 'dinosaurs'[62] or have effaced in favour of other forms of geographical knowledge. It is to these books that we must turn for the evidence concerning what we are obliged to believe about the character of geography, c.1650–1850, in the passage of differences which constitutes geography's historical identity. The number and interrelation of these texts constitutes a textual tradition, sufficiently distinct as a universe of discourse to act as an historical individual.

The task of the historian of geography in this era, then, is to recover the meanings (*inter alia*, political) of texts. In this context, historians have become increasingly sophisticated in the inferences they draw from textual material about the society which produced it as they have recognised that the meaning of a text is not just contained within its content, but is instead:

> defined by three poles ... first, the analysis of texts, be they canonical or ordinary, to discern their structures, their themes and their aims; second, the history of books and, beyond that, the history of all objects and all forms that bear texts; third, the study of practices that seize on these objects and these forms in a variety of ways that produce differentiated uses and meanings.[63]

Instead, then, of simply looking at the text's content, literary historians, historians and bibliographers have focused on the production of meaning as a negotiation between texts, reading, and book production.

At the level of *texts*, the study of content is by no means simple. To read a text in ways sensitive to history as a passage of differences requires careful analysis of the historically differentiated and emergent meanings of terms and concepts, something which Quentin Skinner and J.G.A. Pocock have discussed.[64] Skinner's approach has much in common with Oakeshott's in its dismissal of historical mythologies: of fatuous notions of historical 'relevance'; of prolepsis – the retrospective ascription of significance to a work; and of an approach to the past which simply familiarises it in the light of the present.[65] Skinner is sensitive to the central problem I have identified as bedevilling intellectual histories of geography: 'the ... mythologies I have discussed derive from the fact that an historian of ideas will unavoidably be *set*, in approaching any given writer, by some sense of the defining characteristics of the discipline to which the given writer may be said to have contributed.'[66]

For Skinner, these problems can be avoided by recovering the range of historical meanings any given utterance, sentence or concept could have had, and then trying to home in on which in that range of meanings an author was invoking, in the light of a sensitivity to the generic conventions within which they worked.[67] In the case of geography, its continuity as a textual tradition 1650–1850 allows us to deploy the same method, trying to elucidate what geography meant at this time and what sorts of information and concepts the genre of the geography book could communicate.

The role and status of a text, however, has to be negotiated with *readers*. This is important for two reasons. First, 'reading has a history. It is not always and everywhere the same. ... Interpretative schemes belong to cultural configurations, which have varied enormously over time'.[68]

As such, the ways in which a text could be meaningful change over time with the reading community, and any historical study of a textual tradition needs to beware essentialising its own reading strategies and predilections, just as much as of transposing its own conceptual lexicon into another era. Secondly, readers, of course, create new meanings in texts by misreading, but as McKenzie points out, these in turn become 'new *bibliographical* fact; and it is these facts which constitute the primary evidence for any history of meanings'.[69] In other words, such inflections of meaning as are performed by readers can themselves be recovered, providing historical evidence about the nature of the discourse under discussion. In the case of geography, we will glean valuable information about its character in general, and its political character in particular, by looking at the response of readers to individual texts and to geography as a genre.

The third focus of attention is the *production* of texts and the category of the *author*. The notion of the author as an independent creator derives from the Romantic period, and its projection backwards in publishing history is liable to mislead.[70] More generally, throughout print history, the writer has always been closely connected with the exigencies of publishing, and it is this relationship between a writer, publisher and printer, not the figure of the autonomous author, which has driven the production of texts: 'Authority is a social nexus, not a personal possession ... A hypnotic fascination with the isolated author has served to foster an overdetermined concept of authorship, but (reciprocally) an underdetermined concept of the literary work'.[71]

The exigencies of publishing are important because, as Chartier comments, textual formatting could alter the meaning of a text: 'books are objects whose forms, if they cannot impose the sense of the texts that they bear, at least command the uses that can invest them and the appropriations to which they are susceptible'.[72] Prior to the disciplinary institutionalisation of geography, we shall see that the concept of geographical 'authorship' was very different from that of today, and tended to link geography to politics in historically specific ways.

The remainder of this book attempts to describe an Oakeshottian trajectory, analysing the history of geography and, specifically, the political dimension of that history, c.1650–1850. I start in Chapter 2 by using the methods just described to discern what geography meant in this period in the passage of differences which has constructed its identity, how this related to other forms of geographical knowledge, and why geography *sensu stricto* was, both intellectually and textually, linked to political discourse as this was constituted at the time. Having thereby elucidated the 'historical situation' connecting geography and politics, the bulk of the book is a detailed analysis of incidents within that situation for both geography and allied fields of geographical knowledge (Chapters 3–12), conducted primarily in terms of the textual intentions of geographical authors, but also in terms of readers' responses and the conditions of textual production. In the final chapter, I seek to move beyond the self-understandings of contemporaries to a broader characterisation of the lineaments of the historical situation I have anatomised in terms of the historical covering descriptions of the English and Scottish Enlightenments, their respective politics and their interaction. I also try to move from a study of the historical situation to a broader conception of the historical event of early-modern English-language geography, c.1550–1850, by drawing out from my detailed study the factors which created this situation and drew it to a close.

2
The Sphere of Geography and the Realm of Politics in Britain, c.1650–1850

To understand the 'politics of geography' in the period 1650–1850 in a manner accordant with Oakeshott's characterisation of historical individuals as passages of differences, we need to develop an understanding of the contemporary definitions of the spheres of politics and geography. With respect to geography, meshing the approaches of Skinnerian contextual history and the rubrics of historians of the book suggests three angles of approach: first, analysing geographical *texts* minutely; secondly, looking at the intended and actual *readership* of those texts; and finally, investigating the careers of the *authors* who produced those works, and the *print culture* in which they operated. Each of these approaches will help to elucidate not only the definition, nature and status of geography, but also why and in what ways geography was interwoven with contemporary understandings of the realm of political discourse. Before doing this, however, we need briefly to characterise the constitution of the political sphere in the seventeenth and eighteenth centuries, elucidating its differences from present-day conceptions.[1]

The realm of the political

Vast quantities of work have been done to elucidate the nature and operation of political discourse in seventeenth- and eighteenth-century Britain; it is cited throughout this book. We can at this stage, perhaps, pick out three salient general characteristics of political discourse, which not only highlight the historically-specific constitution of notions of 'politics', but also determine the ways in which contemporaries could have understood geography books to be politicised, something which previous approaches to the 'politics of geography' have ignored.

First, in the period 1650–1800, 'politics' meant what we might now see as 'high' politics, excluding the cultural or social politics which modern analyses of ideology – what might be called the 'hermeneutics of suspicion' – seek to uncover. The term politics was used to refer to discussions of dynastic legitimacy, of representation and of the constitution. If we are to look to contemporary self-understandings of politics as a realm, these are the issues which must be central. As such, for contemporaries to recognise geography books as politicised, they had to discuss these issues, and it was on this basis that geography books could in principle have been seen as 'Tory', and the like.[2] The nature of politics, then, necessitates an analysis of the textual functioning geography books in the period if we wish to uncover the politics of geography, and that politics is to be sought not at the level of ideological presuppositions, but of explicit discussion, unlike the late nineteenth-century fieldwork conception of geography as a practical and implicitly political contribution to the imperial nation state.

Secondly, political argument relied on historical example to a far greater degree than in the present day. England developed its pattern of argumentation from the perceived operation of the common law by precedent, and the tradition of civic humanism inspired by Roman history.[3] This mode of argument rose to prominence in the Elizabethan era, and remained vital long into the eighteenth century.[4] Political argument, therefore, focused on how the constitution had functioned from time immemorial, how the relationship between people, parliament and monarch had operated in previous reigns, and the like. Furthermore, this did not just apply to British history, ancient Greek and Roman examples being central to the construction of political positions.[5] For this reason, the historian of political ideas and languages must be sensitive to the ways in which historical evidence was deployed in the prosecution of political arguments. This was *not* politics by other means, as history was recognised as a sphere of political theorisation. To the extent that geography books touched on history, then, they would unavoidably become a politicised genre.

Thirdly, arguments about desirable forms of political structure could not be readily divided from those about ecclesiastical organisation and doctrine. The established church and the state were seen as a coherent whole. It followed from this that argument in one realm had implications for the other. Defenders of the Anglican church saw it as purified to a more primitive (and therefore more valid) state from the Henrician

Reformation onwards, and as connected with the monarchy from that time. Such a position, frequently drawing on Romans 13, argued that the Christian church had to submit to political powers and preach subordination to them. By contrast, many Dissenters saw worship as an intensely individual matter and established churches as barriers to faith. Those anti-Trinitarian dissenting groups, moreover, which saw Christ as a moral example rather than as divine, saw church structures in a very different way from Trinitarians, for whom the church was Christ's mediating body between God and humanity.[6] As with history, then, theology and ecclesiology were not politics by other means, but rather they were central to the conception of the scope of politics. Indeed, historical and religious argumentation were interwoven: which historical incarnation of the church was viewed as most pure varied according to the commentator's ecclesiological and theological beliefs, which were themselves a gauge of their political stance in the present day. As such, church politics was fused with church history and contemporary politics. If the generic conventions of geography books, then, connected them with ecclesiastical issues, these conventions would bring geography within the realm of political debate.

Geography, geography books, and the realm of the political

The 'Cambridge School' approach to politics has emphasised the need to focus on genres, 'not just on the text to be interpreted but on the prevailing conventions governing the treatment of the issues or themes with which the text is concerned'.[7] This helps to avoid an historically impoverished conception of a discipline's identity drawn from the modern division of knowledge misleading historians as to the character of intellectual enquiry in previous eras. If the accepted boundaries of what a discipline or genre treats change over time, the historian must follow those changing contours. Just as the links between theology, history and politics differed in the early modern era from today, so the nature of geographical texts differed from present generic expectations, as did the relationship between geography proper and allied realms of geographical knowledge. Moreover, the conventions in which geography operated linked it to the conception of politics outlined above.

Texts: defining geography as a body of knowledge

Geography was part of a hierarchy of investigations organised according to the spatial scale of their focus. This set of delimitations was crucial to the definition and functioning of geography vis-à-vis other forms of spatial study, which constituted distinct types of geographical knowledge. Geography was a body of knowledge at a particular spatial scale, that of the earth in general, and this distinguished it from related enquiries.[8]

At a larger scale than geography was 'cosmography', defined by Blount in 1661 as 'the description of the world in general' (a definition closely akin to that given by Cawdrey in 1604). Cosmography studied the 'world' as opposed to geography which studied the earth or terraqueous globe. Where modern usage frequently conflates globe/earth with world, until the later eighteenth century the primary meaning of 'world' was, in Johnson's words, 'the great collective idea of all bodies whatever', which we might call the universe.[9] The understanding of cosmography itself changed over time: Blount had concluded his definition by saying cosmography 'is best handled by *Pliny* in his Natural History', where most eighteenth-century geography books viewed Newton's cosmography, as set out in his 'System of the World' in the *Principia*, as the only one worthy of consideration.

The cosmographical analysis of the system of the world was divided into two parts: 'Astronomy, which shews the structure of the heavens and the disposition of the stars ... And geography, which shews those of the earth.'[10] Johnson's *Dictionary* made the distinction between the arts of the geographer and the cosmographer clear by defining 'cosmographer' as 'one who writes a description of the world; distinct from geographer, who describes the situation of particular countries'. As such, the relationship of geography and astronomy to cosmography was that of parts to the whole. Geography and astronomy were themselves closely allied in their mathematical delineation of the earth: to understand the so-called 'general' part of geography, the mathematical delineation of the earth, required some grounding in astronomy. Both areas of enquiry were classified as 'mixt mathematicks', which dealt not with abstract speculation about number, but with 'quantity as subsisting in material beings'.[11] Moreover, given that both subjects analysed situation, they could demonstrate much of their content by means of terrestrial and celestial globes, and most geography books included sections on both types of globes.[12]

As geography was distinguished from cosmography as part to whole, so it was distinguished from chorography and topography (the politicisation of which we turn to in Chapter 6) as whole to part: '*Chorography* is distinguished from Geography, as the description of a particular country is from that of the whole earth.'[13] Topography, in turn, tended to discuss an even smaller portion of land, being 'the description of a particular place or places, be they Towns, Cities, Shires or Counties'.[14] The relationship between geography and chorography which was established in dictionaries between Cawdrey's in 1604 and Webster's in 1828, came from Ptolemy. Book 1.1 of Ptolemy's *Geography* discussed 'In what Geography differs from Chorography', and argued: 'the end of Chorography is to deal separately with a part of the whole, as if one were to paint only the eye or the ear by itself. The task of Geography is to survey the whole in its just proportions, as one would the entire head'.[15]

There was some change in the definition of chorography over time: Dyche and Pardon noted it related 'only to the Description of a particular County or Province, wherein natural History, and the particular Incidents of the Place are supposed to be recited'.[16] The emphasis of this definition on natural history would have been inconceivable when Camden wrote *Britannia,* the pioneering British chorography of 1586 (for more on which, see Chapter 3), a focus on the natural world only developing in seventeenth-century regional description.[17]

Geography's relationship with cosmography, astronomy, chorography and topography helped to determine its nature until the end of the eighteenth century. Clearly, it was restricted to the scale of the earth as a whole, since the earth's relation to the universe was the province of cosmography, while smaller scales were dealt with by the more detailed enquiries of chorography and topography. It is indeed the case that most geographical compendia dealt with the entire earth, rather than being restricted to a single continent or country. Chorographies provided a mass of detailed information, which the geographer synthesised into a general picture. One reason why geography texts remained so alike in the period was that the scale they treated was hemmed in by conventional relationships with other subjects. For while in some details the definitions of cosmography and chorography shifted over time as the practice of those enquiries altered, the interrelationship between them, and therefore the intellectual space for geographical enquiry, remained constant. J.N.L. Baker argued that: 'in the seventeenth century there was much argument as to what geography meant, how it was related to cosmography, whether chorography was or was not geography [and so on]'.

Cormack's analysis suggests that Baker's picture was exaggerated for the seventeenth century, and it certainly does not capture the very stable system of intellectual relationships in which geographical enquiry was enmeshed in the eighteenth century.[18] Being expected to treat the whole of the earth helps to explain why geography was primarily a textual practice until the end of the eighteenth century, rather than being a realm of fieldwork, which constituted a separate sphere of geographical knowledge: there was no way a geographer could have first-hand knowledge of the entire globe, yet his task as an author was to describe just that. The intellectual space mapped out for geography, therefore, defined it in a way that tended to force it into the study rather than the field.[19] I argued at a methodological level in Chapter 1 that the evidence we have concerning geography's historical character is primarily textual; it is now clear, furthermore and at an historical level, that this evidence obliges us to believe that geography's early modern historical character was textual.

Most of the definitions of 'geography' itself given in geography books and dictionaries were brief. Thomas Salmon, for example, prefaced his *New Geographical and Historical Grammar*, saying that: 'By Geography is understood a Description of the Surface of the natural Terraqueous Globe, consisting of Earth and Water, which is represented by the artificial Globe.'[20] Some definitions formally recognised that geography had two traditions, a mathematical or general tradition and a descriptive or special tradition.[21] The twofold nature of geography was latent, however, in almost all definitions of geography as the activity of 'describing' the earth.[22] Description, of course, had both a mathematical and a prose sense, as in Phillips's definition: 'Describe, to Write, or set down in Writing; to draw the Form of, to Represent, to Explain: In *Geometry*, to draw a Line, a Circle, &c.'[23]

As such, the definition of geography in terms of its descriptive function in no way restricted geography to its special as opposed to its general aspect. General geography was centred on acquiring knowledge about the situation of places in terms of their longitude and latitude, a task which was normally viewed, following Ptolemy, as a mathematical one. Within a situation, special geography (which is the focus of the present study) then acquired further useful information about the natural and human worlds in the manner exemplified by Strabo.

Geography's task, as represented in these definitions, was to determine relative location upon the earth and to describe the phenomena to be found in those locations. This understanding of the role of geography appears to have remained constant throughout the seventeenth and eighteenth centuries and into the nineteenth century, having its roots in the understanding of the purpose of geographical enquiry developed in the classical era, a debt which many definitions explicitly acknowledged.

Those writing geography books did add some further details about geography's scope in the course of defining their subject matter, which give important insights into the role and status of geography in the self-understanding of those actually engaged in writing it. First, while discussing the whole terraqueous globe, geography was said to focus only on its surface. Thus Patrick Gordon's immensely popular *Geography Anatomiz'd* said: 'we intirely restrict the Science of *Geography* to the *exterior Part* or *Surface* of the earthly Globe'.[24] Salmon (cited above) repeated this view in the middle of the eighteenth century, and it took on a new significance in the late eighteenth century, which saw the increasing formalisation of geology as a realm of enquiry built around an interest in rock formations in the earth's crust.[25] In this situation, geography and geology remained distinct enterprises due to geography's traditional emphasis on the earth's surface. Thus Webster in 1828 could define the geographer as 'one who describes that part of the globe or earth, which is exhibited upon the surface' while 'geology' was 'the doctrine or science of the structure of the earth or terraqueous globe, and of the substances which compose it'.[26]

Authors had no illusions concerning the status of geography as an enquiry. Focusing on situation on the surface of the terraqueous globe, geography was mundane in more ways than the literal. As Guthrie's *Geographical Grammar* put it: 'The knowledge of the world, and of its inhabitants, though not the sublimest pursuit of mankind, it must be allowed, is that which most nearly interests them, and to which their abilities are best adapted. And Books of Geography ... promise the best assistance for attaining this knowledge.'[27]

That this was all that was expected of geography can be seen by the criticism of those geography books which tried to move beyond their own sphere in order to render them more intellectually stimulating. Barrow criticised posthumous revisions of Guthrie's *Grammar* itself as being: 'contaminated by an infusion of the modern philosophy in politicks and religion. ... It may cease to be geography, but cannot become history. As far as geography is a science, it is ... easy and pleasant'.[28]

As we shall see, this suggestion that geography books were either normally or ideally free from politicisation in the period was wholly misleading, cutting as it did against the textual construction of geography books, against their intended and actual readership, and against the output of geographical authors.

Although not highly challenging, geography was seen as 'scientific', a fact which a number of the above-cited definitions make clear. Gordon argued 'that *Geography* doth merit the Title of *Science* ... is a

Truth so universally granted' as not to require further discussion on his part. The sense in which geography was seen as a science in the period, however, has to be closely investigated, as it was not the most commonly accepted modern sense. For while Gordon thought the scientific status of geography to be self-evident, he also argued that geography was not concerned with the nature of the earth, its materials or processes, 'designing thereby to distinguish it from *Natural Philosophy*'. Furthermore, while concerned with the mathematical location of places on the earth's surface, geography was not concerned with 'an actual *Survey* of Mensuration ... which the Science of *Geography* presupposeth'.[29] The meaning of 'science' to which Gordon referred in characterising geography is in fact the older one in the English language, and the one which remained dominant in the eighteenth century, namely science as knowledge (the first sense of the term in Johnson's *Dictionary*) as opposed to science as 'certainty grounded on demonstration' (the second sense in the same).[30]

Gordon was by no means exceptional in simultaneously viewing geography as scientific but not demonstrative: indeed his view can be found articulated by the pioneers of English-language geography textbooks. Cunningham, who claimed to be 'the first that euer in oure tongue have written of this argument [of geography]', argued in his *Cosmographical Glasse* (1559) that geography did not have to comprehend the demonstrations of astronomy about the earth, only to be cognisant of its findings.[31] Peter Heylyn was more explicit on this subject in the seventeenth century, distinguishing geography from natural philosophy as Gordon was to do in the next century: 'We take not the Earth simply in its own nature as it is an element, for so it belongeth to *Philosophy*, but for the terrestrial Globe, so it is the subiect of *Geographie*'.[32]

Geography was a coherent body of knowledge about a clearly-defined object, namely the situation of places on the earth and the content of those places in natural and human terms. It was, then, the character of geography as knowledge rather than the way in which that knowledge was verified, which led to the epithet 'scientific' being applied to the subject.

This view of geography as scientific because of its coherent organisation of information helps to explain the relationship between geography and travel writing as a form of geographical knowledge which a number of standard geography texts disclose, and which, as we shall see in Chapter 8, travellers endorsed. Travel writing was anecdotal and lacked the organisational control of geography, such that: 'as for other

Narratives relating either to Countries themselves, or their Inhabitants, and which commonly swell up Geographical Tracts, we reckon them ... rather the *Fringes* of Geography, than its real or essential Parts'.[33]

The geographer and the traveller were by no means synonymous although the geographer did rely upon the traveller for raw material.[34] This relationship was made clear by Guthrie as he explained why his was the first geographical compendium to discuss the present state of learning in each country:

> neither is the lack of this [in previous geographical compendia] to be altogether imputed to the fault of geographical writers. The greater part of travellers, acting solely under the influence of avarice ... were at little pains, and were indeed ill-qualified to collect such materials ... The geographer, then, who could only employ the materials put into his hands, was not enabled to give us any important information upon such subjects.[35]

General educational tracts similarly suggested that geography and travel were related rather than identical. Barclay wrote that travel narratives 'are very agreeable to children, and of great use in improving their geography', but did not suggest the two things were the same.[36]

Most geographies organised their material on any country under a series of standard headings, which was not often the case in travel narratives (although it is clear they were frequently composed in that fashion), and thus geography books did have a distinctive mode of presenting knowledge, which in contemporary terms justified viewing it as a science taking its observations from travel books. In generic terms, then, as well as because of the spatial scale of its enquiry, geography, was seen as a textual tradition of scholarly activity, largely involving the collation of material gleaned from others. In contrast to nineteenth-century geography, which defined itself as a masculine discipline based around fieldwork (a process which we shall witness under construction in Chapter 10),[37] its seventeenth- and eighteenth-century predecessor shows no sign of this conceptualisation.

Geography books, then, were mainly vast compendia, gazetteers or dictionaries with information ranged under a series of headings. The description of each country routinely included sections on law, religion, the constitution and history. As such, the generic conventions of geographical descriptions brought them within the seventeenth- and eighteenth-century conception of the scope of politics. If we agree with Skinner that: 'the essential question which we ... confront, in studying any given text, is what its author, in writing at the time he did write for the audience he intended to address, could in practice have been intending to communicate',[38] we must accept that geography books could not but be intended to make political statements, given the intersection between the conventional subject matters of politics and geography.

Geography books only had space to offer abridgements concerning national histories, constitutions and church establishments, but the manner in which these summaries were constructed gave ample scope to direct the text towards a specific political viewpoint. As in historical work at this time, the claim of impartiality was routinely made, but must be seen simply as a rhetorical commonplace empowering the perspective adopted.[39] Many geography books not only legitimated specific political perspectives, but also had an overarching argument about the course of providence and the ideal earthly church and state polity, which the lessons of history and geography taught the reader with the help of the author's perspicacity. A historicised approach to the politics of geography can only be effected by the collation of texts with their sources in previous geography books and the works they epitomise, the analysis of mutations and continuities of various editions over time, and the close mapping of the political languages of geography books onto the more general languages of politics.

It should now be apparent that geography was defined very tightly in seventeenth- and eighteenth-century Britain: geography was the verbal and mathematical description of the earth; it was a scientific subject, but not in the modern sense; it was restricted in its spatial scale to general discussions of the earth as a whole; it was distinguished from travel writing by its ordered and detached presentation of evidence; it was created by textual collation rather than fieldwork; and its conventional contents interwove it with political discourse. Hartshorne *et al.*'s notion that geography was in a pre-critical stage would have bewildered anyone from the period: if to lack an agreed definition of a subject is to make it pre-critical, twentieth-century geography is a much stronger candidate than its seventeenth and eighteenth century equivalent. Equally, geography *sensu stricto* was poles apart from the diffuse geographical knowledges studied by Livingstone, Withers *et al.* What is discussed under this rubric is material eighteenth-century writers categorically separated from geography *per se*.

Readers: the consumers of geographical texts

The structure of both late-humanist and commercial education in the period meant that geography was not studied for its own sake, but as a propaedeutic subject. Geography books in the seventeenth and eighteenth centuries responded to this in stock discussions of the groups for whom they were intended. Geography's intended readers were exclusively defined in terms of the larger scholarly and practical pursuits in which those readers were engaged, which accepted by implication that geography was a means not an end. Furthermore, it was a means to ends which were political.

The commonplaces about geography's audience were elegantly expressed by Heylyn in his *Cosmographie* (2nd edn, 1657),[40] the politics of which we turn to in Chapter 3. On the scholarly side, geography was above all an essential foundation for history in a humanist education, as practised in grammar schools and the universities: '*History* without *Geography*, like a dead carkasse, hath neither life, nor motion at all ... *History* therefore and *Geography*, like the two Fires or Meteors which Philosophers call *Castor* and *Pollux*, if joyned together, crown our reading with delight and profit; if parted, threaten both with a certain shipwrack'.[41]

A variant on this became a commonplace, geography and chronology being termed 'the eyes of history'.[42] Two branches of historical enquiry in particular, scriptural and classical history, were specified independently as relying on geography. Thus for Heylyn, himself an Anglican clergyman, '*Geography* ... is exceeding useful to the reading of the holy Scriptures, as in discovering the situation of *Paradise*, the bound and borders of those Countreys which are therein mentioned; especially with relation to the travels of the *Patriarchs*, *Prophets*, *Evangelists*, and *Apostles*, yea of *Christ* himself'.[43]

Edward Wells, another Anglican minister and an admirer of Heylyn's *Cosmographie* (on whom we focus in Chapter 7), wrote his *Treatise of Antient and Present Geography* in order to illuminate the geography of the classical world for those reading ancient histories, this being one of many such works in the eighteenth century.[44] Given, as we have seen, that the practice of history was inextricably bound up with political discourse in the period, geography's role as the 'eye' of history made it a facilitator of politics.

If we turn from geography's self-proclaimed scholarly audience to its actual readers, we find that the grammar schools did teach geography as part of a curriculum centred on the classics. It was argued that to read the classical historians profitably required an understanding of the location of the places they referred to. If geography were to be fixed in the pupil's mind, the teacher should 'enliven it with short, agreeable histories, and curious events'.[45] The pupil was recommended, then, that 'of whatever place the mention occurs in his ordinary reading, recourse should immediately be had to his map, to learn not only its relative situation upon the earth, but whatever else has rendered it remarkable in history'.[46] This comment suggests that geography books were referred to in the context of history, which was a political arena. Furthermore, it was argued by some proponents of grammar schools that:

> Mathematical geography, or that part of it which is connected with astronomy, may be deferred ... [As] The drawing of maps, and other minute labours in the pursuit of geography, may be desirable to a person who is designed for some employment connected with surveying or navigation, but are an unnecessary toil to the liberal scholar.[47]

Clearly the grammar schools were still seeking to educate gentlemen in the 'liberal' arts rather than in 'mechanical' subjects, thereby preparing them to take on public and political duties as citizens in the tradition of civic humanism, and in this context it was only descriptive geography which was required.

The grammar schools were preparatory to a continuing classical education at Oxford and Cambridge. The statutes of neither university emphasised geography, but both allowed for geographical education via classics, scripture and history. Daniel Waterland, tutor and then Master of Magdalene College, Cambridge, distilled his experience in *Advice to a Young Student*.[48] Contrary to the image of a Newtonian Cambridge, Waterland asserted that 'the generality of Students are intended to be clergymen, and as such must take the Arts in their Way. They must be acquainted with Mathematicks, Geography, Astronomy, Chronology, and other Parts of Physicks ... and all these are preparatory to *Divinity*, or subservient to it'.[49]

Given the interconnection of church and state, and of political and religious argumentation, the reading of geography was again implicitly being placed in the context of a political education by Waterland. Waterland's Cantabrigian structure of geographical education was closely akin to Oxford's for the simple reason that both universities recommended the same texts: Dionysius's *Geography* in the first year; Keill's *Introduction to Astronomy* in the second year; and Newton's *Principia* for those taking their studies further, together with Salmon's *Geographical and Historical Grammar*. The interchangeability of this curriculum between the universities was emphasised by the reprinting of Waterland's *Advice* at Oxford.

Geography's intended readership was practical as well as scholarly, serving two groups in particular according to Heylyn. First, '*States-men*, who from hence draw their knowledge of the nature and disposition of those people with whom they are to negotiate.' Heylyn here repeated the common defence of geography as an aid to statecraft which was one of the keys to the incorporation of geography into education.[50] Geography also served '*Merchants, Mariners* and *Souldiers*, the several Professors of which kinds of life, find nothing more necessary'.[51] Here, again, we can see that geography's intended readership made it a participant in the political realm, albeit more directly than in the scholarly case.

Actual practical readerships also demonstrated that geographical reading was politicised. There were a number of academies for business education in the eighteenth century, the best known of which was Thomas Watts's. For Watts, geography was 'worthy the prime Man of Business' for reasons that blended the traditional scholarly justifications of the subject with its significance to merchants: 'GEOGRAPHY not only teaches the *Situation* and *Distance* of Places and Countries, the *Knowledge* of the *Earth* in general and particular, but is ... the very Inlet to *History*; which informs us of the *Laws*, *Manners*, and *Customs* of *Nations*, their Advantages and Disadvantages; from whence we may be enabled to form a right Judgment to promote the Interest of our Country, and to lay down such Rules as may be useful in our *private* Concerns.'[52]

Martin Clare's academy for business education at Soho Square had almost the same justification for the study of geography.[53] For both, then, geography was part of the political education of the merchant, as it taught them about the laws of other nations, and, as in civic humanism, allowed the merchant to promote the public good. Unlike the 'liberal' arts approach, however, such knowledge was also vital for private gain. Most curricula for a tradesman's education linked geography and navigation, the sailor needing to know the 'Length, Distance, and Safety of Voyages', and the merchant using this knowledge to time his shippings to coincide with favourable conditions in overseas markets.[54] In this context, it is hardly surprising that the Chelsea Naval Academy also taught geography as part of the theory of seamanship.[55]

The practical inculcation of geography was not just by reading, but also by listening: closely linked to courses of instruction given for aspirant businessmen were public scientific lectures.[56] From the early eighteenth century, public lecture courses acted as experimental expositions of the Newtonian cosmology and mixed mathematics which dominated the Royal Society.[57] These courses tended to include a lecture on the shape of the earth, the use of globes and the traditional material of general or mathematical geography. They were also tied to the violent Whig versus Tory polemics of early-eighteenth century England.[58] As an example, Benjamin Martin's definition of geography eliminated the descriptive, Strabonic element of geography, where the traditional scholarship of the Ancients was still of interest, his view being aggressively Modern:

> GEOGRAPHY *is employ'd wholly in philosophical Speculations, relating to the* Form *or* Figure, *the* Magnitude, *the* Motions, *or* Gravity *of the Earth; the* Variations *and* Vicissitudes *of the* Seasons; *the Division of the Surface into* Land *and* Water: *And many other Affections of Moment can only be well understood from a philosophical Account thereof.*[59]

To say geography was 'wholly' employed in this manner was provocative, given that the grammar schools and universities taught it in a different manner for a wholly unrelated purpose. Clearly, different visions of the nature of geography were in collision here, and this also reflected different political agendas.

In these practical academies, lectures and texts, we see the pressures of commercialisation on the British educational structure. Geography was practised in new sites of learning, and new textual connections were made with other subjects, which highlighted geography's practical utility to an unprecedented extent. Being linking with commercial and naval activity, geography was, of course, being interwoven with the expansion of British power and empire, and was thereby 'politicised' in the modern sense of being an ideology in the service of the state, but it could also be viewed by contemporaries as within the ambit of political discourse as then defined for the reasons outlined.

Lest it be thought these ideas about geography's readership lost popularity, their continuity can be seen in Büsching's *New System of Geography* (1762), which was written and translated into English almost exactly a century after Heylyn's *Cosmographie*. While the facts in Büsching's *New System* were indeed in many cases freshly gathered, the system itself was anything but new.[60] Büsching's introductory section, 'Of the Utility of Geography', has close parallels with Heylyn's: he discusses in familiar terms why geography is useful to the divine, the statesman, and the merchant. The tradition was not entirely static, however. Büsching emphasised that the 'principal use' of geography was to arrive at a knowledge of God. This was not, as for Heylyn, by illuminating the locations referred to by scripture, but because 'our Earth', as described in geography, 'is ... sufficient to impart to us that knowledge of GOD which is attainable by considering the works of nature'.[61] This reflected the shift in the eighteenth century towards natural religion and rational proofs of God, as opposed to the scriptural route to faith via the word of revelation which had been favoured by seventeenth-century High Churchmen such as Heylyn. As Newton's work altered the understanding of cosmography, so it affected the justification of geography in this case. It is the continuity, however, rather than the alteration of justifications of geography's role in intellectual life which is most apparent. Geography texts had a stable and not inaccurate conception of their audience, which recognised that such readers would turn to geography books with political motives in mind. Where the definition of geography and its relationship to other spatial scales of enquiry was heavily influenced by the Ptolemaic heritage, these discussions of the utility of geography derived from Strabonic origins. Strabo (Book 1.1.22–23) emphasised the connection between his historical works and his *Geography*, arguing both were 'useful alike to the statesman and to the public at large'.[62]

The readership for geography, then, was split in two. First, there was a scholarly community of historians, classicists and theologians, in other words, the community of humanist scholars. Secondly, there was a community of those needing a practical education. It is misleading to emphasise one tradition at the expense of the other, or to ignore that both communities had political agendas and looked to geography books for both facts and narratives which bolstered these agendas.[63]

Authors and publishers: the production of geography books

The third angle of approach is the authorship of geography books. Examining who wrote geography books and the milieu in which they were produced sheds further light on how geography operated as a textual tradition.

The term 'geographer' is certainly defined in all the dictionaries I referred to earlier, but in the modern sense being a geographer was if not impossible at least improbable in the seventeenth and eighteenth centuries. Those who wrote geography texts were not geographers in the sense we might understand the term today, as authors of a string of intellectual explorations of the linked themes of space, place and the environment.[64] Rather, most authors who compiled information about the earth's surface at a global scale were either impoverished Grub Street journalists or wealthier scholarly historians. Most of the so-called geographers in seventeenth- and eighteenth-century Britain, be they poor or rich, were in fact authors in numerous genres and rarely of great intellectual distinction. The truth of this observation is hard to avoid if the briefest analysis of the careers of geographers is undertaken.

Let us look first at the poor geographical authors on Grub Street. Grub Street was a topographical reality in eighteenth-century London, being a street in the Cripplegate area. More importantly, it was a name associated, as the area was in fact, with cheap publications by hack writers. The title of Richard Savage's poem, *The Author to Let*, neatly summarises the status of the denizens of Grub Street, willing as they were to turn their pen to a variety of compositions in order to eke out a living. They were immortalised in Alexander Pope's poem, *The Dunciad*, their status as dunces epitomising the esteem they were held in by contemporaries producing more innovative literature. Johnson defined Grub Street as 'Originally the name of a street in Moorfields in London, much inhabited by writers of small histories, dictionaries, and temporary poems; whence any mean production is called *grubstreet*.'

Many have pointed to Johnson's ironic gesture in including dictionaries in this list, but from the present perspective, the mention of 'small histories' is more interesting. Histories, as mentioned earlier, were closely linked with geographies, and this extended to their production, with geographical compendia and small histories being comparable publishing ventures. Such compendia were clearly within the ambit of Grubstreet. Indeed, it has been argued that 'there was ... a broad field of annals-compilation which was the special preserve of Dunces'.[65] As an example of a Grub Street geographer's career, we can take William Guthrie, whose *Geographical Grammar* was the most popular and, as we shall see in Chapter 9, one of the most intellectually ambitious, of the late-eighteenth century compendia of geography. His career is typical of a Grub Street writer. Guthrie was a miscellaneous writer for the *Gentleman's Magazine*; he went on to produce one of the 'small histories' Johnson saw as typifying Grub Street productions, in his case called *A General History of the World* (1764–67). Guthrie received his education at Aberdeen University and was described by Johnson as 'a man of parts. He has no great regular fund of knowledge; but by reading so long, and writing so long, he no doubt has picked up a good deal'.[66] This was a charitable version of the Popean image of the Dunce, as Johnson was recollecting a friend, not satirising a foe. Equally typical of a Grub Street career was that of Thomas Salmon. He went on Anson's voyage round the world (1739–40), but his writings show little evidence that this made him a geographer in a modern sense, his *Geographical and Historical Grammar* (which is analysed in Chapter 7) being the result of collation, not fieldwork. Salmon's works were in the main historical, with geography being seen as contributing to history. Both poor Grub Street authors, then, developed *oeuvres* where geography was linked to the politicised realm of historical writing, their prime aim being to write enough to earn a living.

Of course, some British geographers were above the ranks of Grub Street in terms of wealth, but their publishing careers could still be worthy of a Dunce. Like their fellow authors on Grub Street, they also published geographies as part of a broader *corpus*, normally centred on historical enquiry. An example of this is Laurence Echard (discussed in Chapter 7), whose career was not unusual for a university student of the eighteenth century, being made a fellow at Christ's College, Cambridge, before moving into life as a parish priest in Lincolnshire. His writings include popular geography books, but seen in the context of his *oeuvre* it is clear he saw geography as subordinate to history, both secular and ecclesiastical, and to classics. He published translations of Lucian and Terence's comedies, as well as a *General Ecclesiastical History* and a *History of England*. Edward Wells, whose geography books have been cited earlier (and whose work is also analysed in Chapter 7), also fits this category, having been a student at Westminster and Christ Church, Oxford, before his clerical career. Wells's career is exactly one of those mapped out for a wealthy but talentless author in *The Dunciad*, where Pope speaks of Christ Church finally falling into line with the reign of idiots on the Isis.[67] His geography books once more are part of an *oeuvre* including classical translations, scriptural histories and Anglican polemics. Clearly he was writing geography books to mesh with a humanist education in the classics and Christianity. Hearne's assessment of Wells's writing career was the same as Johnson's of Guthrie's, without the charitable lens of friendship to soften the tone:

> [Wells] writ & scribbled many books, on purpose to scrape up money tho' he was a single man, & was otherwise very rich. The truth is, he was a man of great industry, but the books he published, as they are very many, so they are inaccurate, & contain very little that is curious.[68]

Wells and Echard, in the eyes of their contemporaries, may have differed in their social, financial and geographical situation from Grub

Street authors, but not in the scholarly mediocrity which led them to write hack geography books. A further similarity with Grub Street hacks was that Echard and Wells both produced geography books as part of a wider output of works in the political genres of history, the classics and religious debate.

Geography was clearly not the career of these authors, be they poor or wealthy. They compiled geographical information to facilitate the real business of scholarship as they saw it. This may help to explain something which has always annoyed modern geographers who have read any number of the works under discussion. They are flagrant plagiarisms of one another, often being copied word-for-word.[69] Looking first to the context of Grub Street, and of seventeenth and early eighteenth century publishing more generally, originality had never been the main aim: publishers in that milieu sought to compile marketable products and paid by the page, not according to the quality of the work. Authors had a positive incentive to copy material from other sources, as this allowed them to maximise their productivity and therefore their income. Given that poverty and Grub Street were synonymous, this was the prime consideration for these authors. The same held for Grub Street publishers. Even after the institution of the 1710 Licensing Act, which in theory gave publishers only a limited duration of copyright over books, tacitly, until a landmark legal judgment of 1774, publishers still saw themselves as having perpetual ownership of 'copy'. Particularly in the seventeenth century, the publisher (or 'Stationer', in contemporary terms) owning copy could alter, epitomise and reprint titles, such that word-for-word copies of earlier geography books might not be viewed as plagiarism, if both works were under the same publisher.[70] Beyond this, the emergent conception of the author as owner of his work was seen by many in the eighteenth century as not holding for 'tasteless compilers' of the sort who produced geographical compendia.[71] Geography books, then, were a genre where usurpation and the breach of copy were the norm, making it a realm of textual transmission where the function of authorial expression was relegated, although it was far from absent, as will become apparent through the biographical approach to the politics of geography adopted throughout this book.

Wealthier authors such as Wells and Echard had less need to write to keep themselves from starving, but they had no more incentive to break the generic norms of geography books than the pens for hire on Grub Street. They essentially reproduced the humanist model of education in which they had been formed, seeing the role of geography in a very traditional and therefore subordinate light. The innovations they introduced were anything but revolutionary, taking the form of improving aspects of the information geography books conveyed, while continuing to copy most of the standard material.

To see plagiarism as a weakness of the geography of the period, and to expect authors whose self-image was as a geographer, is to demand that geography conform to subsequent modes of academic organisation and print culture which had no meaning until the nineteenth century. In an educational system where, as we have seen, geography was a body of knowledge with a subordinate function, the prestige of authorship as a self-conscious social function lay in other realms. Many have been sensitised to the historical evolution of concepts of authorship by Foucault's 'What is an Author?' He argued that texts we might call scientific in modern parlance, a category in which he included geography texts, began to lose their need for the authority of an author in the eighteenth century, this being replaced by 'the anonymity of an established or always redemonstrable truth'.[72] As has been pointed out, geography as defined in the eighteenth century did not adopt the dynamics of what we now call the sciences, being seen as scientific in a different sense and being connected to the workings of humanist scholarship. For this reason it is unsurprising that geography did not see this drive to authorial anonymity. Truths were not redemonstrated as would be demanded by scientific method, they were simply copied and repeated. Equally, the sanctity of the creative author, which Foucault sees 'literary' discourses as developing in the eighteenth century, does not capture the dynamics of eighteenth-century geographical writing. From the evidence of geography books, Eisenstein's chronology of the development of authorship appears more satisfactory. For Eisenstein, the invention of the printing press made the notion of an author possible, but the first centuries of print history were dominated not by patrons or the market sponsoring the creative author, but by 'the more plebeian and prosaic early capitalist entrepreneur who hired scholars, translators, editors and compilers'. In the context of British print culture, this view is reinforced by Johns, who shows that it was the Stationers not authors themselves who drove the functioning of book production. The authors such a system produced were 'familiar with print technology and commercial trade routes' rather than being secluded intellectuals, and their works were often accordingly less exalted in their intellectual intentions.[73]

Because of this conception of authorship, geographical authors were frequently also important participants in political discourse. Or, to be more accurate, their geographical authorship was a small part of an *oeuvre* whose aims were historical, classical, theological, and therefore, by definition in contemporary terms, political. This applied to wealthier authors, as their ambition was to contribute to the realms of enquiry which humanism valued, namely theology, classics and history. It also applied to poor authors, since the works most in demand, and therefore most frequently sought by publishers from hack writers, were in the same genres, given that the market was largely constituted by those educated in a humanist framework. Repeatedly, as this work develops, we shall see writers of histories and religious tracts transferring their political languages into their geographical epitomes and abridgements. The functioning of authorship, then, given the commercial operation of British print culture in a society educated according to the values of humanism, linked the writing of geography books with the production of works in political genres.

Plan of the book

We have looked at geography books from three angles: texts, readers and authors. In each case we have seen that geography was interwoven with politics as defined by contemporaries. Given that I have argued that the political languages of geography books can only be disclosed by minute analysis, some breadth of coverage is inevitably sacrificed, throwing into question the extent to which the sample I analyse represents the population of geographers in the period. I have chosen to examine the intellectual biographies of those writers whose publications *qua* geography were the most historically significant, the most popular, the most frequently cited and the most revealing about the processes of political language usage and appropriation. It is hoped that this minimises the dangers of inherent bias towards any particular form of politicisation. The writers selected come at 20 to 30 year intervals, and the interrelationships between them show they were very conscious of writing within a coherent geographical discourse which had consistent patterns of interconnection with political idioms. Alternating with this study of the politics of geography taken *sensu stricto* is a series of chapters which look at the politics of the areas of geographical knowledge which we have seen were viewed as proximate to geography in contemporary divisions of knowledge: maps, chorography, travel writing and fieldwork.

My analysis starts with Peter Heylyn's *Cosmographie* (1652), which Sitwell calls 'the first large-scale special geography by an English author'.[74] As Sitwell's assessment makes clear, Heylyn is the logical starting point for any analysis of this genre. But Heylyn is not only the *origo*, he is the *fons et origo*: given the central importance of plagiarism, both structural and textual, to the production of later geography books, Heylyn became the source of much of their patterning and of their political language. As we shall see, Heylyn's geography defended the church-state vision of his great patron, William Laud. Being written during the Interregnum, this was, of course, a profoundly subversive political move in defence of a disestablished establishment. This political positioning, defending an 'established' church and state structure, while attempting to elide the ambiguities in the legitimacy of that establishment, is one repeated in different ways throughout the 'long' eighteenth century.

If Heylyn is viewed as a geographer of the Interregnum, the subject of my next chapter, John Ogilby, geographer to Charles II, is a purveyor of geographical knowledge in the era of the Restoration. Ogilby's road maps, as embodied in his *Britiannia* (1675), both culminated a career which repeatedly linked geography to the restored monarchy and constituted the benchmark against which road maps were measured for over a century. *Britannia*, then, amounts to one of the most significant products in the cartographic realm of geographical knowledge which was produced in the 'long' eighteenth century. Ogilby allows us to investigate the politico-geographical iconography of maps, an iconography which was also manifest in his design of the triumphal procession which greeted the restored Charles II.

Reverting to geography *sensu stricto*, I focus on Edmund Bohun to analyse the processes of textual transmission and intertextuality in the political language of geography books in the era of the Glorious Revolution, an event Ogilby did not live to see. Bohun's *Geographical Dictionary* was far less important in terms of subsequent editions and influence than Heylyn's. Yet Bohun self-consciously modelled his personal politics and the political language of his *Geographical Dictionary* on his image of Heylyn. In two ways Bohun shows us that political languages were unstable and liable to appropriation. First, he appropriated Heylyn's political language of geography in the contexts of both the defence of James II and the defence of the events of 1688/89. Secondly, the textual ambiguities of Bohun's appropriation were in turn appropriated by critical readers to align him with Jacobitism. Attachment to a political language, then, could either create or destroy authority, depending on the interface between authors and readers. The analysis of Bohun shows some of the political dynamics of the three-way negotiation between authors, readers and texts which historians of print culture have theorised.

If Bohun's Stuart links led to accusations of geographical Jacobitism, my next chapter, which turns to chorography in the form of Edmund Gibson's 1695 and 1722 editions of William Camden's *Britannia*, shows how geographical knowledge could be used to model a fidelity to the altered monarchical succession established in 1688/89 by a group formerly loyal to the Stuarts. Camden's *Britannia* remained the revered text of chorographical knowledge until the last decades of the eighteenth century, when the weight of new archaeological information overtook it. As such, its significance to the dissemination of geographical knowledge in the eighteenth century is hard to overestimate. In this chapter we see how the humanist procedures of textual editing and scholia could be used to create a political persona. Gibson *et al.* in the 1695 edition modelled Camden as an Anglican who would have abhored James II's Catholicism and went on in the 1722 edition to annotate *Britannia* so as to articulate an historical justification of the Hanoverian succession. The change of monarchs in 1714 led to a change in textual strategies, but the common thread between the two editions was an anti-Jacobite loyalism. This did not go unnoticed by Gibson's readership, his High Church Anglican opponents drawing *Britannia* into their polemics over church government in the period between Gibson's two editions.

Geography *sensu stricto* for the first half of the eighteenth century is treated via brief sketches of three writers, Laurence Echard, Edward Wells and Thomas Salmon. These authors have been chosen for their popularity, each having written more than one geography book which went through a number of editions. Echard's and Wells's geography books were popular at grammar schools and universities, being recommended in numerous educational treatises. Echard and Wells produced very brief treatments of geography, but their politics can be interpreted in the light of the historical writings of Echard and the theological and ecclesiastical polemics of Wells. Thomas Salmon's *New Geographical and Historical Grammar* (1749) was more popular than the geography books of Wells or Echard, going through eighteenth separate versions. Sitwell's assessment is accurate: 'a typical special geography, except that the historical section on England includes a chronology that is unusually detailed'.[75] Salmon is analysed, then, both because his work was popular, and because, given the importance of images of history to eighteenth-century politics (see above), his historical section allows his political language to be established more unambiguously from his geographical work itself than is the case with Echard or Wells. The three authors taken together show an attachment to the established church and state, but no detailed agreement as to what that 'establishment' was, or where precisely 'orthodoxy' lay. This shows that if the political languages of geography books were predominantly defences of orthodoxy, not only could they be appropriated by subversive readings (as in the case of Bohun), but also that the conception of 'orthodoxy' was unstable over time. Indeed, 'orthodoxy' was as much a rhetorical term of approbation in the political language of these works as a definable political reality. For geography books, then, as for political language more generally, William Warburton's comment applied: 'Orthodoxy is my doxy; heterodoxy is another man's doxy.'

The following chapter examines the politics of travel writing, which, as we have seen, was closely related to geography proper. As in the previous chapter, we see competing claims to the 'orthodox' high ground, this time in three of the most important travel writers of the eighteenth century, each of whom was a self-confessed Tory Anglican: Thomas Pennant, John Wesley and Samuel Johnson. We will discover that the ways in which these three travellers responded to and described the landscape, government and established church in the regions through which they travelled related to their individual conceptions of a Tory Anglicanism. The significant variable between the three was theological stance, and this drove their differing constructions of travel narratives. Travel, then, exemplifies the point that the textual politics of geographical knowledge was inseparable from the religious sphere in this period. Furthermore, we will see that the audience for travel books was sensitive to this, and responded by engaging with the religious politics which Pennant, Wesley and Johnson had articulated.

For later eighteenth-century geography, I analyse the politics of William Guthrie and John Pinkerton. Guthrie's *Geographical, Commercial and Historical Grammar* (1770) is an obvious choice in terms of its popularity and its innovations in the deployment of political languages in geography books. In terms of popularity, Sitwell comments that 'by almost any measure, Guthrie's *Grammar* was the most popular special geography of them all',[76] the most obvious measure being its 46 editions. Beyond this, Guthrie provided a copy text on which numerous other geography books were based, or, as in Pinkerton's case, the benchmark against which geography books were measured. One element of Guthrie's lasting importance was his substantial reworking of the political language of geography, broadening the treatment from an English to a British context by introducing themes from the political philosophy of the Scottish Enlightenment. Pinkerton's *Modern Geography* (1802) went through six full and three abridged editions in the early nineteenth century, and was unusual in terms of its intellectual rigour. Pinkerton allows us to analyse the response of the political language of geography, as reworked and 'Scotified' after Guthrie, to the events of the French Revolution, which marked the beginning of fundamental changes in the discursive framework of British political languages. Initially these changes were registered in problems in using Catholicism and absolutism as political 'others' to be stigmatised and in an increasingly assertive British nationalism, both themes registered in *Modern Geography*.

In terms of popularity and influence, John Walker's *Elements of Geography* would not be included, but it is highly significant as a rare example of a geography book which, being written from a Quaker perspective, developed a critique of all the varieties of orthodoxy in church and state defended by other geography books. That dissent should be so intertwined with the demands for a more democratic politics is unsurprising given the interweaving of religious and political languages, although the leading element in the dissent-democracy nexus is disputed.[77] Walker represents a path to a more subversive form of textual geographical politics which was not taken up.

In many ways, it is Walker's contemporary, James Rennell, the subject of the next chapter, who represents the path towards a new interconnection between geography and politics which was taken up. Where geography had always simply utilised the findings of fieldwork and travel, Rennell starts to construct the tradition of geography as a fieldwork subject which was to become increasingly dominant as the nineteenth century progressed. Rennell's importance to forging a new type geography has been recognised from the first obituaries on his death in 1830. His career pattern, the texts he produced, and the nature of their politicisation all bespeak a new conception of geography, whose political impact is as much in the field as in the text. The political language of geography is recast as geography is recast, being developed much more at the level of *ideological assumptions* about the legitimacy of empire, European superiority and the like, than at the level of *explicit textual interventions*, as had been the case for the textual tradition originating with Heylyn. Rennell is, however, a liminal figure, engaging with both a textual and a fieldwork tradition of geography.

The effect of fieldwork such as Rennell's on the textual tradition of geography *sensu stricto* is analysed in the next chapter, which discusses

James Bell and J.R. McCulloch. Bell's *System* (1829–32) has been described as 'the last British example of a massive description of all the countries of the world written by a single author' and also as perhaps the best.[78] It is noticeable for its proliferation of statistical information about the world, which bespeaks a new and quantitative approach to geographical description. This is predicated on the expansion of imperial activity and of fieldwork geography of the sort pioneered by Rennell. Bell's *System* acts, furthermore, to internalise this conception of geography within the genre of the geography book, thereby altering the textual character of geography. But Bell continues to rely on textual construction to make his political points, which emanate from the framework of the later Scottish Enlightenment as it faced questions of reform in the early nineteenth century; Bell is therefore intellectually as well as chronologically the inheritor of Guthrie and Pinkerton's Scottish reconfiguration of geography. This is far less true of Bell's successor, J.R. McCulloch. A noted political economist in Ricardo's generation, McCulloch's geographical gazetteers of the 1830s and 1840s continued the quantification of geography seen in Bell, subordinating the old textual tradition of geographical politics to the dictates of political economy. This was a key moment in the reconfiguration of geography and its politicisation. McCulloch shows himself notably uninterested in religious polemics or dynastic issues, being concerned with questions of democracy and industrialisation. In other words, McCulloch's geography books respond to the new conditions for political discourse which were established by the radical changes in British political life instituted in the 1830s and deploy new intellectual frameworks to discuss those conditions. The textual tradition of geographical politics with which this study is concerned has, by the time of McCulloch, became weakened to the point of disappearance, the only remnant being a generic one: the persistence of the gazetteer as the site of politicisation.

In a coda, I discuss the work of Halford Mackinder whose significance as one of the 'founding fathers' of modern British geography is undisputed. In other words, I close where standard histories of British geography, political geography and geopolitics open. Mackinder is contextualised both in terms of the crisis of confidence in Edwardian politics and in terms of the antecedent tradition of the interweaving of politics and geography with which this study has been concerned. Mackinder is shown to have been reacting against the statistical geography of the mid-nineteenth century, and as such reinstituting a textual politics of geography. However, he reversed (at least rhetorically) the lines of force between geography and politics, aiming to show how physical geography drove political events, rather than simply incorporating political argumentation into a geographical text. The result is a textual geography which entirely renounces the generic conventions of the gazetteer. Yet Mackinder is shown in fact to have followed the same procedures as his eighteenth century predecessors to a considerable degree, incorporating pre-existent political positions into his geography. Mackinder, then, broke the last lines of force with the textual connection of geography and politics which had run from Heylyn on, but did not alter the discursive privileging of politics over geography.

3
'Geography is twinned with divinity': the Laudian Geography of Peter Heylyn, 1621–57

Prelude: Camden's politics

From the earliest examples of books of geographical instruction the role of geography in education has been tied to politics. Strabo's *Geography*, which remained so important to the conceptualisation of geography's scope and purpose until at least 1800, opened with a discussion of the utility of geography, which averred its importance to statesmen and commanders.[1] Similarly, and deriving from Strabo, the earliest English-language geography books were also justified in terms of their role in a political education. Thus Cunningham's *Cosmographical Glasse* (1559), which, as we have seen, he claimed was the first geography book in English, emphasised its utility to 'the defence of our Country':

> Cosmographie herein do so much profite, that without it both valeaunt Corage, Policy and Puisaunce oftentimes can take no place. For by her we are taught whiche way to conduct most safely our ooste, where to pitch oure tentes, where to winter: yea, and where most aptlye to encounter with them in the fielde.[2]

The geography books spanning the period from the Reformation to the Reform Act by and large take a less militaristic view of the nature of politics, but they continue to argue that geography contributes to a political education, and they also demonstrate their authors' political identities by the languages and arguments they deploy.

This cannot be seen as a deviation from some putatively 'objective' geographical tradition: geography, like other areas of enquiry in the humanist educative structure, was designed to prepare the student for public life. This applies to the English geographical tradition as a branch of humanism, which received its first canonical expression in William Camden's chorographical work, *Britannia* (1586). *Britannia* was the single most quoted source in geography books when discussing Britain for the next two centuries, and elements of its political project resonated with future generations in very different political contexts.

Camden's own position built on a patriotic belief in England's unique status as inscribed in its geographical location: 'It is certainly the master-piece of Nature, perform'd when she was in her best and gayest humour; which she placed as a little world by it self, upon the side of the greater, for the diversion of mankind.'[3] Politically, this patriotic pride in *Britannia* derives from the English church and state system in the aftermath of the Reformation. In terms of the state: 'The King ... hath soveraigne power and absolute command among us, neither holdeth he his empire in vassallage, nor receiveth his investure or enstalling of another, ne [*sic*] yet acknowledgeth any superior but God alone'.[4]

Connected with the absolute independence of the English state was that of the English Christian church, which for Camden had been founded and received its pure form and doctrine prior to the intervention of Rome. While Camden debunked numerous myths about British antiquity, he maintained the traditional tale that the English church had been established by Joseph of Arimathea and was therefore primitive and apostolic.[5] This allowed Camden to view the English Reformation as a revolution in the sense of a revolving of the wheel of history to its starting point, rather than as a radical break in English history. The type of church establishment Camden saw as primitive (and therefore pure) was a *via media* between Roman Catholicism and extreme Protestantism. For this reason, he was critical of those Protestants who showed no interest in the ecclesiastical monuments of Britain:

> There are certaine, as I heare who take it impatiently that I have mentioned some of the most famous Monasteries and their founders. I am sorry to heare it, and with their good favour will say thus much, They may take it as impatiently, and peradventure would haue us forget that our ancestoures were, and we are of the Christian profession when as there are not extant any other more

> conspicuous, and certain Monuments, of their piety, and zealous devotion toward God. Neither were there any other seed-gardens from whence Christian Religion, and good learning were propagated over this isle, howbeit in corrupt ages some weeds grew out over-ranckly.[6]

Camden's politics, then, had much in common with those of his contemporary Richard Hooker, as expressed in his *Laws of Ecclesiastical Polity*, defending the church and state by law established as legitimate developments grounded in England's original political and ecclesiastical independence, while not denying the Roman Catholic influence within that developmental trajectory. Both emphasised that their position was a *via media* between the extremes of licentiousness and tyranny, and between Presbyterianism and Roman Catholicism, the two spectra being interrelated, in that both extremes of religion were seen as indiscriminately encouraging tyranny or anarchy.[7] Both saw England as an 'empire' in that term's older sense: as a contiguous unit ruled by one sovereign power. England as a geographical 'world by it self' was for that reason a world apart in church and state, an empire over itself, from which no recourse could be had to outside powers. The Reformation merely reinstated the empire which geography and history had initially installed.[8]

As Camden's remarks on monasteries in *Britannia* suggest, his position on the English church and state did not go unchallenged even within the field of geography. In 1599 George Abbot, future Archbishop of Canterbury, published his *Briefe Description of the Whole Worlde*. Abbot was far more closely associated with the Calvinist tradition than Camden. Agreeing that the English church had been founded independent of Rome, his image of the Roman Catholic corruption of that primitive and catholic church was far more uncompromising than Camden's, something which was revealed in his attitude to monasteries. Admitting the 'fayrenes of our greater and larger Churches', Abbot continued that the beauty of monasteries concealed 'those which liued there, [who,] giuing themselues to much filthyness, and dyuers sorts of vncleannesse, did so poure downe the vengeaunce of God vpon the places, that they were not only dissolued, but almost vtterly defaced by K. *Henrie* the eight'.[9]

As should be clear, the earliest English-language geographies and chorographies were not divorced from political debate, engaging with the major issues concerning the church and state. Camden's influence on subsequent geographical writing was far greater than Abbot's, and this helps to explain the High Church inclination which geography displayed until the accession of George III. *Britannia* was based on original research in a way Abbot's *Briefe Description* was not, and it was also a monumental folio by the time of his death, where Abbot's book was a modest octavo.[10] This does not explain why Camden's own political and ecclesiological proclivities, suitably updated, remained so important, given that his information could have been fitted into entirely different political projects, and indeed was in Gibson's editions of *Britannia*, as we shall see in Chapter 6. Much of the explanation for the persistence of Camden's political predilections in later geographical work must lie with the generation which followed him and constructed the Laudian church-state nexus from the 1620s. For the Laudians also made a significant contribution to the geographical tradition which ensured a continuation of Camden's politics of geography into Charles I's reign and beyond as part of their more general promotion of scientific learning.[11] Their geographical contribution was masterminded by Peter Heylyn.

Claiming Camden's mantle: Heylyn and Laud

Peter Heylyn (1599–1662) attended Hart Hall, Oxford from 1615, being elected to a fellowship, in which capacity he was obliged to give lectures on cosmography.[12] In 1621 these lectures were published as *Microcosmus* by Oxford University Press in an octavo of 418 pages. Thirty years later, a much expanded edition of this work was published as *Cosmographie*, the second (1657) edition of which covered 1100 folio pages.[13] Heylyn clearly wished to place himself and his geographical writings in the tradition of the previous generation of scholars, led by Camden and Sir Robert Cotton. In an era when classical scholarship remained the *ne plus ultra*, Heylyn paid Camden the ultimate compliment of describing him as 'the *Pausanias* of the *British* Islands' (*C.*, p. 304; see also p. 28). Further, just as Camden had recovered British history and geography as a source of patriotic pride, so, for Heylyn in the next generation, Camden himself was now a source of patriotism: 'one *Cotton* for the Treasury, and one *Seldon* [*sic*] (now Mr *Camden* is dead) for the study of the like antiquities; are worth all the French'.[14] Heylyn placed himself as a successor to Camden as part of his creation of a public persona of loyalty to the monarch and adherence to a catholic vision of the Anglican church. Camden's *via media* in church and state could be appropriated and modernised in the 1630s, the context in which Heylyn entered the political arena. Where Helgerson has argued that Camden's work could be subversive to monarchy, creating a national identity built around the land rather than the sovereign,[15] Heylyn did not countenance such a possibility. It would appear that from Heylyn's perspective to link himself with Camden was to assert his orthodoxy.

In his role as a lecturer in Oxford during the 1620s, Heylyn was already establishing his political identity via his geography. *Microcosmus* is dedicated to Charles, Prince of Wales, who within four years would accede to the throne and who in November 1621 'received verie graciously' a copy from Heylyn.[16] Heylyn's dedication, in the Strabonic tradition of geography as advice literature for princes, hopes that his book will 'be somewhat seruiceable to Princes which deale in the affaires of many nations' (*M.*, ¶2v). Such a profession of loyalty, especially coupled with the book's defences of church ornament, attacks on Presbyterianism and distancing of the English Reformation from European Calvinism (*M.*, p. 243 and pp. 248–90), was also bound to appeal to William Laud.[17] Laud was president of St John's College, Oxford, an increasingly influential dispenser of patronage within the university and was appointed Bishop of St David's in the year of *Microcosmus*'s publication. The precocity of Heylyn's Arminianism should be noted: arguing that the English Reformation was a native growth 'bearing respect neither to *Luther, Zwinglius,* nor *Calvin*' (*M.*, p. 249), *Microcosmus* went against the Calvinist orthodoxy which prevailed in Oxford until at least 1630, pitting an insular picture of the Reformation against the pan-European Calvinist model.[18] The book, then, made a twin profession of loyalty to the future king and to the Arminianism of Laud from within the Oxford context. The political position of *Microcosmus* as a text, then, together with its context of publication and dedication make it a precocious part of the emergent Laudian and Caroline project.

Before discussing Heylyn's geographical works in detail, it is worth sketching some of the salient points of the broader political and ecclesiological framework in which they participated. Laud was appointed Archbishop of Canterbury in 1633 on the death of George Abbot. Abbot's churchmanship, as we have already seen from his geography book, the *Briefe Description*, was more Calvinist than Camden's. Laud and Heylyn, by contrast, swung the pendulum back towards the position of *Britannia*, defending the church ornament and liturgical formality which Abbot and others dismissed as Popish corruptions of Christianity, arguing they were vital generators of a suitably reverent attitude to the church as an establishment and of spirituality within that establishment. Laud ran an aggressive campaign defending his views, which can be summarised as follows:

> continuity with and descent from Christ and his Apostles; the central importance in the life of the Church of the episcopacy; a deep concern that the worship of the Church should be of prime importance in the life of the Church, and should be conducted with reverence and awe; a focus on the altar, in churches furnished and adorned in such a way as to enhance the beauty of holiness and stimulate worship ... and an affirmation of the English Church as part of the historic Church, joined still, in spite of outward division, by the one Catholic faith.[19]

Accompanying this ecclesiastical politics was a view of the monarch as an absolute power, divinely appointed. As a consequence, Laud stressed obedience to the state and the monarch as a Christian prerequisite. He viewed the 'incipient egalitarianism of Calvinism'[20] as potentially dangerous, undermining awe for the monarch. Just as episcopalian hierarchy and ornament were essential to preserving reverence for the church, so they buttressed the unassailable position of the monarch. In church and state, then, the Laudian project was committed to defending a hierarchical society as essential to the creation of deference, whence sprang social stability.

Heylyn had signalled his affinity with the Laudian position long before it gained the ascendancy, and continued to defend it through all vicissitudes. Both of his geography books articulate this position, *Microcosmus* in the era of its genesis, and *Cosmographie* in the midst of the Interregnum which followed its overthrow. These geography books must be seen as part of a wider publishing campaign in which Heylyn established a consistent political position on church and state through historical, theological and travel books.

Laudian geography

A convenient entry point into the politics of Heylyn's geographical writings can be found in his patriotism. In the tradition of Camden, Heylyn sees Britain as geographically designed to be exceptional: 'Environed with turbulent Seas, guarded by inaccessible Rocks: and where those want, preserved against all forein invasions by strong Forts, and a puissant Navy' (*C.*, p. 294).

British exceptionalism and independence are validated by the facts of geography, and contrasted with other nations, which are seen as inferior. As such, Heylyn is troubled by those who travel to the continent, as they thereby endanger their Englishness:

> I have met with some Gentlemen, who upon the strength of a little travell in *France*, have grown so *un-Englished*, and so affected or besotted rather on the *French* Nation, that they affirm the *English*, in respect of the *French*, to be an heavy, dull, and *Phlegmatick* people ... A vanity, meriting rather my pity, than my anger. (*C.*, p. 300)

But while all this is entirely comprehensible within the tradition of Camden, the bases of Heylyn's patriotic defence of England are more precise, and belong to the generation of Laud and Charles I. These bases can be found in the detailed narrative of Heylyn's geographical works, and they are the nature of the Church of England and the structure of the English constitution. In a sense, this is unsurprising, given that national identity in the early modern period tended to be founded on church, state, monarchy and legal system, rather than race and empire as in the nineteenth century.[21] Heylyn's specific visions of the English church and state, however, delimit his patriotism as Laudian.

Looking at the state first, English geographical exceptionalism is matched for Heylyn by an exceptional and ideal constitution:

> The Nobility of this Countrey is not of so much unlimited Power, as they are (to the prejudice of the State) in other Countries: whereas in other places they have some absolute, some mixt government; so that upon any little distast, they will stand on their own guard, and slight the power of their Soveraign. And on the other side, the Commonalty enjoy a multitude of Priviliges above all other Nations (*C.*, p. 300)

If Heylyn seems to be advocating a balance of power, this is anything but his intention. While he wrote of balancing the three estates of the realm, his political theory was not of a government of checks and balances. For Heylyn, the nobility and the commons should have liberties within a system of absolute monarchical power.

This vision of an absolutist balance of powers, which sounds paradoxical after three centuries of political theory associating balance with checks to absolutism, receives its fullest expression in Heylyn's narrative of his French travels:

> They therefore which have written of Republicks do most applaud and commend the mixt manner of rule, which is equally compounded of the Kingdoms and the *Politeia* ... In these there is referred to the King, absolute Majesty; to the Nobles, convenient authority; to the People, an incorrupted liberty: all in a just and equall proportion ... A rare mixture of Government, and such at this time is the Kingdome of *England*.[22]

For Heylyn, balance could not be produced without an absolute monarch. This related back to the concept of sovereignty which persisted from Elizabethan times and argued that absolute power must finally rest in one discernible location in the polity. The time Heylyn spoke of when England achieved this 'rare mixture' was the 1620s; by the time his travels were published in 1656 the meaning of the term 'republic' had taken on a new resonance it would never lose in England, and the absolute majesty of monarchy had been irrevocably damaged.

Heylyn continued to defend this vision in the Interregnum, and drew support from his predecessors including, perhaps most interestingly for the present argument, Camden:

> the King established is an absolute *Monarch*, from whom the meeting of the *three Estates* in Parliament detracteth nothing of his Power and Authority Royal ... And *Cambden*, as renowned an Antiquary as any of the Age he lived in, hath told us of the King of *England*, supremam potestatem & merum imperium habere ... To prove this last, he cites these memorable words from *Bracton* an old *English* Lawyer, *omnis quidem sub* Rege, & *ipso sub nullo sed tantum sub Deo*, that every man is under the King, but the King under none saving only God.[23]

Seamlessly combining the language of common law with that of absolutism, Heylyn, then, did not see the king as one estate of the realm, but as absolute ruler over these estates.[24] It is noticeable that Heylyn's views were republished in the midst of the Exclusion Crisis of 1679–81, in which the legitimacy of excluding James, Duke of York from acceding to the throne at the death of Charles II was debated. As Heylyn drew on Elizabethan sources to defend absolute *jure divino* monarchy, so he was in turn drawn on as a figure whose historical status made his views more, not less, pertinent in a subsequent political context. As we shall see in Chapter 5, Heylyn's geographical writings were also reprinted and his geographical mantle was taken up by Edmund Bohun in the same period, again in an attempt to recast his Laudian synthesis in the altered circumstances of the 1680s. In both cases we see the legitimating force of historical example in political argument in the 'long' eighteenth century.

In a significant and subversive posture for the 1650s with Charles II in exile, a belief in absolute monarchy permeates the texture of Heylyn's discussion in *Cosmographie*. This manifested itself in several ways. First, Heylyn's description of William the Conqueror in his section on the history of England is highly pertinent:

> *William*, surnamed the *Conqueror*, after the vanquishment and death of *Harald*, acknowledged and Crowned King, altered the antient Laws of *England*, and established those of *Normandy*, in place thereof: governing the people absolutely by the power of the Sword. (*C.*, p. 318)

Whatever the factual accuracy of this view, in the seventeenth century it amounted to a significant defence of absolute monarchy. In this period, defenders of the rights of parliament who wished to constrain the power of the monarch argued that the constitution predated the monarchy, and that successive monarchs had conformed to the liberties granted time out of mind by the Ancient Constitution. Within this argument, William I could not be seen as a conqueror, because that would amount to an absolute break in the continuity of constitutional government, and the subsequent legal and constitutional structure would be the grant of an absolute sovereign. As J.G.A. Pocock points out:

> once men had appealed to the immemorial, the laws must be either absolutely immemorial or subject to an absolute sovereign ... A polemical situation could therefore arise, in which to put forward any theory as to the origin of the English law at a time within recorded human history could be interpreted, and even intended, as an argument in favour of absolute monarchy. We shall see this happening in 1681.[25]

As stated above, many of Heylyn's political tracts as well as the *Cosmographie* itself were reprinted at just this juncture. Moreover, given his numerous defences of a Laudian conception of balanced absolutism, it is unlikely that he did not intend his brief comment on William the Conqueror to be read in this light, especially as he repeated that William was a law giver to England in his French travels.[26] Within the binary structure of immemorial rights or absolute monarchy which the historical bent of political argumentation established, Heylyn opted for the latter alternative in the Interregnum, where Parliamentarians invoked the former option, arguing they were defending ancient liberties.

Further evidence in *Cosmographie* also shows it to contain a defence of absolutism. Heylyn praises James I and Charles I (see *C.*, pp. 354 and 342), both of whom were frequently seen as flawed monarchs because of their tendency to absolutism. It should be emphasised that for Heylyn absolutism could not be equated with tyranny or despotism. Indeed, *Cosmographie* pictures English monarchs as absolute in power, but not tyrannical in its deployment, which is contrasted with the French monarchs, who are tyrannical precisely because they do not receive the absolute submission that should be accorded to a sovereign. His summary of French government sketches this position:

> As for the *Government* of these Kings, it is meerly *Regal*, or, to give it the true name, *Despotical* ... As for the *French* Parliament or Assembly of the three Estates, which heretofore were of great credit and renown, and looked on as the principal Bulwark of the Publique liberty... [Louis XIII] finding them to retain something still of their antient stomack, and apt enough to clash with that absolute Soveraignty which his Predecessors had attained to, he resolved to make no more use of them ... The power of the *French* King over his Subjects being so *transcendent*, it cannot be, but that his *Forces* must be very great. (*C.*, pp. 236–7)

Where the monarch is not accorded unquestioning obedience, Heylyn sees all parties as losers: parliamentary representation is denied by the king who fears for his position; despotism replaces benevolent kingship; and 'miserable is the condition of the common Peasant, partly by reason of the intolerable Taxes laid upon him by the King [to fund the armies which enforce his power], and partly by those great but uncertain Rents which are set upon him by his *Landlord*' (*C.*, p. 175). It becomes apparent why for Heylyn submission to the monarch is a duty[27] and why his comment that 'the Kings of *England* are more absolute Monarchs than either of their Neighbours of *France* or *Spain*' is said as a compliment.[28] Clearly, Heylyn's *Cosmographie* had a part to play in the prosecution of this argument: by a comparative analysis of the countries of the world, Heylyn believed he could show the baleful effects of falling away from the balanced absolutism that the England of Charles I had established.

If *Cosmographie* defended a Caroline monarchical absolutism, it also advanced a Laudian position by its attitude to the visible church and the history of Christianity in England. Looking first at church history, Heylyn follows Camden in accepting that 'the *Christian Faith* was first here planted, as some say, by S. *Peter* and *Paul*; more probably (as others say) by *Joseph* of *Arimathea*' (*C.*, p. 301). The Roman Catholic church was an imposition which eventually perverted the primitive Christian tradition, and the Reformation simply returned the church to its traditional status as 'consonant to the Word of God, and most pure Antiquity' (*C*, p. 302). Heylyn expands on the nature of the English Reformation in a passage which makes his affinity with Laudian High Churchmanship transparent:

> [C]orruptions ... had been brought ... by the power and tyranny of the Church of *Rome*. ... [The Reformation] being in most other Countries received tumultuously, by the power of the People; was here admitted upon mature deliberation, by the authority and consent of the Prince and Prelates: the *Architects* in this great work, without respect unto the Dictats of *Luther*, or *Calvin*, looking only on Gods Word and the Primitive Patterns, abolishing such things as were repugnant unto either, but still retaining such *Ceremonies* in Gods publick worship, as were agreeable to both, and had been countenanced by the practice of the Primitive times. A point wherein they did observe a greater measure of Christian prudence and moderation, than their neighbour Churches; which in a meer detestation of the See of *Rome*, allowed of nothing which had formerly been in use amongst them, because defiled with *Popish* Errors and abuses ... Whereas had they continued an allowable correspondency in their extrinsicals of Religion, with the Church of *Rome*; their party in the world had been far greater, and not so much stomached as it is. (*C.*, p. 302)

The above extract summarises most of the elements of Heylyn's Laudian programme, and is in no sense an attempt to relate an impartial narrative of the English Reformation. This can be seen from several elements of the description. Unlike most Protestants, Heylyn implicitly allows that the Roman Catholic church is a true church and not the devil's work. Accepting that the Roman Catholic church was corrupt, Heylyn argues Calvinism and even Lutheranism were irrational in jettisoning elements of worship sanctioned by the Bible or the early church fathers (i.e. by 'primitive patterns'). The appeal to the Fathers is significant, as it shows Heylyn does not advocate a *sola scriptura* approach, but rather one which grants that the early church's direct links with the practice of Christ and the Apostles gives it canonical status. It is this which makes Heylyn's vision of a pure church a Laudian one, and which distances him from Scottish and English Calvinists, as well as from much of the European Protestant tradition.[29] The *via media* between Calvinism and Roman Catholicism is the line of true Catholicity for Heylyn as for Laud.

This image of a true church is reinforced at many points in *Cosmographie*, Heylyn's argument recapitulating that of his more openly controversial theological and historical works. First, monarchs are assessed according to the degree to which their religious policies encouraged a Laudian conception of the church. Thus Henry VIII was correct to 'pare away the excresencies' of Roman Catholicism, but the '*Avarice* of some great persons in Court and State' skewed the Henrician Reformation (*C.*, pp. 300 and 302). Where Edward VI and Mary I move to the extremes of Calvinism and Roman Catholicism, Elizabeth I finds the *via media* once more.[30] Secondly, having mentioned the 'architects of the Reformation' he goes on to name them as John Jewell, Richard Hooker, Lancelot Andrewes and William Laud (*C.*, pp. 302–3). This is not so much a catalogue of the Reformation as of the High Church conception of the Reformation.[31] As in Heylyn's historical writings more generally, so in his geographical work the aim is to prosecute an argument, not to give an impartial narrative.

The geographical format of *Cosmographie* does enable Heylyn to modify the form of his argument. By looking at the Protestant church in other countries, Heylyn reinforced his claim that only England had reformed its church rationally. The other Reformations, being products of the violence of extreme Calvinism, were unstable and threatened the harmony of the constitution. The French Huguenots are a case in point: 'being grown too insolent by reason of so great a strength, and standing upon terms with the King as a *Free Estate* (the *Common-wealth of Rochell*, as King *Henry* the fourth was used to call it) they drew upon themselves the jealousie and fury of King *Lewis* the thirteenth' (*C.*, p. 176). Heylyn has no sympathy for the Huguenots,[32] and his reference to 'the Commonwealth' gives a clue as to why this is the case. He believed that the English civil war was at least in part the result of Scottish Calvinists foisting their irrational Reformation on the Catholic Reformation of England:

> [The Reformation] was here [in Scotland] made by a strong hand, according to the judgement of *Knox* and others: not taking counsel with the *Prelates*, nor staying the leisure of the Prince, as they did in *England*; but turning Prince and *Prelates* out of all autority [*sic*]; ...[They] have endeavoured ever since by practices and correspondency with that party here, and finally by force of Arms, to thrust their own Constitutions and Forms of Worship on the Church of *England*. (*C.*, p. 332)

Given the harmonious balance in church and state which Heylyn portrayed for England, it is unsurprising he saw the importance of a multiple kingdoms explanation for the Civil War. If England had achieved the rare and ideal mixture of power and liberty in church and state facilitated by an absolute monarch and a Catholic church, all groups should have been content with their situation. Heylyn would be loath to admit the English empire unto itself was riven by faction in the 1630s, and so the slide away from the ideal polity had to be ascribed to the incursion of a geographically exotic influence with alien ideas of church and state government.

Heylyn was equally critical when his geographical survey covered Roman Catholic countries. But his anti-Catholicism did abate interestingly in his discussion of France because 'the *Gallican Clergy* stands more stoutly to their naturall rights, against the usurpations and encroachments of the see of *Rome*, than any other that live under the Pope's authority' (*C.*, p. 176). Papal power in France had been curtailed, such that Heylyn could see here some echoes of the Catholic but independent conception of the church that Laud modelled for England.[33] Heylyn and the Laudians were frequently accused of being closet Roman Catholics by their Calvinist opponents precisely because their views on Rome were not unrelentingly negative. Most famously, Gilbert Burnet asserted that Heylyn's history of the Reformation, *Ecclesia Restaurata*, converted James II to Roman Catholicism. Similar suggestions of a greater tolerance towards Roman Catholicism than Calvinist Protestantism might have been levelled at *Cosmographie*. More impartially it can be said that in constructing his descriptions of national religions, Heylyn's evaluative criterion was proximity to the Laudian conception of true catholicity in church doctrine and government. As such, the patriotic message that the reign of Charles I had seen an ideal polity could not but be reinforced by the way in which Heylyn constructed his geographical survey.

Geography is better than or better with divinity?

It should be apparent that Heylyn's geography books are tied to his broader theological and political programme. Yet most commentators, contemporary and modern, have assumed a distance between Heylyn's geographical writing and his other works. Thus Vernon argued in his 1682 biography that during the Interregnum Heylyn 'took care that what he writ should be beneficial to Royal Government'. But equally, he suggests that *Cosmographie* is the exception to the rule, as Heylyn undertook this 'to divert his mind from the sad complexion of the times'.[34] Likewise, E.W. Gilbert, one of the few modern geographers to write about Heylyn, argues that in the Civil War Heylyn 'returned from theology to the study of geography,' and that this 'must have given him great solace'.[35] Gilbert's piece is entitled 'Geographie is better than divinity', and refers to an incident recorded in the 'Preface' to *Cosmographie* where Heylyn, buffeted by defeat in the 1640s, is advised by a 'tall big Gentleman' to return to geographical studies and end his career as a political controversialist. The problem with this categorical separation between geography and divinity is that any special geography of the variety which Heylyn wrote had to treat political and ecclesiastical history, and therefore could not but engage with debates about the structure of the church and state.

The unavoidable interrelationship between geography, divinity and polity can be seen in Heylyn's writings by focusing on the recurrence of a simile which appears in *Cosmographie*. As we saw in Chapter 2, in discussing the utility of geography, Heylyn wrote: '*Geography* without *History* hath life and motion, but very unstable and at random; but *History* without *Geography*, like a dead carkasse, hath neither life, nor motion at all' (*C.*, p. 19).

Linking geography to history gave it an inescapably political role in a culture where historical evidence was seen as decisive in the political sphere: 'The early modern mind perceived "history" as a means of access to religious and moral rectitude ... in an age of religious disunity, the past, and the presentation of the past, became a displaced crucible for ideological dispute.'[36] This can be traced through Heylyn's recurrence to the simile of the dead carcass. In his *Full Relation*, Heylyn argued that any political analysis must 'begin with the person of the

King, without whose influence and presence, the Court is but a dead carkasse, void of life and Majesty'. Later in his discussion of the French state, he similarly argued that 'Religion is as the soul of a State, Policy as the body; we can hardly discourse of the one, without a relation to the other ... The Common-wealth without the Church, is but a carkasse, a thing inanimate.'[37] Heylyn refers, of course, to the trope of the state as the body politic, but the repeated use of this simile does highlight a series of connections in the modes of argumentation used by Heylyn and his contemporaries. The body of geographical information must be understood to be a competent historian, and rational views on the bodies political and ecclesiastical are founded on a grasp of history. If this held generally for seventeenth-century political discourse, it applied *a fortiori* for Heylyn: his Laudian disposition made the appeal to primitive practice a decisive one for present policy. To give just one example, Heylyn attacked the Huguenots in the following terms: 'as are their Churches, such is their Discipline, naked of all Antiquity, and almost as modern as the men which imbrace it'.[38] The lack of antiquity or the presence of modernity is enough to discredit an ecclesiastical body *eo ipso*. The term 'modern' is not a neutral one to Heylyn, but carries a pejorative force, distancing a church from its true form which is its primitive form.[39]

In fact, the 'Preface' to *Cosmographie,* which has been cited to imply the division of geography from divinity, argues quite to the contrary. Establishing his persona as an author, Heylyn wrote that: 'as I have taken on my self the parts of an *Historian* and *Geographer*; so have I not forgotten that I am an *English-man*, and which is somewhat more, a *Church-man*' (*C.*, A3r).

Looking across space at the nations of the world, the task Heylyn set himself in *Cosmographie* was to affirm the values of Caroline monarchy and Laudian ecclesiology. As such, his geographical writings, just like his historical writings, were persuasive rhetorical deployments of information to argue a political case.

There is a sense, however, in which Heylyn's tone in *Cosmographie* is indeed quietist in the midst of the more aggressive political argument he develops in other genres. Looking across history and geography, the framing argument of *Cosmographie* is one of Christian humility at odds with the Laudian proselytising zeal of the 1630s:

> though I cannot tell what effect the reading of this following Book may produce in others, yet I can warrantably say thus much of my self, that the observation of the fall of so many puissant Empires,

> the Extirpation of so many mighty and renowned Families, the desolation of so many flourishing *Christian* Churches, as the composing of this Book did present me with ... did more conduce to the full humbling of my soul under the mighty hand of God than either the sense of my misfortune, or any other morall consideration which had come before me (*C.*, A5r)

This comment was made in close proximity to a reference to the 'Tragedies of blood and death which have been lately acted on the Stage of *England*' (*C.*, A4v). The Civil War, when Heylyn took a broader perspective, was but one further example of the rise and fall of empires, and his Christianity suggested this should direct the reader beyond history and geography. Having pointed to the moral of his *Cosmographie* in the 'Preface', Heylyn reverted to it in concluding his massive folio. If the Caroline–Laudian synthesis had been an ideal earthly polity, and it had been destroyed by the vicissitudes of time, the workings of providence demanded a transformative view which moved from the mundane facts of geography and history to a world transcending time and space:

> as much as the most flourishing Country which is here described, doth fall short of the unspeakable glories of that *Paradise* wherein God placed our Father *Adam*, so much and infinitely more did that Earthly *Paradise* fall short of the unspeakable glories of the Kingdom of *Heaven*. To the diligent and carefull search of which *Heavenly* Kingdom I heartily commit the Reader: not doubting but the Works of GOD which are here presented, and that *vicissitude* of Humane affairs which is herein touched at, may prompt him to some serious thoughts of that mighty GOD who made all these Works, in whom is no shew nor *shadow of change*. (*C.*, p. 1105)[40]

While the framing rhetoric of *Cosmographie* is quietist and humble, the argument is sporadically more assertive. In many ways this is an adequate description of the experience of reading the book. The enormous bulk of information tends to anaesthetise the reader to the book's polemical argument. But in the 1650s, there was no other safe way for Heylyn to write. Contemporary biographies of Heylyn describe the persecution he suffered during the Interregnum, and they all dwell on one intriguing incident, when Cromwell's men found the house in which he was staying. The house had a priest hole:

> which Room formerly was supposed to be made for the hiding of Seminary Priests and Jesuits, because the House heretofore belonged to a Papist family ... In which room, instead of a Papist, a right Protestant Doctor [i.e. Heylyn], who was a professed Enemy both to Popery and Puritanism, was now secured from the rage and violence of the Soldiers, who sought after him with no less eagerness, than if he had been a Heretick, followed by the *Spanish* Inquisition.[41]

Heylyn's *Cosmographie* was something of an intellectual priest hole, in which persecuted elements of the author's doctrine could be both expressed amongst and concealed by the bulk of other material. The 'tall big Gentleman' who told Heylyn that geography was better than divinity did not have the same historical mentality as Heylyn. He would thus attack Heylyn's theological, political and historical works with the 'rage and violence' of Cromwell's soldiers, but leave Heylyn's Laudian doctrines secure from that rage in a geography book because he assumed it to be outside the sphere of politics. For Heylyn, by contrast, geography was twinned with, not divorced from, divinity and politics thanks to its role as the vital force enlivening history: subversive monarchical and ecclesiastical doctrines could be hidden from Parliamentarian and Puritan rage in the intellectual priest hole which geography provided for bodies of theological, historical and political argument.

4
John Ogilby and the Iconographic Roads to a Restored Royalist Geography, c.1660–75

Fabulous priest holes: circuitous routes to royalism

Looking back from the vantage point of 1670, John Ogilby argued that it was the Civil War which had made him a writer: 'in the first Fluctuations of the late Grand Rebellion, I being left at leisure from former Imployments belonging to the quiet of Peace wherein I was bred, in stead of Arms, to which in parties most began to buckle, I betook my self to something of *Literature*'.[1] This does not mean that Ogilby's response was that of the quietist, any more than it had been Heylyn's, but that for both literature became the buckler they donned in the Royalist cause. Both Heylyn and Ogilby participated in a more general movement defending monarchy and Laudian Anglicanism in the Interregnum, the 'golden age of High Anglican theology and apologetic'. Further, both were at the 'lower' end of spectrum of literary aspirations in this golden age, contributing primarily to the 'host of popular controversial works, now forgotten, but effective propaganda in their day'.[2] But where Heylyn's contributions came in the form of a massive folio geography, Ogilby's contribution came in Aesopian fables.[3]

Ogilby published the first edition of his *Fables of Æsop* in 1651. A number of the fables, as he rendered them in verse, contained fairly direct political messages with a Royalist edge. Fable 12, 'Of the Frogs desiring a King', was one such example:

> Since good *Frogmorton, Jove,* thou didst translate,
> How have we suffer'd, turn'd into a State!

In several Interests we divided are;
Small hope is left well-grounded Peace t'obtain,
Unless again
Thou hear our Pray'r,
Great King of Kings: and we for Kings declare.[4]

The 'translation' of the first line clearly referred to the notion of Charles I as a King and Martyr, translated to heaven at his execution. This was a link which had already been made in the context of Ogilby's writings by the dedicatory poem prefixed to his 1649 translation of Virgil by Henry Proby: 'Yet all Translations are not Falls, for men/Say they're Translated that are gone to heav'n.'[5] Ogilby as a translator of both Virgil and Aesop was reminding his audience of the more recent translation of an executed king, thereby keeping the memory of the royal line alive in the public sphere as surely as did Heylyn.

The reference to the 'translation' of the king turning the nation into a state was one which was taken up in the most explicitly politicised of the fables, Fable 40, 'The Parliament of Birds'. At the end of the golden age, 'Civil War turn'd Kingdoms into States', and at this point:

The *Birds* reduc'd thus to a Popular State,
Their King and Lords of Prey ejected, sate
A frequent *Parliament* in th'ancient Wood,
There acting daily for the Nation's Good.

The fable goes on to depict a disagreement in the parliament as to whether flax trees should be cut down to prevent men from making ropes to catch the birds. The birds vote to allow the retention of flax trees, the result being that they are destroyed by fowlers. The conclusion was a bleak one for a kingdom such as Interregnum England, which had been reduced to a state:

Thus soon this rising State was overthrown,
And Man e're since did Rule the Earth alone:
When this sad Ditty, silver'd o're with Age,
A Captive *Stare* sang in his woful Cage;
When Civil War hath brought great Nations low,
Destruction comes oft with a Forein Foe.[6]

The foreign foe in question was most probably the Dutch, with whom Cromwell was about to engage in a naval struggle which would continue to dominate English foreign policy into the 1670s. But the message of Ogilby's fable was also designed to bolster the spirits of defenders of monarchy, who could see no signs of the hoped-for restoration in 1651: as the birds had bickered and eventually destroyed themselves, so would the parliamentarians. While the politics of the *Fables* could be read for support by each party in the Civil War, an ambiguity which rendered them a safe priest hole for Ogilby, their 'widest moral' was one which was fundamentally congenial to the dispossessed Royalists: 'the hope of leniency and better times'.[7]

Coronation routes: restored Royalist geography

'The Parliament of Birds' proved to be prophetic rather than fabulous, with Charles II returning to the throne in 1660. At the Restoration, the careers of Heylyn and Ogilby both intertwined and diverged as they emerged from their intellectual priest holes to the light of a re-established monarchy and Anglicanism. Their careers diverged in that Heylyn received no promotion within the church despite his years of trenchant pamphleteering. Heylyn was too closely associated with the Laudian position, while Charles II always sought a more balanced ecclesiastical settlement on his return, as signalled by the tolerant line taken by the Declaration of Breda in April 1660. Thus, on appointing new bishops in 1660, 'very few of the new nominees had held positions of authority in the church when Archbishop Laud had been at the peak of his power. ... The neglect of Peter Heylyn was particularly striking'.[8] By contrast, Ogilby benefited from his conspicuous literary loyalism after the Restoration: 'from the time of his appointment as Master of the King's Revels in Ireland [in May 1661] until he became Royal Cartographer (later Royal Cosmographer), a title conferred near the end of his life, Ogilby was a steady petitioner for and recipient of royal favour'.[9]

Ogilby's ascendant career intertwined with Heylyn's quiescent career in the coronation of Charles II in 1661. Heylyn's part was enacted on the 23rd April in Westminster Abbey, where he had been a prebend from 1631. As part of the coronation ceremony, it fell to Heylyn to carry the sceptre with cross to the altar.[10] By this time, Ogilby's role had already been successfully completed, for he had designed the iconography of the triumphal arches through which Charles had processed the day before. This procession saw a significant political use of geographical themes in the public sphere, as the monarchist and loyalist credentials of geography could at last be openly displayed, freed from constraints of textual disguise which had proved necessary in the Interregnum.

Coronations and processions served political purposes, in that they 'reinforce[d] specific socio-legal norms – the domination of the City by the Crown, of the polity by neofeudal principles, and of the social order by aristocratic blood'.[11] But this is not something which modern critical analysis has had to tease out of the available material. On the contrary, processions were explicitly understood by contemporaries to have political meaning, none more so than that which Ogilby helped to design in 1661. As Edward Hyde, Earl of Clarendon, reported, the coronation had been designed:

> That the Novelties and new Inventions with which the Kingdom had been so much intoxicated for so many Years together, might be discountenanced and discredited in the Eyes of the People, for the Folly and Want of State thereof; his Majesty had directed the Records and old Formularies should be examined, and thereupon all Things should be prepared, and all Forms accustomed be used, that might add Lustre and Splendour to the Solemnity.[12]

If this was the message of the coronation ceremony, the procession could enforce the same themes by similar methods to a wider audience on the streets of the metropolis. As Pepys recorded, 'the way is so thronged with people to see the Triumphall Arches that I could hardly pass for them'.[13] As with the coronation, the procession was said to be enacted 'according to the Ancient Custome',[14] but in fact it encoded rather more contemporary political messages, notably through the deployment of geographical imagery, to which I now turn.[15]

Long before Ogilby was appointed His Majesty's Cosmographer in 1674, he had been involved in interconnecting monarchy and geography, starting with the Coronation Procession in 1661. Ogilby was in charge of the iconography of the triumphal arches, or, as the King's warrant put it, of 'the Poetical part thereof, consisting in Speeches, Emblemes, Mottoes, and Inscriptions'.[16] The arches of the procession created an entire geographical schema at three scales – the international arena, the multiple kingdoms of Great Britain, and the metrocentric interrelation of London and England – which together acted to envisage Charles II as an absolute monarch originating policy from the geographical heart formed by London, and thereby determining the operation of the national body politic and its extended imperial manifestations.

Looking first at the widest scale, the international, the second arch in Ogilby's procession treated Britain in the context of European rivalry with the Dutch for dominance of the seas, commerce and empire. For just as the Dutch were the foreign foes of Ogilby's Æsopian fables in the 1650s, so they remained the chief threat at the Restoration. The second arch was erected in Leaden-Hall Street, outside the offices of the East India Company, which was battling for access to the lucrative markets of Asia with the Dutch East India Company, which had claimed a monopoly.[17] This was reflected in the entertainment which received Charles II, which included a poem of geographical and imperial import:

> For You [i.e. Charles] have out-done *Solomon*, and made
> Provision for a more than *Ophir* Trade;
> Among Your first of unexpected Cares
> Enlarg'd our Charter, and dispel'd our Fears
> Of the incroaching *Holland's* Rival Force.
> Nor can we doubt, but by the bounteous Source
> Of Your Successful Right, not only We,
> But all the Merchants of Your Realm shall see
> This *Empory* the *Magazine* of All
> That's Rich, from *Phoebus* Rising to his Fall;
> And Your *Imperial Title* be the same
> In Deed, which *Spain's* proud Crown vaunts but in Name.
> (*Relation*, p. 10)

The entertainment at this arch was not designed by Ogilby, as he made plain (see *Relation*, p. 9, note), but it did fit into the geographical schema of the whole procession, which Ogilby masterminded. Furthermore, Ogilby's own entertainment at the third arch which the monarch encountered included a song by three 'sailors' which conveyed the same message of imperial triumphalism, albeit couched in a more hearty tone: 'King *CHARLES*, King *CHARLES*, great *Neptune* of the Main! / Thy Royal Navy rig, / And We'll not care a Fig / For *France*, for *France*, the *Netherlands*, nor *Spain*' (*Entertainment*, p. 92). The anti-Dutch theme was one which Ogilby also carried into his later compilations of travel accounts, commenting in *America* (1671) on 'how dangerous Intruders the *Dutchmen* are generally upon other Princes [*sic*] Dominions'.[18]

It was not only the spoken entertainments at the arches which betrayed an anti-Dutch rhetoric and an aggressive espousal of a British empire gained by domination of the seas. The second arch itself had an inscription (presumably designed by Ogilby) pertaining to this theme: 'Neptuno Britannico, Carolo II, Cujus Arbitrio Mare Vel Librum, Vel Clausum.' (*Entertainment*, p. 51; see Figure 4.1). This inscription was strongly anti-Dutch, as was suggested by Ogilby in the accompanying text, which justified the claim that Charles II was the arbiter of whether the seas would be open or closed as follows:

> And sure, if any Nation may plead Prescription for the Title, the *King* of ENGLAND may, having had a longer uninterrupted Succession in the Dominion of the BRITTISH [*sic*] *Seas*, than the ROMANS in the *Mediterranean*, or any other Nation, that *History* ha's [*sic*] acquainted us with. The Antiquity whereof being purposely, and at large declared by Mr SELDEN. (*Entertainment*, p. 52)

The arch and its accompanying explanation refer to John Selden's treatise *Mare Clausum* (published 1635), which had responded to the argument of the Dutch jurist Hugo Grotius in *Mare Liberum* that the North Sea was a zone of free access to all nations, which could not be owned by any one country. This debate was of direct relevance to the battle between the Dutch and English East India Companies, with the English demanding free access to the Indies and the Dutch declaring the area a closed sea, such that both parties were reversing their positions from those held about the North Sea.[19] The second arch, outside the East India Company offices, then, was declaring that the naval might of the restored Charles II would cut through these intellectual skirmishes, leaving England to decide whether the seas were open or closed.

It has recently been argued that the Anglo-Dutch wars were about far more than an economic rivalry, ideological disagreements being pivotal to the language of debate and to the pretexts for conflict.[20] The political languages evoked by the iconography and entertainments Ogilby designed clearly support such a conclusion. Pincus argues that the English feared that the Dutch sought, through their maritime predominance, to gain universal monarchy, thus replacing the Spanish in that role. This fear comes through in the coronation procession: the poem spoken outside the East India offices clearly envisages a contest for universal monarchy, not simply a commercial rivalry, going further than Pincus's analysis in its overt desire that England and Charles II take over the mantle of universal power from the Spanish rather than simply thwarting the Dutch.[21] Again, the description of the second

Figure 4.1 The Second Arch in John Ogilby's Coronation Procession for Charles II, from Ogilby's *Entertainment* (1662)

arch revolves around the same themes, asserting that it is England which by right of precedence and antiquity has dominion over the seas. The legitimating language for imperial aspiration is a quasi-legal one from Seldon, English precedency creating a property right which rendered Dutch claims a violation of common law.[22]

In the political language of the time, the Spanish and the Dutch were modelled as the same type of enemy, who 'Popishly' sought to usurp legitimate English power over the seas which had been established by common law precedent (itself often figured as the legal equivalent of Protestantism): 'These people do arrogate to themselves St Peter's power on the Seas ... it is mare liberum in the British Seas, but mare clausum on the coast of Africa and in the East Indies.'[23] Of course, by claiming that the English had such absolute power over the seas, Ogilby's procession placed Charles II in the role of the absolute ('Popish') monarch, something we will see again at the other geographical scales on which the procession operated. The absolutist tinge to Ogilby's procession would have been clear to any of its spectators who recalled the arch at Temple Bar designed for Charles I by Inigo Jones (1636–38). Jones had connected maritime power and the monarch by the use of a triumphal arch, this time defending the levying of ship money, which was seen by many as marking Charles I's descent into tyranny.[24]

At the scale of world geography, then, Ogilby's procession made strong claims for the absolute power of the monarch, advocating that Charles's Britain succeed to Spain as the universal monarchy and oust the upstart Dutch. While this was akin to the Popish absolutism of the Spanish and the Dutch, it was also figured in the political language of the procession as different because it was legally sanctioned by the antiquity of the British control of the seas. Just as the return of the monarch was figured in the legal fiction of an unbroken occupancy of the throne which ignored the Interregnum, so British control of the seas was an uninterrupted succession (*Entertainment*, p. 52).

The second geographical scale at which the processional arches operated was that of the multiple kingdoms which comprised Great Britain. While it has been argued that most people did not think in terms of multiple kingdoms in the era of the Interregnum, however important modern historians view that context as being to the nature of the Civil War,[25] it is clear that Ogilby, as a Scot, saw the loyalty of Ireland and Scotland as vital to the unity of Britain at the Restoration. This theme was explored at the first and third arches in the king's coronation procession. The description of the iconography of the first arch makes this most apparent: 'On the *South* Pedestal is a Representation of BRITAIN'S MONARCHY, supported by LOYALTY, both Women; *Monarchy*, in a large Purple Robe, adorn'd with Diadems, and Scepters, over which a loose Mantle, edg'd with blue and silver Fringe, resembling Water, the Map of *Great Britain* drawn on it, on her Head *London*, in her right Hand *Edinburgh*; in her left *Dublin*' (*Relation*, p. 2).

The third arch reinforced the message with the mythological figure of the Geryon from Hesiod, a three-bodied creature '*in his three left-Hands the three Escutcheons of* England, Scotland, *and* Ireland: *before him the King's Arms with three Imperial Crowns; beneath, in great Letters*, CONCORDIA INSUPERABILIS' (*Entertainment*, p. 114). Further, at the third arch a song echoed the theme of the arch in referring to Charles II – 'three Nations glory' – as the man 'Who fixt with gentle Smiles / Two fluctuating Isles' (*Relation*, p. 27). The mere presence of the king was seen, then, as overcoming the problems of uniting three kingdoms into one nation state, the solution being grounded in loyalty to the anointed monarch. As with the international scale, the absolute power of the monarch was a legal right which should be willingly acceded to.

The third and final scale of geographical images in Ogilby's procession was that intertwining the nation and the region, the iconography of which was encoded in the second arch, and which enforced a metrocentric message about the pivotal status of London. Of course, the mere fact of having the coronation procession in London made it symbolically 'a microcosm in which the joyous reception of the monarch by the City represented the ideal relationship for the nation'.[26] The omphalatic status of London, drawing together the nation in loyalty to its restored monarch, was made clear by symbolism on a stage at the second arch: 'a person representing the River *Thames*; his Garment Loose and Flowing ... his Head Crown'd with *London*-Bridg ... Four Attendants in White, representing the four fresh Streams, which fall into the River *Thames*, viz *Charwell*, *Lea*, *Colne*, and *Medway*' (*Relation*, p. 11).

All England, then, was tied to London according to this fluvial symbolism. London was a crowned figure, greeting the return of a crowned monarch. The centre which was London symbolised Charles II, and both drew in the people to pay tribute as naturally as the rivers of southern England joined as tributaries to the Thames.

The geographical iconography at three scales analysed thus far was also bolstered by the symbols of geography *sensu stricto* at the second arch (see Figure 4.1, above). Over the cornice was the figure of Atlas, bearing the terrestrial globe. In Ogilby's description, Atlas was thus represented by the ancients 'from his admirable knowledge in the motions of the Heavens, and the nature of things here below' (*Entertainment*, p. 92), these earthly things 'here below' presumably being geographical, given the relationship between geography and astronomy outlined in chapter 2. The arch also had four niches, containing allegorical figures representing arithmetic, geometry, astronomy and navigation, all staples of mathematical geography. The figure of Geometry made the geographical theme particularly apparent, holding a compass, her niche being inscribed: '*Descripsit Radio Totum Quae Gentibus Orbem*' (*Relation*, p. 13). Finally, the second arch had personifications of the four continents, which stood for the contents of descriptive geography. Given that the second arch contained the symbolism of Britain's international dominance and its internal coherence, it is unsurprising that the traditional geographical personifications were invoked at this point. The suggestion was clearly that Charles II and Britain grounded their unity and their expansionism on a solid grasp of geography. Geography was loyal to Charles II and would be a foundation stone for his reign at all spatial scales.

Exactly what was 'restored' in the period 1660–63 has been the subject of much controversy among historians. Was the Laudian church system reinstalled or jettisoned in favour of a more moderate system?[27] To what extent did Charles II regain the absolute power Charles I had wielded in the personal reign?[28] As Seaward has suggested, the deepest instincts of parliament and the nation were conservative, fearing the expansion or the diminution of royal power. As such, the aim was to defend the legally-established ancient constitution, with the king at its head.[29] The geographical symbolism of Ogilby's 1661 coronation procession recognised this, creating a tiered or hierarchical absolutism grounded in loyalty rather than power. Reversing my order of presentation, the king was envisaged by the procession as based in London; London drew England together as both a microcosm and a commercial and political heart; England was further united with Scotland and Ireland by loyalty to the legitimate monarch; and, finally, Britain would take on all-comers to defend and enlarge its international standing, which was itself legally justified by its prior claim to dominion over the seas. At whichever spatial scale, then, the king was at the heart of affairs, his power deriving from both the affective ties of loyalty and the prescriptive ties of legality.

Britannia: 'Regular and absolute' routes

John Ogilby's other great geographical project, *Britannia* (1675),[30] was the first road map for which accurate measurements were made. As

Figure 4.2 Wenceslaus Hollar's Frontispiece to John Ogilby's *Britannia* (1675)

with his coronation procession, so the iconography and construction of *Britannia* fitted into debates over the politics of absolutism which bulked large in later Stuart political life. The difference was that *Britannia* appeared long after the euphoria over the Restoration had been replaced by a fear of monarchical power. By about 1663, the acclamation of the king had been replaced by hostility, although it should be emphasised that this was only a hostility towards the extension of the king's powers, not towards the king *per se*.[31] The major fear, and a fear which grew in the 1670s, was that the king was trying to fashion an absolutism of the variety which Louis XIV enjoyed in France. Whether there is any compelling evidence to substantiate this fear is disputed,[32] but it was without doubt a key component in the political language of the 1670s, and one with which Ogilby engaged in *Britannia*. Another important shift was in the nature of patriotism in the political language of the period. According to Pincus, with the close of the Anglo-Dutch wars in 1668, national identity increasingly turned in on itself, coming to mean 'defence of a peculiarly English culture not commitment to a particular religious position or a purely monarchical form of government'.[33] As we shall see, Ogilby's *Britannia* reflects this shift in the scale of patriotic identity towards the national and the internal, but not the shift in the sources of that identity away from the monarchical.

Ogilby's *Britannia* was explicitly written in the context of peace with the Dutch after twenty years of intermittent conflict. Peace was said by Ogilby to '*have Laid Open to Us all those* Maritin [*sic*] Itineraries' for international commerce.[34] Symmetrical with this, and in accord with a patriotism increasingly insular in its geographical scope, Ogilby designed *Britannia* to '*Improve Our* Commerce *and* Correspondency *at Home*' (*Britannia*, Ar). In other words, *Britannia* focused on the smallest of the three scales with which the triumphal arches of the coronation procession had been concerned. Yet just as the Anglo-Dutch wars had created a patriotic identity which was simultaneously commercial and political, so *Britannia* retained such a twofold patriotism. Commercially, *Britannia* was of 'infinite Advantage ... in Reviving and Propagating the great Soul of the World, *Commerce* and *Correspondency*' (*Britannia*, Bv). Politically, the book, dedicated to Charles II, gave '*a True* Prospect *of This Your Flourishing* Kingdom' from which '*may be Taken, Pregnant Hints of* Security' (*Britannia*, Av). These two species of patriotic purpose to which *Britannia* contributed were also captured in the frontispiece by Wenceslaus Hollar (see Figure 4.2, opposite), which pictured a commercial landscape of agriculture, shipping, roads and coaches, all being centred politically on a palace, flying the royal standard, wherein courtiers surveyed the whole prospect.

Above the prospect in Hollar's frontispiece hover three cherubs, the central one of whom holds a map of London. This captures another important aspect of the political geography of *Britannia*: its focus was not only national, but metrocentric, drawing a British identity together around London, just as the personification of the River Thames did in the coronation procession. In the interim between the Restoration and the publication of *Britannia*, the symbolic significance of the capital had increased thanks to the Great Fire of 1666 and the subsequent rebuilding. Ogilby was quick to realise this, figuring London as the heart of an empire (as Dryden did contemporaneously in *Annus Mirabilis*): 'no Work appears more Perspicuous than that Stupendous Miracle! the Raising from a Confused Heap of Ruines ... Your Imperial City ... hereafter to be the Business of Foreign Nations to See and Wonder at'.[35]

Even more than in 1661, London was a 'type' of the monarchy, both being restored after disaster. This was something to which Ogilby recurred at a number of points in *Britannia* and in his *London Survey'd* (1677), a text which glossed a map of London he had constructed with William Morgan. Addressing the king, London was described as '*this Great* Emporium *and* Prime Center *of the Kingdom, Your Royal* Metropolis' (*Britannia*, Av). The capital had seen 'a continual Succession' of bishops for ten centuries, a legal fiction equivalent to the uninterrupted succession of monarchs on which the Restoration had been predicated.[36] But London was not simply symbolically royalist; the capital was also a loyal garrison. London's militia was said to be 'a Force sufficient to Oppose any Enemy either Forreign or Domestick'.[37]

Given that London was symbolically and actually pictured by Ogilby as the centre of royalism, the fact that *Britannia*'s entire construction was centred on London was not politically neutral; it was a gesture of loyalty to Charles II as assuredly as the dedication of the book to 'his most serene and sacred Majesty'. All the measurements for *Britannia*'s itineraries started from Corn Hill in London, which had also been the site of the second (and most geographical) arch in the coronation procession. All the roads measured were defined by Ogilby according to their relationship to London:

> We have distinguisht ... into *Direct Roads* and *Cross Roads*, calling such *Direct* as proceed Directly from this Our *Prime Center* and *Grand*

> *Metropolis*, LONDON, to the less *Central Cities, Capital Towns*, or other Eminent Places of the *Kingdom*; and calling such *Cross* as lead from some of the said *Lesser Centers* to another like *Capital Town*, or Place of Eminency. (*Britannia*, Bv)

Having been measured thus, the itineraries were presented accordingly: *Britannia* starts with 'independent' routes directly from London; then moves on to 'dependent' roads, being those which branch from the direct London roads; before finally dealing with the 'cross' routes unconnected to the capital. As such, the whole construction of the route maps and their presentation radiated out from London, which was thereby symbolically and geographically constructed as the centre of the kingdom. London was a symbolic type of the king and in actuality his loyal garrison, and this was matched by the construction of *Britannia* as a text dedicated to the king and centred on London.

Looking beyond the construction and format of *Britannia* to its content, we find that loyalty to the monarch was evidenced in both the route maps and the surrounding text. The route maps had royal crests, as can be seen from the first road depicted, that from London to Aberystwyth, the cartouche to which had the royal crest and motto (see Figure 4.3). The same motto was contained in an emblem at the head of the sparse accompanying text (*Britannia*, p. 1 – see Figure 4.4). The text was mainly a listing of the distances between places, but it did elaborate some form of loyalism, in that it always mentioned towns which had been loyal to the monarchist cause in the Civil War in glowing terms. Thus Bacon Street in Lichfield was mentioned as 'the chief Street in the Town till burnt in the late Troubles, in which this City appear'd so Eminently Loyal' (*Britannia*, p. 44). Similarly, Colchester was said to have 'suffer'd greatly under the late Pressures of prevailing Disloyalty' (*Britannia*, p. 38). On the other side, displays of Parliamentarian bravery went unrecorded, Ogilby preferring instead to attack the 'Fanatick Zeal' of the usurpers (*Britannia*, Ev), and taking apparent pleasure in recording an historical example of the quelling of disloyalty in his account of Norwich: 'suffer'd much ... by that Arch Rebel *Ket* ... but [Kett] being Subdu'd, was at last Hang'd on his own Oak of Reformation' (*Britannia*, p. 152). Here *Britannia*'s stance came together with that in Ogilby's *Fables*, with which we started. Ogilby continued to expand his fables in the 1660s and 1670s, including more direct attacks on the Parliamentarian cause than he had dared to in the

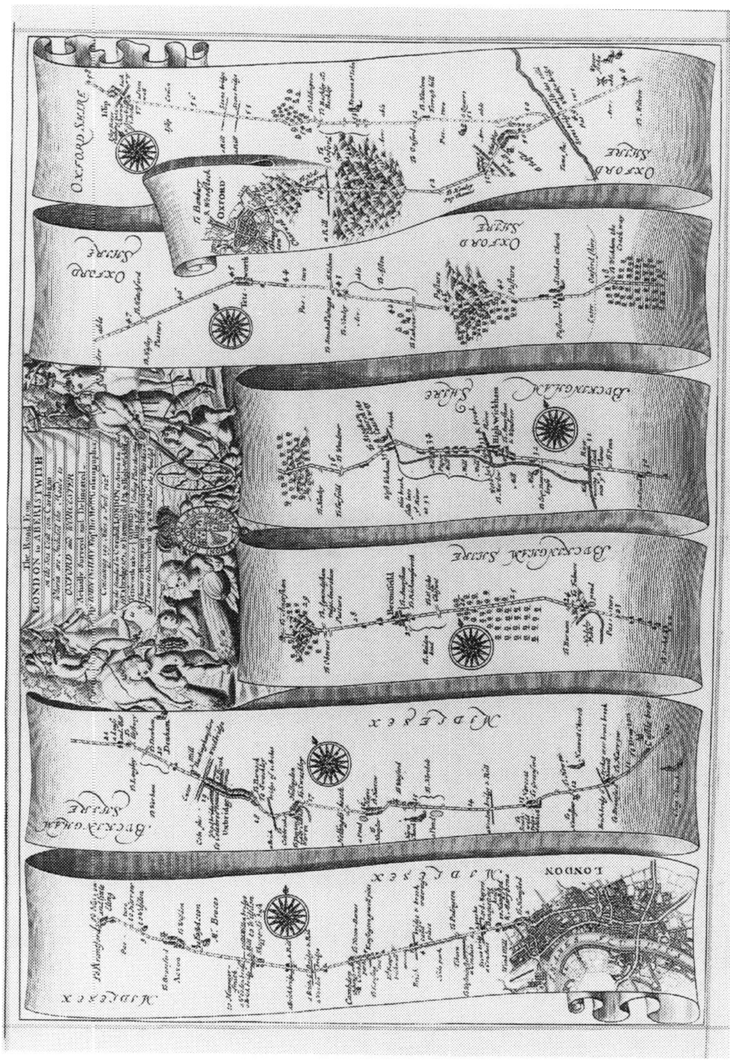

Figure 4.3 London to Aberystwyth route map, part 1, from John Ogilby's *Britannia* (1675)

Figure 4.4 Emblem at the Head of the text to John Ogilby's *Britannia* (1675)

Interregnum, lambasting the hypocrisy of those who switched to support Charles II in his tale of the 'Antimonarchical Republick Stork'; attacking the Fifth Monarchists who had caused such a scare to the Royalists in Venner's Revolt of 1661 in his fable 'Of the Ant and the Grasshopper'; and lamenting the 'bad work' of the 'Covenanting Kirk' in 'Of the Satyr and the Sword.[38]

The mensuration, structure, maps and text of *Britannia*, then, conveyed Ogilby's continued royalism. As with the coronation procession fourteen years previously, Ogilby centred loyalism on London, the intervening years having only strengthened the symbolic intertwining of the restored Charles II and the reconstructed metropolis in his geographical iconography. Centred on London, *Britannia* was designed to give a 'prospect' of the nation which allowed for efficient governance, the route lines radiating as lines of information into London and Charles, just as the lines of the garden at Versailles centred on the Sun King. As Ogilby himself put it:

> Now, the Methods made use of by *Geographers* in the *Description* and *Illustration* of *Kingdoms* and *Countries* are various, Some following the Natural Traduction of *Rivers* and *Mountains*, Others the *Distinction* of *People* and *Inhabitants*, Others again more frequently, the *Politique Division* of *Princes*; but *Antiquity* and the Practical Succession of *Geography* has more especially commended to Us the *Itinerary Way* as the most Regular and Absolute. (*Britannia*, Br)

The reference to the itineraries of antiquity linked *Britannia* as a project with the routes mapped by the Greek and Roman emperors, Ogilby himself citing Alexander the Great, Julius Caesar and the Antonine Itinerary of Britain (*Britannia*, Av). As with the coronation procession, classical precedent was invoked to model Charles II as an imperial ruler, with *imperium* over both an expansive overseas empire and over the 'empire' of Britain, the latter being based on accurate geographical knowledge and the ability to deploy the forces garrisoned in London wherever they were needed which Ogilby's route maps offered. Describing the itinerary as 'absolute', Ogilby meant that it was the most mathematically precise method of geographical description, but in his hands the 'absolutism' of the itinerary was clearly both mathematical and political.

Britannia down the road: the politics of eighteenth-century road maps

In terms of accuracy, *Britannia* was a pioneering project. For this reason, the influence of Ogilby's road maps was long lasting: indeed, 'not until the early nineteenth century were their comprehensiveness and innovations wholly superseded'.[39] While this has been recognised for road maps as a genre, it is also the case that the political meanings Ogilby incorporated into *Britannia* resonated down to its eighteenth-century derivatives in the form of an assumed loyalism to church, state and monarchy.

The most important successor to *Britannia* in the early eighteenth century, Emanuel Bowen's *Britannia Depicta: or Ogilby Improv'd*, not only took on Ogilby's royalism, but in one way amplified it.[40] The route maps themselves were taken from *Britannia*, but they were adorned with a whole series of crests which Ogilby's originals had not included. If we look at the first sheet of the route from London to Aberystwyth (see Figure 4.5; compare with Figure 4.3), Bowen's version had crests for the boroughs of Uxbridge, Southwark, London, Westminster and High Wycombe, and the arms of the Bishopric of London and the Deaneries of Westminster and St Paul's. The effect of this was to surround the route maps with the symbols of church and state. The crests of the boroughs emphasised local governmental processes in a way Ogilby had not, his work being published at a time when Charles II was contemplating attacking borough statutes to draw the localities into line with his aims. This apart, the iconographic effect of Bowen's work was akin to Ogilby's, creating a pervasive reminder of church and state. That this was even more insistent in Bowen than in Ogilby suggests that road maps underwent a development which was the reverse of that traced by Helgerson.[41] Where Helgerson shows the gradual effacement of symbols of royalism from county maps, national identity becoming connected to the land rather than an individual, road maps show the proliferation of symbols of authority, which obscure the landscape on the maps, and make the human construction of authority and identity ever more insistent.

The road maps of the eighteenth century were predominantly small, cheap, commercial ventures, designed to be of use to tradesmen. Yet they still retained the hierarchical and humanist elements which Ogilby had incorporated into *Britannia*, a folio produced under royal

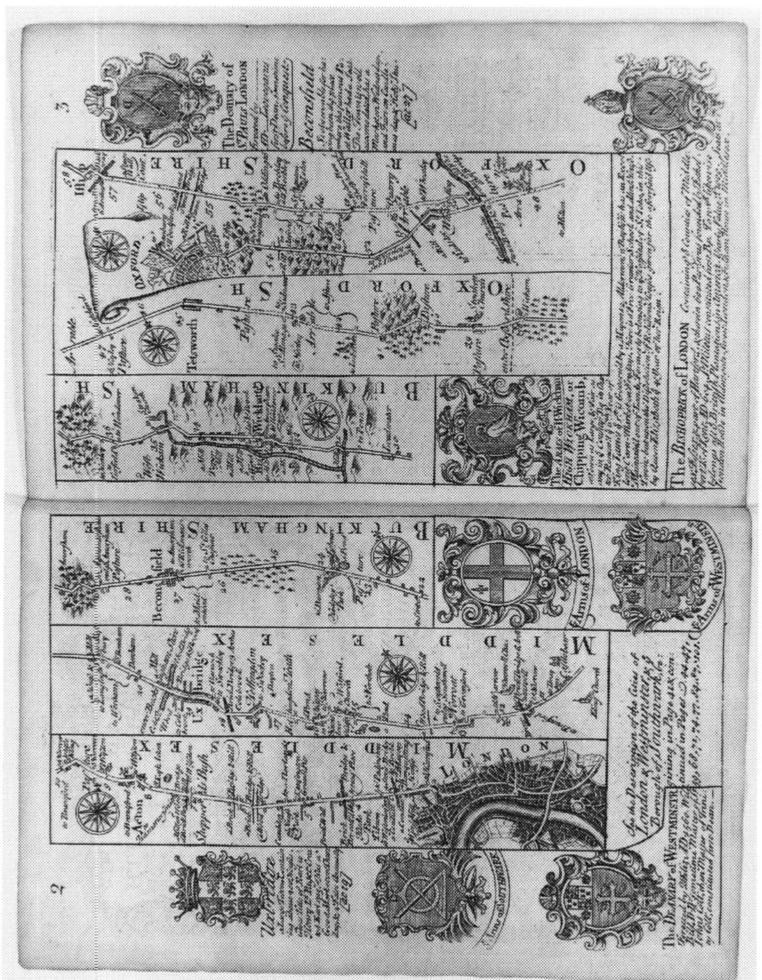

Figure 4.5 London to Aberystwyth route map, part 1, from Emanuel Bowen's *Britannia Depicta* (1720)

patronage. Thus, amidst details on postal services and prices, Dodsley's *New and Accurate Description* (1756) also found space for a section on Roman roads, almost certainly taken from John Horsley's *Britannia Romana*, which continued the link between modern road maps and the Antonine Itinerary which Ogilby had established. The legitimacy of the form of geographical knowledge embodied in road maps, then, continued to be strongly dependent on classical precedent.[42] Furthermore, even road books with the scantiest textual information, such as John Potter's *Traveller's Pocket Book* (1780), continued to reinforce an aristocratic and royalist perspective by their choice of material such as the following: 'Near Basingstoke is Basing, where formerly was a fine Seat belonging to the Family of Powlett, now Duke of Bolton. It was a Garrison for his Majesty in the Civil Wars, and held out a Siege two Years but was at last taken by Storm and burnt.'[43] Even John Cary, whose work is seen by Fordham as finally superseding Ogilby's, still included an 'Index to the Country Seats'.[44]

It is clear that a diffused Royalism permeated eighteenth-century road books thanks to their derivation from Ogilby's *Britannia*. As with Ogilby, in these books commercial utility and hierarchical loyalty to the powers by law established coexisted iconographically and textually. In one key way, however, all the eighteenth century books of road maps differed in their royalism from that in *Britannia*: they were loyal to an altered monarchical succession, not to the Stuarts. Ogilby, as part of his loyalty to Charles II was also loyal to his brother, James, Duke of York, the future James II. Indeed, *Britannia* said that the city of York was 'Honor'd by giving Title to several Princes of the Blood, Particularly the most Illustrious Duke of York Sole Brother to his Present Majesty' (*Britannia*, p. 14). This theme was also taken up in the frontispiece and the engraving on page 1 (see Figures 4.2 and 4.4, above): in both cases a flag was held (by a cherub and a mermaid respectively) which centred on York as the second city in ecclesiastical terms and as giving the immediate successor to the throne his title. Ogilby did not live to see the turmoil of James II's reign, but the problems he might have faced in developing a Stuart royalist geography of the sort he had crafted under Charles II become apparent in the work of Edmund Bohun, to whom I now turn.

5
The Political and Geographical Appropriations of Edmund Bohun, 1684–1710

Bohun and Heylyn

In May 1686, Edmund Bohun noted in his diary that he had agreed to 'draw a smal historical, geographical dictionary' and 'to fill up and continue the historical passages of Heylin's *Cosmographie*'. His motive was not intellectually exalted: 'the reason why I took up this was, because I found my estate would hardly support me and my family'.[1] Whether Bohun approached the bookseller (Charles Brome) or vice versa is not clear, but in either case it will become apparent that Bohun was a politically appropriate individual to update Heylyn's project of a geography in defence of the established church–state nexus.

The political context Bohun faced was very different from Heylyn's, given that the Restoration of the Stuart monarchy and the accession of the Catholic James II had intervened between Heylyn's edition of *Cosmographie* and its proposed revision. Yet Bohun faced persecution as Heylyn had before him. Indeed, the reason that Bohun needed to scrape together money from various pieces of hackwork for the London press was his uncompromising defence of the Anglican church in the face of a Catholic monarch whose policy with respect to English religious life seemed to many more extreme than that of the Puritans. He wrote of his ostracisation at this time that:

> in this year [1687] the struggles grew very great between the popish party and those of the church of England; and I being engaged in it to a publick disputation with one of the priests belonging to Whitehall, I treated his reverence with so little respect that I was, for

it, turned out of the commission of the peace for the county of Suffolk; and continued so till the abdication of king James the second.[2]

Bohun's persecution was obviously less life-threatening than Heylyn's, but the parallel I have drawn between the situations the two men faced is not wholly fanciful, since Bohun at several points linked himself and Heylyn as belonging to the same political tradition. For example, in defending the absolute divine right of monarchs, Bohun reprinted Sir Robert Filmer's *Patriarcha*, and added in an appendix a letter from Heylyn praising Filmer.[3] Moreover, Bohun's correspondence shows him seeking further information about Heylyn from his surviving relatives.[4]

Bohun's Anglican Royalism, 1684–88

While Bohun might choose to frame his contributions as in the tradition of Heylyn, what did this mean in the changed political situation he faced 25 years after Heylyn's death, and how was this manifested in his geographical writings? Bohun's position has been described as an Anglican Royalism, and the nature of this position must be sketched before the politics of his geographical work can be understood.[5] Bohun defended the absolute power of the sovereign against the notion of an ancient constitution, as Heylyn had in his comments on William the Conqueror, and drew on the same juridical tradition to do so:

> Go to *Spelmans* Collection of Ancient Saxon Laws, and see if you can find one, that was made by the People without the Prince. So that this is a very silly Proposition, and the allegiance which we owe our Soveraign, and our Ancestors ow'd to his, is the cause of our submission to the Laws, and not the Laws, the cause of our submission to him, and them.[6]

Just as with Heylyn we saw what to the modern mind seems the paradoxical defence of a balanced absolutism, so in Bohun we find that the absolute power of the monarch creates rather than destroys liberty for subjects: 'the greatest Liberty in the World (if it be duly considered) is for a People to live under a Monarch, it is the *Magna Charta* of this Kingdom, all other shews or pretexts of Liberty, are but several degrees of Slavery, and a Liberty only to destroy Liberty'.[7]

Twinned with this belief in an absolute monarch is the argument that usurpers must be resisted, and particularly that 'this was the Doctrine of all Christians, till *Pope Zachary* revived the old Heathen Doctrines'.[8] It is here we start to see why Bohun, despite professing unshakeable loyalty to the crown, was ostracised by James II. For Bohun intimates that it is Catholics and Presbyterians, not Anglicans, who countenance resistance to a monarch, because they have departed from the example of the primitive church. Thus, while Bohun defended James's right to succeed in the Exclusion Crisis, he argued in his *Diary* that both this crisis and the Civil War showed the impossibility of loyal behaviour outside the Anglican communion:

> we may see where our puritans sharpened the shears with which, at pleasure, they sever the bonds of all our oaths ... But they are the sworn enemies of the papists ... It is not of their dogmas: of these they are the copyists and most strenuous defenders. For both have derived from hence the arguments by which they promoted the late rebellion, from hence the license of lying, the devices of disloyalty and its justification, contempt and insolence towards kings.[9]

Such views can hardly have been popular with a Catholic king, and they were expressed just as stridently in Bohun's published writings, notably his *Apologie for the Church of England* (1685).[10]

In the period 1684–88, then, Bohun's political tracts consistently labelled him as an Anglican, and argued that the stability of the church and state was only safe in such hands. His publishing also consistently linked him to Filmer and Heylyn, and thus to the absolutist tradition of Anglican Royalists. Bohun sought to drag the political arguments of the Interregnum (which themselves looked back to the Laudian and Caroline church–state nexus, and even to Jacobean England in Filmer's case) into the debates about the same issues of monarchical power which raged from the Exclusion Crisis until James II's flight in 1688. Regardless of the intervening years, by editing Filmer and adding a laudatory letter from Heylyn, Bohun clearly tried to position himself as the third figure in a triumvirate. In this context, it can be suggested that Bohun's agreement to update Heylyn's *Cosmographie* was part of the creation of the same political persona. The reprinting of Heylyn's *Historical and Miscellaneous Tracts* in 1681 had not been a coincidence, but a contribution to the debate about James II's succesion; likewise, Bohun's publication of *Patriarcha* in 1685 was designed to contribute to the debate about which parties demonstrated loyalty to James II when installed. In this context, there is no reason to think that when Bohun agreed to reissue a work by Heylyn in the 1680s, he did this without the aim of asserting the message of absolute royal power and the superiority of the Anglican church which Heylyn had written into *Cosmographie*, while at the same time reinforcing his textual affinity with that position.

It was not the updated *Cosmographie*, but Bohun's *Geographical Dictionary* which was the first of his geographical works to be published in 1688. Its texture shows elements of the Anglican Royalism Bohun had defended in the years leading up to its publication. In the 'Preface' Bohun both distances the *Geographical Dictionary* from politics and at the same time signals that such a separation cannot be rigidly maintained:

> I have as much as was possible endeavoured to please all the World, and declined saying any thing that might give any persuasion of Men an offence, my Observations being purely Geographical, and Chronological, and if any Moral reflection has in any place broke loose, I hope it will appear as Innocent, and Inoffensive, as Necessary and Loyal.[11]

The most obvious political position the *Geographical Dictionary* adopts is of overt loyalty to James II. Thus in the article on 'Lo stato di Modena' Bohun makes one of the deviations from the geographical and the chronological to the loyal that he had signalled in the preface, praising James II's queen, Mary of Modena: 'the Noblest Branch it [the Royal family of Modena] ever produced, is *Mary* the present Queen of *England*, whose Virtues are too great, and too many to be here inserted' (*GD1, sub.* Modena). Perhaps the most extreme statement of loyalty, however, came in the discussion of Bridgewater:

> on *Sunday, July* 5. 1685 the late Duke of *Monmouth*, Natural Sone to *Charles* II. of ever blessed Memory, was intirely defeated, being then in Rebellion against His present Majesty *James* II upon a *Moor* near this place, 2000 of his Army being kill'd, though he had above double the number of His Majesties Forces, and the advantage of a surprize by Night, both of which were over-rul'd by the Providence of God. (*GD1, sub.* Bridge-Water)

True to his principles, Bohun refuses to countenance the Protestant Monmouth, as this would entail disloyalty to the divinely-appointed monarch.

This loyalty is undermined to a certain extent, however, as can be seen in what might be expected to be the acme of loyalism, the article on England:

> It has also the best Government, and the best Religion, of any Nation in the World, and as much Learning, Civility, Arts, and Trade, as any other. Our Fleets excel at Sea, and our Foot at Land, those of all other Nations. In short, we want nothing to make us happy, but gratitude to God, and Union amongst our selves. (*GD1*, *sub.*, England)

The praise of the nation has some disjunction from praise of the monarch, because the 'best religion' is that by law established, the Anglican communion of which James II was not a member. Seen in this light, it is hard to see how union amongst the English can be achieved without some alteration either in the established religion or in James's attitude towards it. This view is reinforced when we consider that Bohun in this period attacked both Roman Catholicism and Presbyterianism as divisive forces which could resist the monarch and destroy English unity and happiness. The point was made in the continuation of the article on England:

> the late Rebellion in 1640 ended in 1660 by the Restitution of *Charles* the Merciful and Just. So that the Common-wealth the Factious would have erected here in the State, and the Anarchy they settled in the Church, were of no great duration, though the Miseries that brought them in, the Calamities that attended them, and the Judgments that have followed them; may be eternal Monitors to *English* Men to be Loyal to the King, and stedfast to the Church. (*GD1*, *sub.*, England)

Once again, the tension in Bohun's position is apparent: what should an Anglican do when remaining loyal to the king is at odds with being steadfast to the church?

Bohun's Anglican Royalism, post-1688

As has been shown already, Bohun's agreement to publish a revised version of *Cosmographie* was not a politically neutral act. As for Heylyn, so for Bohun *Cosmographie* was designed to be a book written from the political wilderness, asserting Anglican monarchy in the midst of a victorious denominational opposition (led in this case, to complicate matters, by the monarch himself). Between agreeing to the project in 1686 and its publication in 1703, however, James II was forced to leave England and William III and Mary II acceded to the throne in 1688/89. Given this, the loyalist persona Bohun hoped to create by his geographical work had to be retailored to the new circumstances ushered in by a Protestant monarchy.

In one sense, the events of 1688/9 might seem to diffuse the tension I have pointed to in Bohun's political and geographical career until that time between loyalty to the monarch and steadfast Anglicanism. But equally, where previously his loyalty to the church might be seen as subversive of his adherence to the monarchy, to many his espousal of the divine right of kings and the necessity of overthrowing usurpers made his adherence to the monarchy equally precarious after its alteration in 1688, given that James II was an exiled monarch whose accession Bohun had defended in print from the Exclusion Crisis until the *Geographical Dictionary*, the latter published on the eve of William of Orange's arrival.

In fact, Bohun's political career after 1688 has great symmetry with that before, in that he published political tracts in defence of the absolute right of monarchs, but the grounds of that defence in the end led to his exclusion from political preferment. Bohun's defence of the events of 1688/9 was couched in Tory dynasticism, not in Whig contractarian language.[12] For Bohun, Anglicans had not ousted James II, but rather he had deserted the country when they had refused to cooperate with the changes in the constitution of church and state he had been seeking to implement. This was not resistance to the monarch, which Anglicanism could not countenance, but non-assistance of the monarch.[13] Moreover, William of Orange, as a sovereign himself, had pursued a legitimate grievance against James II in the only way available, and had conquered him in a just war. As such, the allegiance of the English people, while remaining absolute, had to be transferred to William and Mary.[14] As Bohun put it:

> It is denied that King *James* his Subjects were bound to stand by him and fight for him. He had notoriously invaded and destroyed all our Civil and Religious Rights and Liberties, and designed the Ruine and Destruction both of them and us ... and therefore if it were unlawful to resist him, it was also as unlawful to assist and enable him to destroy the true Religion, the English Liberties and Immunities; nay the very Nation.[15]

Two points about this passage are of particular importance. First, it maintains the doctrine of unquestioning Anglican submission to the monarchy which Bohun had taken from Heylyn and Filmer. Bohun, in short, is asserting the continued political relevance of Heylyn; to edit his *Cosmographie*, therefore, was not simply an antiquarian or scholarly enterprise, but a political statement. From being a defence of Anglican royalism in the wilderness as it might have been envisaged in 1686, the editing of *Cosmographie* could be reinterpreted as reinforcing a Tory version of the events of 1688/89. Secondly, the passage renders explicit the implicit tension in Bohun's article on England in the first edition of the *Geographical Dictionary*: the vision of church and state to which the article was a pæan was one which James II was subverting, as is here accepted.

Bohun's *Cosmographie* came out in 1703, long after the immediate urgency of the argument over how to characterise the events of 1688/9 had subsided. As so often in his career as a political thinker, Bohun had been on a losing side. His conception of 1688 as a desertion by James II and a conquest by William of Orange, while popular in the immediate aftermath, lost out to the Whig image of 1688/89 as establishing a King-in-Parliament rather than replacing one absolute monarch by another.[16] Yet *Cosmographie* continued to defend some version of Bohun's own absolutist argument, as might be expected in a continuation of Heylyn. While Bohun's continuation of the history of England to the present was cautious, it was also revealing:

> from this Marriage [of James I to Mary of Modena] the 11*th* of *June* 1688 there was pretended a Spring, a Prince of *Wales*, which is still living in *France*. *William* III Prince of *Orange*, and *Mary* Eldest Daughter of *James* II, Succeeded in the Abdicated Throne of *England*.[17]

This potted summary was supportive of a dynastic reading of the events of 1688/89. Denying that James II had a legitimate successor (the notorious 'warming pan' myth), Bohun suggests that James II voluntarily surrendered the throne, and that the next in line succeded. For a more complete rendition of events, *Cosmographie* directs the reader to the *History of the Desertion* (*EBC*, p. 286), Bohun's account of the voluntary desertion of James II and the conquest of England by William III.[18] Given this, Bohun is still in a position to deny the argument that government is a human creation. In *Cosmographie*, Bohun does just this, using a strategy gleaned from Heylyn himself. Having surveyed the nations of the Old World, Bohun came to America, where he added 'An Advertisement to the Reader'. The lack of arts and good government in the New World was alleged to give geographical sanction to Bohun's assertion which ran as follows:

> that Learning and Arts, Religion and Government, are not the Productions of Time, and that they came not into the World by humane Inventions ... The whole World might have been to this Day, as Barbarous and Ignorant as they now are, if the Divine Providence had not drawn it out of its *Chaos* of Moral, as well as Material Confusion. (*EBC*, p. 941)

Cosmographie, then, continues to assert Bohun's affinity with Heylyn and Filmer on the divine sanction of the state.

Was Bohun a geographical Jacobite?

Edmund Bohun has often been accused of political venality, in that he shifted from passive obedience to one line of monarchs to passive obedience to another. Yet as Goldie points out there is at least one fundamental continuity which links his political thought before and after the Glorious Revolution: 'He regarded kingly government, prerogative, obedience and the Church as the pillars of the social order. Stuart monarchy was the keystone of this edifice but only for as long as it served to maintain that order.'[19] Given that unswerving loyalty to the incumbent monarch is the Ariadnian thread in Bohun's opinions, it might seem that the last charge he was vulnerable to was that of Jacobitism. Yet it was precisely such an allegation that Bohun had to refute in 1690–93, in a controversy which revolved around his *Geographical Dictionary*.

As previously mentioned, the first edition of the *Geographical Dictionary* was published before the arrival of William of Orange, but this was followed by a second edition allegedly published in 1689 when the new church and state settlement was in place. It was this second edition which sparked the accusation of Jacobitism. The details of the allegations are only known through Bohun's rebuttals. He prefaced his *Three Charges* (1693) with this disclaimer:

> The only Reason of the Publishing these Discourses at this time, is the confuting a common Report spread and believed amongst them that know me not, that I am a *Jacobite*, or a person disaffected to the

> present Government. And this for want of a better foundation is grounded upon some passages (for the most part unknown to me) pretended to be found in the second edition of my *Geographical Dictionary*... I am informed there were some few things added, and upon them chiefly this Calumny is built; As the *Boyne*, which River was not in My first Edition, to be sure there is nothing of his Now Majesties Passing it.[20]

The suggestion was clearly that some of the entries in the '1689' edition encouraged loyalty to the Stuarts. Bohun took the allegation seriously enough to also add a disclaimer addressed to geographical readers in the form of 'A Letter to Mr Laurence Eachard, The Author of the Compendium of Geography', prefaced to Echard's popular gazetteer.[21]

Bohun's rebuttal of the charge of Jacobitism was that he was not responsible for the second edition of the *Geographical Dictionary*, this being an edition printed by the publisher Charles Brome without Bohun's assistance. The accusation had been directed at Bohun because, like so many pieces of editorial hackwork, the revised version continued to bear Bohun's name, and the preface had not been altered in any way to suggest the assistance of another hand.

If we separate off the question of authorship for the present from the accusation about the textual tendency of the book, we find that plausible grounds could be adduced for seeing the second edition of the *Geographical Dictionary* as Jacobite. To understand this, it must be understood that Jacobite rhetoric was often encoded in subtle ways in literature.[22] This means that most of the evidence about Jacobitism, be it true or false, takes the form which we can see the debate over the *Geographical Dictionary* adopting:

> Jacobites may be identified by what they said and did, and by what others said about them. Unfortunately, most of the surviving evidence falls into the latter category – it consists of accusations, allegations and scraps of information from which inferences may be drawn. These types of material must by very carefully weighed, and few solid conclusions can be reached from them.[23]

Turning to the *Geographical Dictionary* with this in mind, plenty of seditious innuendoes could be found. Thus the article on Bridgewater continued to assert that James II defeated the Protestant Monmouth 'by the Providence of God'; and Mary of Modena, while no longer described as present queen of England, was 'Queen Consort to King *James* II' (*GD2, sub.* Bridge-Water and Modena). The article on the river Boyne in the second edition showed by its reference to the Battle of the Boyne that, the title page notwithstanding, the *Geographical Dictionary* must have been published in 1690. As Bohun had asserted, the article on the Boyne could be read as entirely innocuous:

> *Boyne, Bouinda*, a River in the Province of *Leinster* in *Ireland*, hard by *Drogheda*; where King *James* II and his Army, being about 25000 men, encamped on the South Side of the River, received the Defeat of Jul. 1 1690 by K. *William* in person. (*GD2, sub.* Boyne)

Equally, a more suspicious reading could have pointed out that the '1689' edition of the *Dictionary* had the exiled James II winning a victory at Bridgewater by providence while William III merely won. Moreover, the Boyne article, like those on Modena, Bridgewater and St Germain-en-Laye, the exile of the Stuarts, titled James Stuart King James II at a time when 'the names of King James II and his heirs were officially anathematized'.[24] To carry the argument one stage further, to show how the politics of allusion could be read into this entry, its reference to Drogheda could lead the reader to that article. The second edition altered the entry under 'Drogheda', in what might appear provocative ways:

> This Town falling into the hands of the Duke of *Ormond* in 1649 and being taken soon after by storm by *O. Cromwell Septemb.* 11 of the same Year; he put all the Garrison which was about four thousand, to the Sword; and as to the Inhabitants, he spared neither Sex nor Age, poor nor rich, but initirely ruined all before him ... This being the first place the Tyrant took in that Island, this cruelty was intended to strike a terror into the other places, that he might with the greater fidelity reduce the rest of that Kingdom under his Dominion, and accordingly he had great and indeed incredible success in all his future attempts. {Since then the Town is well recovered ...} **Which town surrendered to K. William, in two days after the fight at the Boyne**. (*GD1 and 2,* sub. Drogheda)

The passage in curled brackets was removed from the second edition and replaced by the words in bold type. This could be seen as a way of aligning William III with Cromwellian violence and tyranny; both in such a reading would be usurpers who sought dominion in a nation over which they had no *jure divino* right, and used the same strategy of terrifying the mainland by violence in Ireland.[25]

At this point, we must return to Edmund Bohun, whose work, edited by another hand, exposed him to all these seditious readings. For while the reader may find evidence for a Jacobite reading of the *Geographical Dictionary* once sensitised to the possibility, Bohun's explanation is far more convincing, given the nature of geographical publishing in the period. In his own defence, Bohun pointed out he could not be convicted of Jacobitism as the only edition of the book he was responsible for came about before the change in monarchs, when it was not possible to be a Jacobite. Moreover, as he said, the second edition was advertised 'as *Corrected* and *Enlarged*; when in truth it is neither'.[26] Bohun's point, which is entirely accurate, is that the second edition is not so much a new edition updated in the light of political affairs, but a reprint to increase sales, which therefore does not represent its nominal author's current political outlook. Bohun had explained in his post-1688 pamphlets why loyalty should be transferred to William, but was being attacked by the scrutiny of political opinions he no longer held and which had been reprinted in new and inappropriate circumstances.

This process of attacking Bohun after 1688 through his professions before that time was one which was paralleled outside a geographical context in the sphere of public life. Under William III, Bohun's loyalty was rewarded by his appointment as licenser to the press. As licenser, Bohun was unpopular with those who sought greater freedom of the press, and a tactic used to discredit him was to paint him as a Catholic sympathiser and Jacobite. As with his geographical writings, the justification for such allegations was to point to his pre-1688 writings advocating absolute submission to James II.[27] The accusation that Bohun was a geographical Jacobite, then, appears most plausibly to be the product of Grub Street print culture rather than doctrinal conviction. It can also be seen as part of a campaign to disqualify him from public life. In both cases, it appears that the ability to embarrass Bohun was at least as significant as the possible truth of the accusation. As Bohun had sought to appropriate the Anglican Royalism of Filmer and Heylyn under James II, so this project was appropriated to attack him under William III. The accusation of Jacobitism saw Bohun's Anglican Royalism as tipping over into an essentially Roman Catholic defence of tyranny, just as Heylyn had been accused of being too Catholic to be a good Anglican by Gilbert Burnet. If Bohun sought to appropriate and update Heylyn's position, his critics updated Burnet's sceptical readings of Heylyn's Laudian political language to accord with altered political circumstances.

Was the second edition of the *Geographical Dictionary* Jacobite?

One further question arises about the second edition of the *Geographical Dictionary*. Accepting that Bohun was not a Jacobite, was he the victim of sloppy Grub Street editing or of provocatively inappropriate editing? The nameless editor of the second edition was identified on the title page of subsequent editions as John Augustine Barnard. Barnard's identity as editor and almost as much information about his character as is known today was known to Bohun from at least early in 1693, when Charlett wrote to him: 'Pray let me heare after from you, you know, who Augustin Bernard is, that writ the Preface to ye Dictionary agst you, a Convert [to Roman Catholicism] in K.J. time, Moral Philosophy Lecturer by mandate, then assumed the name of Augustine, since that, tis said, he is reconverted.'[28]

Barnard, then, had become a Catholic and been appointed to his post in Oxford by James II's intervention, joining the Jacobites in Ireland after James's flight, before being reconciled to the Williamite succession, c.1690.[29] This being the case, it is at least possible Barnard *did* edit the *Geographical Dictionary* in 1689/90 in such a way as to support the Jacobite cause. Bohun's accusers, then, may have stumbled onto a textual truth, despite the fact that their aim was merely to besmirch Bohun's reputation by reading past political professions in altered circumstances.

Looked at again, the key entries in the second edition cannot simply be dismissed as sloppy editing. Given that most spurious new Grub Street editions were straight reprints of earlier editions, and given also that most entries in the *Geographical Dictionary* were entirely unaltered from 1688 up to and including the 1710 fourth edition (which even failed to adjust the article on Scotland to take into account a political event as momentous as the Union of 1707), it is not without significance that entries such as that for 'Drogheda' were altered in the second edition, but their references to the Stuarts were kept in place. The adjustments made show that editorial attention had been paid to these sensitive articles, but they retained laudatory references to James II which were evidently inappropriate after the Glorious Revolution. Where Bohun's accusers, therefore, seem to have deliberately ignored his (first) edition to substantiate the charge of Jacobitism, if we ignore

Bohun and look to Barnard as the editor, it is precisely in the comparison with the first edition that the possibility of Jacobitism gains credence.

One other comparison can be used to interrogate the suggestion that the second edition of the *Geographical Dictionary* was Jacobite. It could be argued that if geographical gazetteers were designed to convey historical and political information, most of the references discussed as potentially Jacobite in the *Geographical Dictionary* were generically unavoidable. A comparison with Herman Moll's contemporaneous gazetteer, *A System of Geography* (1701), only gives partial support to such a suggestion. Moll's entry under 'Modena' is far longer than Barnard/Bohun's, yet he makes no reference to the family connection with James II as the *Geographical Dictionary* did. Similarly, where the *Geographical Dictionary* pointed out that James II lived in exile at St Germain, Moll has an article on St Germain, but makes no reference to what was probably the most interesting fact about it to Englishmen. The two gazetteers only agree where the *Geographical Dictionary* is more moderate, as in the entry under Boyne where Moll also mentions the river as the site of a recent battle, but gives no details.[30] Moll's *System*, then, tends to confirm that direct references to James II were not conventional in geography books after 1688. That Moll adhered to such a policy in the rather less fraught circumstances a decade after the exile of James II, where Barnard failed to in the year following the event suggests a certain recklessness on Barnard's part.

The possibility (and it is no more) that Barnard edited Bohun's *Dictionary* in a Jacobite manner raises one final point. Barnard, as the fourth edition of the *Geographical Dictionary* stated, was the grandson of Peter Heylyn and the son of one of Heylyn's biographers (see *GD4*, A3r). He may, therefore, have been showing another trajectory that Anglican Royalism could take after 1688, prioritising loyalty to James II rather than loyalty to Anglicanism as Bohun had. This would make a fascinating series of textual appropriations between Heylyn, Bohun and Barnard. As Bohun wished to use Heylyn's *Cosmographie* to articulate his political position while outside the sphere of preferment, so perhaps Barnard used Bohun's *Geographical Dictionary* for the same purpose. If so, Heylyn's Laudian intellectual priest hole, the geography book, had been appropriated in one way by Bohun, who consciously modelled his position on Heylyn's, and Bohun in turn had been appropriated by a Catholic with another lineage to Heylyn. Perhaps Heylyn, Bohun and Barnard had all found ways in which to mobilise the geographical compendium to express subversive and marginalised politico-theological positions.

The patterns of transmission and translation in the political languages of the three men show how complex the politicisation of geography books could become. Each appropriated established authorial reputations in altered political contexts, and thereby exposed themselves to attacks which appropriated their language and updated standard patterns of criticism in new circumstances. Geography books normally played a small part in such politico-discursive manoeuverings because of their creation by process of epitome and plagiarism, but it was these very qualities which brought the *Geographical Dictionary* onto centre stage. As its editing failed, by chance or design, to keep pace with political realities, the dynamics of transmission, translation and criticism operated to render a piece of geographical hackwork unusually controversial.

6

Edmund Gibson's Editions of *Britannia*: Loyalist Chorography and the Particularist Politics of Precedent, 1695–1722

Introduction

As we saw in Chapter 2, where geography books normally treated all the countries of the world, they were complemented at a smaller scale by 'chorographies', which collected and arranged natural, historical and antiquarian information topographically. In the British context, as noted in Chapter 3, the founding chorographical text was William Camden's *Britannia* (1586), which presented information by county, those counties being arranged according to a framework dictated by the Saxon kingdoms.

The information collected and conveyed by Camden never left the public eye, being continually translated, reprinted and plagiarised, but some hundred years after the first edition of *Britannia* a substantial new scholarly edition was undertaken which aimed both to give a faithful English translation of Camden's text and to update the work in the light of a century of topographical and chorographical research.[1] Coordinated by Edmund Gibson, this edition was conducted by a group of scholars, most of whom Gibson had come to know at Oxford. Gibson acknowledged drawing on the expertise and regional gleanings of over 25 scholars in the preface to his edition of *Britannia*, published in 1695, such that its *authorship* could not reflect a uniform political position, unlike Camden's original.[2] Thus contributors to *Britannia* included Ralph Thoresby the dissenter and Robert Sibbald, a temporary convert to Catholicism, as well as the Anglican Gibson. It also included 'trimmers' ambivalent about the Revolution of 1688/89, notably John

Evelyn, Robert Plot and the work's 'greatest promoter', Arthur Charlett (*1695*, sig.a.r). The political diversity of the contributors was probably wider still, as there are several names not listed among the contributors whom it is unlikely Gibson did not have recourse to, perhaps most notably the nonjuror George Hickes, who had inspired Gibson to, and assisted him in, his earlier Anglo-Saxon antiquarian work at Queen's.[3] Yet the 1695 *Britannia* did have a coherent political position *as a text*, and one which was the loyalist antithesis of Hickes's stance as a non-juror. This position can be characterised as supportive of the Protestant succession established in 1688/89, and the text's political coherence, despite multiple contributors with varied personal politics, can be ascribed to two factors. First, and most importantly, Gibson's hand was dominant in the *Britannia*: in addition to penning the prefatory material, we can assume from the absence of other named contributors in his acknowledgements that Gibson was responsible for the notes to some 16 counties out of the 38 in the English section of *Britannia*.[4] Given this, the political position he chose to impress left a deep and indelible mark. Secondly, a number of the other contributors broadly shared Gibson's religious and political sympathies. Looking at the contributors, who include White Kennett, William Nicolson, Thomas Tanner and Jonathan Trelawny, it is noticeable that they went on from their work on *Britannia* in the 1690s to clerical careers at the heart of political life in the period 1690–1750. While their precise allegiances diverged, all supported the Revolution of 1688/89 and the Hanoverian succession in 1714, whether Whigs like Gibson, Tories like Trelawny or first Whig then Tory like Nicolson, and were therefore 'Whiggish' when compared to Heylyn, Ogilby or even Bohun. As Douglas summarises matters:

> As the years elapsed it began to appear as if historical erudition was coming to be held as an indispensable qualification for elevation to the new episcopate, until at last within two years of the accession of George I, Anglican policy was largely directed by a group of men who, closely knit together by their political opinions, were further united by the fact that their special achievements had been in the sphere of historical scholarship.[5]

This connection of politics and historical scholarship is one which Douglas recognises had a more general application in the period: 'historical contemplation became a mainspring of political action, and the work of the scholar supplied the weft on the loom of English political life'.[6]

Given that Camden's original editions of *Britannia* had political purposes (see Chapter 3), that historical scholarship was a realm of political debate, and that the revisers of *Britannia* were starting careers in public life dedicated to the defence of a Protestant settlement which excluded the Jacobites, it is surprising to find that the possibility that Gibson's 1695 edition of *Britannia* and its 1722 update were politicised projects has not been canvassed. In fact, the scholarly consensus has held quite the reverse view of Gibson's *Britannia*, seeing it as an apolitical project, wherein the contributors earned their scholarly credentials before moving into public life. Thus the modern biographers of Edmund Gibson and White Kennett, while chronicling their political and ecclesiastical careers, exempt their chorographical work, seeing it as 'in happy contrast with the bitterness of their incursions into the world of politics'.[7] Further, despite arguing (as we have seen) that historical scholarship was politicised and that Gibson *et al.* became political actors, Douglas exempts *Britannia*, writing that the contributors formed a 'distinguished and disinterested company'.[8] While there can be no doubt that this company was distinguished, I will argue that it was in no way politically disinterested in the production of *Britannia* in 1695. Rather, the contributors' shared affirmation of the Glorious Revolution informed their edition of Camden, rendering their chorographical labours supportive of a loyalist position which effaced detailed questions concerning the justification of the events of 1688/89, demanding instead loyalty to William and Mary simply by virtue of their being ensconced. Establishing political and scholarly credentials were in no way separate and sequential elements in the lives of this group, as has been suggested, with the former following on from the latter. In fact, they were intertwined. This will then be further evidenced by an analysis of the ways in which *Britannia* and the scholarly methods deployed therein were drawn into controversies in ecclesiastical politics in the period 1696–1722 by Gibson's circle and its opponents, notably Francis Atterbury. Finally, I will show that in the aftermath of these controversies, Gibson's 1722 edition of *Britannia* became far more overtly politicised, the 1695 edition's diffuse calls to loyalism being rendered more focused and explicit in the face of Jacobite threats to the Protestant Hanoverian succession.

The *Britannia* of 1695: Williamite Monarchism

In the course of the 1680s, the future contributors to Gibson's edition of *Britannia* had been as loyal to the Stuart succession as Ogilby or

Bohun. Indeed, this had even manifested itself in a geographical context in the case of William Nicolson. Nicolson, who contributed notes on Northumberland and York to *Britannia*, started his scholarly career by writing for Moses Pitt's *English Atlas* (1680–83). The *English Atlas* was dedicated to Charles II, had an engraving of him as the frontispiece to the first volume, and proudly declared that 'his most Sacred Majesty has been Graciously pleased, for the promoting of this Design, to permit that his Collections of Maps and Descriptions of Countries may be perused, that such of them as have not yet been published, and are thought proper for this work, may be taken into it'.[9] Nicolson wrote the section on Germany for the second volume, which also flaunted its Stuart loyalism in the form of an engraving of Charles's queen, Catharine of Braganza.[10] Moses Pitt was one of John Ogilby's main competitors, and it is clear that Nicolson was contributing to a project which supported the Stuart monarchy as much as Ogilby's *Britannia* and more than Bohun's *Geographical Dictionary*. In this Nicolson was not alone among the future contributors to the revised Camden: Kennett had been loyal to the Stuarts in the 1680s and had scruples about the oaths of allegiance to the Williamite succession which he had to take as an ordained minister. Similarly, Gibson himself had scruples as a young man about taking the oath of allegiance, which delayed the conferment of his BA from Oxford for almost a year.[11]

As we have seen, Bohun became enmired in accusations of Jacobitism thanks to his geographical expression of Stuart loyalty, and this damaged his political credibility when the Stuart monarchy was replaced. By contrast, Nicolson, Kennett and Gibson managed to negotiate the change in succession and went on to flourish in public life. Many, of course, had to make such a switch, but Gibson *et al.* chose to do this in part through a geographical (or, more accurately, chorographical) project.

Before 1695 both Gibson and Kennett had already produced scholarly editions of texts which contained geographical and chorographical information: Gibson included geographical notes to his edition of the Scottish poet William Drummond's *Polemo-Middinia* in 1691 and appended an index of places to his edition of the *Anglo-Saxon Chronicle* of 1692; while together Gibson and Kennett had produced an edition of William Somner's chorographical *Treatise of the Roman Ports and Forts in Kent* (1693).[12] Yet the edition of *Britannia* was far more important, given Camden's cultural cachet and the wide readership which a new edition would attract amongst the intellectual and political elite.

It was in such a work that geographical scholarship had to articulate its political loyalty unambiguously if the contributors wished to receive ecclesiastical promotion. The possibility of political adjustments and interventions in Camden's text was signposted by Gibson's *Proposals*:

> The large Discourse of Mr *Cambden* concerning the Government of the *Romans* and *Saxons* in *England*, concerning the Degrees in *England*, and the Courts of *Britain*, will by large Marginal Notes be much Corrected and Improved, and some Original Discourses on that Subject will be added by an Eminent hand.[13]

In fact, however, the politicisation of Gibson's *Britannia* came not in this section, but rather in the prefatory material concerning Camden's life which preceded it, and in the additional notes appended to each county description which followed it. In short, Gibson *et al.* sought to appropriate both the character of William Camden and the text of *Britannia* to articulate a transition from their Stuart loyalism of the 1680s to a defence of the Williamite Protestant succession in the 1690s.

Gibson's 'Preface' to the 1695 edition of *Britannia* emphasised the stature of Camden in a way which echoed Heylyn: 'He was stil'd the *Varro*, the *Strabo*, and the *Pausanias* of Britain' (*1695*, A2r). Similarly, Nicolson was in no doubt as to the importance of Camden's main text – 'His *Britannia* ... may honestly be styl'd the common Sun, whereat our modern Writers have all lighted their little Torches' – adding his own classical comparison to Gibson's list, since '*Camden* has been the same thing as *Homer* was of old to the Poets of *Greece*.'[14] As these comparisons make clear, it was not just the *Britannia* which was important, but William Camden himself as the founder of British chorography. Given this, the ways in which Camden was presented as a person in the 1695 edition have an important bearing on its political message.

Gibson's prefatory 'Life' emphasised Camden's religious orthodoxy: in response to an accusation that Camden was a crypto-Catholic (*1695*, d1v), Gibson argued that on the contrary Camden's career in Oxford had been blocked by 'the Papist party' (*1695*, b1r) and added that his attack on the Gunpowder Plot had been placed on the Index Librorum Prohibitorum (*1695*, c1v–c2r). Lest it be thought Camden diverged towards the other extreme of Puritan 'enthusiasm', Gibson made it clear he was 'known for his fidelity to the Church of *England*' (*1695*, b1r) and that he had infuriated the Puritans by his note on Banbury (*1695*, col. 270, note k and see below, p. 215).

Furthermore, as headmaster at Westminster school, Camden propagated his loyalism to boys who were to become Anglican bishops: by 'forming Youth into Religion and Learning', Camden 'had laid the foundation of those pillars which prov'd so considerable supports both to Church and State' (*1695*, c1v and c2v). Two points are worthy of note here. First, Gibson and many of his contributors were trying to start clerical careers as they compiled the *Britannia*, and to model Camden as a loyal Anglican at the same time as they placed themselves as lineal successors to him in their role as chorographer-clerics was also to imply their loyalty to the new Protestant regime rather than to the exiled Roman Catholic, James II. Secondly, whereas Camden's Anglicanism was explicitly labelled as this distanced him from the Stuarts, the fact that the monarchy he provided the foundation for by his teaching was the Stuart one of James I/VI was effaced in the unspecific covering term 'state'. If Camden supported the pillars of church and state, the Anglicanism of the church was emphasised while the Jacobitism of the state was elided by Gibson, who had himself (together with many of his contributors) been a Stuart as well as an Anglican supporter, and needed to perform precisely the same sleight of hand in his own life as he had in Camden's 'Life', effacing and emphasising the same loyalties.

The 'Life' finally linked the scholarly and political elements of Camden's career – chorography and loyalism – in two ways. First, arguing that local studies rose to prominence as Roman Catholic '*School-Divinity*' lost ground with the Reformation, such that Protestantism and chorography were inextricably bound together (*1695*, b1v). Secondly, linking Camden's antiquarian work and his pedagogy:

> tho' his time was divided, the whole was spent in the Service of the Publick. He was always either exciting the present Age to Virtue and Honour, by representing to them the Venerable Monuments of their Ancestors; or laying a Foundation for the Happiness of Posterity, by forming Youth into Religion and Learning. (*1695*, c1v)

Again, this interweaving of learning and loyalty was clearly a mantle to which Camden's successors and continuers could lay claim. Lest the contemporary relevance of Camden's example was lost on Gibson's audience, the 'Dedication' to John Somers forced the point home. As a dedicatee, Somers had impeccable Whig credentials, having asserted that James II had abdicated and helped to draft the Declaration of

Rights of 1689. Merely to dedicate the *Britannia* to Somers was a clear statement of loyalty to the new regime, yet the 'Dedication' also linked Camden's life and legacy to Somers's politics. As a Junto Whig, Somers's life was said to be 'one continu'd Zeal for her [Britain's] Honour and Happiness' (*1695*, Ar). Somers and Camden, then, were parallel instances of devotion to public life. Furthermore, if Camden had taught a series of future bishops who became pillars of the church and state, Somers's actions in 1688/89 in 'Defence of her [the Church of England's] distressed Prelates' (*1695*, Ar) had prevented those pillars from crumbling in the face of the Roman Catholic onslaught of James II.[15] The true preservers of Camden's twin legacy, then, were those who had ousted the Stuart succession Camden himself had supported and the scholar-clerics who aligned themselves with William and Mary. Transposed by a century, Camden's authority was firmly positioned as anti-Jacobite and anti-Catholic by Gibson and his contributors as they forged just such an identity for themselves.

Turning to the text of *Britannia*, Gibson *et al.* commented upon Camden's original text by means of a series of notes which were printed at the end of his descriptions of each county. The notes were painstaking in their detail, being the substantiation of Gibson *et al.*'s scholarly credentials. This demanded a display of control of archival sources and a scepticism about conjectures not grounded in particularistic topographical information. These notes added new information, correcting and glossing Camden, and thereby acted as a humanist scholia to a classical text by the 'English Pausanias'.[16] The political aims of these notes were less rigidly defined than those of the prefatory material, their cumulative effect being to support a more diffuse monarchical and Anglican agenda in ways similar to those analysed for the text to Ogilby's *Britannia*.

The general monarchism of *Britannia* was articulated not through a defence of the events of 1688/89, but rather through the treatment of the Civil War. On the one hand, the activities of the Parliamentarian side were criticised in the notes, being seen as deleterious to antiquarian scholarship (*1695*, col. 161, additional notes) and as destructive of antiquities themselves, such as Nonsuch Palace: 'there's nothing of all this to be seen, scarce one stone being left upon another; which havock is owing to the late Civil Wars' (*1695*, col. 165, note h; see also col. 638, additional notes). If reverence for antiquities developed a love of country and a passion for public service, as the prefatory material asserted, the Parliamentarians and all those disloyal to the reigning monarch were shown by such notes on their attitude to antiquities to be politically irresponsible. On the other hand, the Royalist side was consistently praised in the notes: as with Ogilby, sites of Royalist victory and bravery such as Worcester and Maidstone had eulogistic glosses appended (*1695*, col. 525, note m; and col. 217, note o). Further, if the Royalists destroyed antiquities such as the walls of Gloucester, this was not viewed as havoc, as when practised by their opponents, but as just retribution (*1695*, col. 246, note l).

The Royalist balance of the notes did not simply construct a political or dynastic argument, but was also given a religious aspect through the deployment of providential rhetoric. Roman Catholic tales of miraculous actions at certain sites in Britain were dismissed as irrational superstitions, perhaps most notably the stories about the healing powers of the well at Holywell, imparted by the martyrdom of St Winifred: 'such frivolous superstitions, are too much honour'd, when we use arguments to confute them' (*1695*, col. 691, note c; see also col. 41, note y).[17] By contrast, miraculous occurrences related to the Royalist cause were reported without scepticism. Thus at Stratton, the Royalist victory in 1643 was 'follow'd [by] a prodigious crop of barley, ten or twelve ears on one stalk' (*1695*, col. 23, note r). This loyalist providentialism was even more apparent in a note on Harewood in Bedfordshire:

> On the west side of this County is *Harewood*, made remarkable for the standing of the river, An. 1399. The very same thing happen'd (as I have been inform'd) the 18. or else the 28. of January, in the year 1648. And as the first was look'd upon to be a prognostick of the Civil Wars that ensu'd [i.e. the Wars of the Roses]; so, as the circumstances fall out, may this be thought of K. Charles I's death; there being nothing so extraordinary on foot at that time. (*1695*, col. 289, note a)

None of this, of course, was directly supportive of the post-1689 settlement, yet it can still be identified as consonant with the more overtly Williamite prefatory material of the 1695 *Britannia*. As recent studies have shown, the Whig defence of 1688/89 was by no means exclusively conducted through a natural rights and contractarian framework of the sort articulated by Locke's *Two Treatises*. In fact, contractarian arguments were notable by their absence. Rather, the 'Whig view of the Revolution was based firmly on English history', seeking in particular to incorporate Charles I as a Whig hero, thus avoiding the slur that Whigs were the successors to the regicides.[18] Within the framework established by the prefatory material, Gibson's *Britannia* sought to align itself with a monarchism which abhorred the events of the Civil War. Furthermore, the events of 1688 were viewed within a firmly anti-Catholic mentality:

> For more than a century, anti-Catholicism had been an indelible characteristic of British society ... Its presence, whether overt or covert, had informed every stage of James II's undoing, just as it underpinned every step of William of Orange's advancement to the throne ... For most of Britain the dynastic issue had become inseparable from that of religion.[19]

As such, the support of an anti-Catholic line within the notes to *Britannia* was covert support for the altered succession established by the events of 1688/89. Finally, political languages after 1688/89 were by no means inimical to providentialist rhetoric: court polemic as developed under William and Mary by the clergymen they promoted 'presented William as a providential ruler who had a divine commission to protect the protestant church in England, and to return the nation to its pristine faith, piety, and virtue'.[20] The court polemicists who preached of William as a Godly monarch were in the same network as the contributors to Gibson's *Britannia*, that network centring on Thomas Tenison and William Wake. *Britannia*'s providential aspect was the deployment in a scholarly context of the language of 'courtly reformation' developed by Gibson *et al.* in the pulpit at William's court. However discretely, then, the Royalism, anti-Catholicism and providentialism of the notes to the 1695 edition of *Britannia* were not a contradiction to, or retreat from, the loyalist politicisation of the prefatory material, but supported it at a more tacit level.

Characterising *Britannia* as a 'tacit' politicisation raises questions concerning the *intentionality* of the textual positioning discussed here. Was *Britannia*'s politicisation inadvertent and/or inevitable in the context of 1695? Certain points are clear. First, Gibson's prefatory material was strongly anti-Catholic which, in the context of the 1690s could not but make it anti-Jacobite. Secondly, it is true that the royalism of the 1695 edition of *Britannia* was vague. Praising the Stuarts in the form of its comments on the Civil War, *Britannia* could be read as criticising only one of their number, namely James II, for his Catholicism, rather than as an attack on the Stuart dynasty *per se*. Perhaps making a denominational and therefore a dynastic choice was inevitable, but the choice Gibson and his main contributors made was anything but inadvertent. As such, it can be said that the 1695 *Britannia* was an intentionally Williamite and Protestant work. Beyond questions of textual intention lie issues of textual *reception*. Here there can be no doubt that Gibson's *Britannia* was construed as Williamite and Protestant, indeed as polemically and Whiggishly so, by Francis Atterbury in the Convocation Controversy, to which I now turn.

Geography, chorography and the Convocation Controversy, 1695–1722

In the years between the publication of Gibson's two editions of *Britannia*, 1695–1722, British public life was riven by the divides of the 'rage of party', as the Whig and Tory parties, with only minor complications caused by the Court-Country divide, came to dominate the political landscape: 'The political vocabulary of the early eighteenth century, then, was based quite manifestly on the assumption that the essence of politics ... was the contention of two great national parties for supremacy.'[21] The division between the two parties mapped onto a divide in ecclesiastical politics as well,[22] which became most pronounced in the Convocation Controversy, which was started by Francis Atterbury's *Letter to a Convocation Man*, written in 1696, and was to have repercussions for the next 20 years or more. The Convocation Controversy centred around the respective rights of the Upper and Lower Houses of the Church of England's supreme decision making body, the Convocation. Atterbury led a High-Church Tory assertion of the rights of the Lower House, wherein his ilk were predominant, by claiming that the Lower House was the twin of the parliamentary House of Commons, and therefore had the right to meet at the same time as the Commons, irrespective of being called by the monarch or the bishops of the Upper House. By contrast, the Whig clergy defended the supremacy of the Upper House (wherein William III's promotions had made the bishops overwhelmingly Whiggish),[23] claiming that the analogy between Convocation and Parliament was an historically-misleading one. The Whig position defended the episcopal hierarchy and the powers of bishops. The contributors to Gibson's *Britannia* joined the Whiggish side of the Convocation Controversy, or rather they forged that position. It is to the lines of force between *Britannia* and these disputes in church politics that I now turn to uncover one specific and one structural connection.

At the specific level, one note in the 1695 edition of *Britannia* was drawn into the Convocation Controversy. This was 'note w' in the Northumberland section, which was written by William Nicolson and concerned the Synod of Twiford: 'At the famous Synod (mention'd by our Author) S. *Cuthbert* is said to have been chosen Bishop. By the account that *Bede* (and especially his Royal Paraphrast) gives of the matter, it looks more like a Parliament than a synod' (*1695*, col. 872).

Nicolson went on to support his contention with an analysis of the meaning of the Anglo-Saxon term, 'witena', which saw it as meaning 'parliament', as it 'signifies *Senators* or *Parliament-men*'. This note was contentious to the High Churchmen of the Lower House as it asserted that Parliament and Convocation were not analogous. Francis Atterbury, the doyen of the High Churchmen, attacked this note head-on as part of his argument that Saxon meetings which were called 'synods' consisted of both temporal and spiritual leaders, and therefore that Parliament and Convocation had the same rights as they derived from the same origin:

> *Which the Learned Mr* Nicholson, *with all his* Saxon *knowledge, seems not to have consider'd where he asserts the Meeting at* Twiford ... *to have been no* Synod, *but a* Parliament. *An Instance, which shews how fit he is for that Office he has taken upon himself, of being an Umpire in this Controversy.*[24]

Atterbury then went on the offensive, seeking to appropriate Camden's authority to his own cause from that of *Britannia*'s editors. As part of his argument that Convocation was the twin of Parliament, and should meet at the same time without let or hindrance, he pointed out that 'Mr *Cambden* knew no better, whose words are, *Synodus, quae Convocatio Cleri dicitur, & semper simul cum Parliamento habitur.*'[25] Atterbury, then, had both disputed the scholarly competence of Nicolson, by extension questioning the authority of Gibson's *Britannia*, and gone on to claim Camden's authority for his position.

The clash between Nicolson and Atterbury was to have practical political consequences, Nicolson as Bishop of Carlisle refusing to install Atterbury as Dean in 1704, and Atterbury leading a campaign against Nicolson's right as bishop to act as Visitor of Carlisle Cathedral in 1707.[26] Yet the principal concern here is with the intellectual response to Atterbury's appropriation of Camden's authority. Nicolson defended himself in *A Letter to the Reverend White Kennett* (1702), arguing that while Bede used the term 'synod', in his own time, 'the word *Parliament* suits better ... and therefore I humbly offer it, as an amendment of my author's text, to put in a word of modern use, instead of one now obsolete, as to the true meaning and import there intended'.[27]

Nicolson, then, reclaimed the intellectual high ground by asserting that his knowledge as a Saxonist allowed him to convey the sense, where Atterbury could only light upon the term itself, being ignorant of its historical mutation. Nicolson's fellow contributor to *Britannia* and the recipient of his published letter, White Kennett, supported his knowledge as a Saxonist: 'I leave Mr *A* to be taught *Saxon Knowledge* by that Reverend Person whom he despises for it.'[28] Both Nicolson and Kennett, then, claimed that a true knowledge of Saxon would have allowed Atterbury to see that synods and parliaments differed constitutionally, and therefore that the Whig line they supported was true.

The ramifications of this exchange could still be seen in two ways in the 1722 edition of *Britannia* which Gibson *et al.* prepared. First, Nicolson's offending note was left unchanged, since to have adjusted it would have been to concede that Atterbury had scored a hit against the Saxon knowledge of Nicolson and, therefore, against the authority of Gibson's edition.[29] Secondly, Atterbury's positive appropriation of Camden's authority was contested. Gibson's translation was faithful to Camden's 1607 original, and therefore had to concur with Atterbury that he said parliament and convocation met at the same time (*1722*, col. cclvi–vii). Yet Gibson added after Camden's description of the ecclesiastical courts that they were not independent, but under the authority of the Archbishop of Canterbury, an annotation which clearly linked with Whig arguments about the supremacy of the Upper House of Convocation (*1722*, col. cclvii). By its glosses and its overlooking of Atterbury's points, then, the 1722 edition of *Britannia* added to the rebuttal of Atterbury's incursion into the realm of chorography which the Whiggish clergy had claimed as their own.

At the structural level, the modes and sources of argumentation which Gibson *et al.* used to defend the Whig position in the Convocation Controversy were exactly the same as those they had developed in their chorographical labours. Above all, the Convocation Controversy was fought out in the realm of historical scholarship, a realm in which Gibson *et al.* had proved their credentials in *Britannia*. Thus, while disagreeing with Atterbury's attempts to parallel Convocation and Parliament, Gibson saw:

> one known custom in Parliament, that may well deserve to be consider'd in Convocation, *viz.* the Recourse they have, and the

> Deference they pay, in all contests about Privilege, to the *Records* and *Journals* of former Parliaments. For both our *Civil* and *Ecclesiastical* Meetings are *Ancient* Constitutions.[30]

It was precisely through such a respect for precedent and the particularities thrown up by historical scholarship that Gibson *et al.* claimed to have reached their position about the supreme authority of the monarch and the Archbishop of Canterbury over the Lower House of Convocation. By contrast, Atterbury was stigmatised as a mere systematist, who did not respect historical evidence:

> It is no new thing among *Speculative* Men, to frame a Variety of Schemes for the more perfect Government of Church or State ... But when we are speaking of the present *Rights* in an Establish'd Constitution; the Question is not, into how many Forms a working Fancy may mould the Constitution, but *where* it is that *Law* and *Custom* have plac'd the Right.[31]

Gibson had attacked exactly the same sort of approach where historians were 'led away with a chain of groundless stories and extravagant conjectures' in the *Britannia* (*1695*, col. 565, additional note).

The danger of a speculative approach, ungrounded in historical evidence, was that it could pull the monarchy and the Anglican church away from a stable *via media*. On the question of the monarchy, for Kennett, Atterbury's views were linked with the 'Republican Principles' which *Britannia* had shown were alien to loyal churchmen, both being ahistorical travesties of the English polity, such that 'A *Milton* could not write of Crown'd Heads with more Irreverence and Spite' than had Atterbury.[32] Further, and on the question of church government, Atterbury's elision of parliament and convocation was 'an *Erastian* Error' which destroyed the church's independence.[33] Gibson elaborated on the danger to Anglicanism, Atterbury's arguments being stigmatised as both Puritanical (since they undermined the episcopal power of bishops) and Roman Catholic (since they denied the independence of the church): 'as these new Claims, are in their *tendency* thus favourable to the Measures of PRESBYTERY; so the Principle upon which they are chiefly founded, gives the PAPISTS a manifest Advantage against our Reform'd Church'.[34]

In other words, just as Gibson's 'Life' of Camden had shown that he trod the Anglican *via media* and was loyal to the established state, so the attack on Atterbury showed that he diverged from such a position in religious and political matters because he had failed to immerse himself in historical scholarship of the sort Camden and his successors, namely Gibson *et al.*, practised.

If we look to the positive side of the Whig message in the Convocation Controversy, there is a further link with the modes of argumentation they had developed in *Britannia*. Just as Gibson and his circle defended the need for an episcopal hierarchy, so they argued that this demanded a geographical delineation of church authority. If Atterbury's ahistoricism threatened to bring chaos by failing to respect the authority of the bishops, so there was a spatial equivalent, as Gibson made clear in a later attack on the Methodists:

> The great Preservative of Religion, and of Order and Regularity in the Exercise of it, is, the Provision that is made for the Performance of publick Offices by Persons lawfully appointed within particular Bounds and Districts; and if these be broken down, nothing can follow but Disorder and Confusion.[35]

Respecting ecclesiastical geography meant, as in the Convocation Controversy, looking at historical records, since that geography had developed as a '*Common Law of the Church*', rather than through any system.[36] When historical records were consulted, it was found that:

> our Ecclesiastical Officers took their limits of Jurisdiction from a like extent of the Civil Powers. Most of the Old Saxon *Bishopricks* were of equal bounds with the distinct *Kingdoms*. The *Archdeaconries*, when first settled into local Districts, were commonly fitted to the respective *Counties*.[37]

This analysis of ecclesiastical geography had close similarities with *Britannia*. As with Camden, the spatial framework was set by the counties, arranged according to the Saxon kingdoms; within each region, local precedent, as discovered by historical enquiry, was the supreme arbiter. Clearly, the skills of the chorographer were an essential prerequisite to the analysis of ecclesiastical geography, and a respect for that geography was in turn essential to the maintenance of order in church and state.

At the outbreak of the debates over Convocation, Nicolson argued that scholarly chorography was divorced from the ecclesiastical and political sphere:

> Amongst the other blessings of peace that we are now encouraged to hope for, surely learning will begin to flourish again; and (instead of

> many rascally disputes about matters of religion, which are now in fashion) the more valuable study of antiquities will grow modish. I am abundantly convinced, that, if ancient history were more heeded, and better understood, we should be much more at quiet than we are.[38]

This putative divorce was disingenuous to say the least, given the ways in which learning was deployed by Nicolson and others in the 'rascally' religious debates of the Convocation Controversy. Indeed, the last sentence sees Nicolson undercut his own position, by asserting that political quiescence was the outcome of scholarly labour. If that were the case, antiquarianism would efface the question of the legality of the events of 1688, which was precisely what adherents to the new succession sought. Geographical argumentation was key to debates over the status of the Convocation, with Camden's own position on this question being fought over.

A great deal of scholarly attention has been focused on the development of what Pocock called the 'common-law mind', and it has been recognised that Gibson and his contributors possessed such a mentality, which looked to precedent and historical argumentation.[39] Less appreciated, however, is that this mind-set was just as much a chorographical one, arranging historical evidence in a spatial framework. It is clear that Gibson *et al.* were interested in geography as well as history in the construction of their arguments in the Convocation Controversy, since they had built their reputations in that field, and were attacked by Atterbury on that subject. More generally, in the post-1688 context, localism was a popular discursive framework for Williamites: 'Perhaps nothing done in the 1680s by Charles II and James II caused so much reaction against them as their interference with local privilege'.[40] The common-law mind as developed by Gibson's circle, then, had a spatial element which might be called the chorographical imagination, both being grounded in a scholarly reverence for precedent and the particular, which formed the basis of a critique of Tory, crypto-Jacobite and High Church arguments as overly schematic and not grounded in the history and spatiality of English governing institutions.

'Hanoverianising' *Britannia*: the edition of 1722

To understand the major shift in the political rhetoric of the 1722 edition of *Britannia*, one further element of the political language of

Gibson *et al.* in the interim years, 1695–1722, needs to be recognised in addition to those discussed in the previous section. Looking at the sermons and pamphlets of Gibson and others, it is apparent that they grew more concerned about the safety of the Protestant succession in the face of a Jacobite opposition as the years passed.[41] Gibson described the events of 1688 as a providential deliverance from a Catholic onslaught in highly charged language: 'Do not many of us *remember* the Day, when we were upon the point to be made Slaves to a foreign Yoke, and every Heart trembled for the Ark of God, for the True Protestant Religion...?'[42] Kennett described the 'Catholic yoke' in still more lurid terms: the 'Stream of innocent Blood' shed in Mary's reign would have been repeated by James II, and while 'we had indeed no *Fires* yet kindled, for the burning of us, they called *Hereticks* ... that Forbearance was chiefly owing to a Repeal, in the former Reign, of the Act *de Haeretico comburendo*'.[43] Similarly, Gibson lambasted those Tories who had contemplated changing the succession back to the Stuarts in Anne's reign, which would have been 'the utter Destruction of our Religion and Liberties'.[44] The defeat of the 1715 Jacobite rebellion against the newly-installed Hanoverian succession gave full scope to the fears of Gibson and others: for Kennett it could only be explained by saying 'there is a great deal of the Hand of God visible in it', while Gibson's entire thanksgiving sermon paralleled the divine favour bestowed on England with Old Testament Israel.[45] The more general trend in the argumentation of adherents to the new succession in the early eighteenth century which Gibson and the contributors to *Britannia* followed here was towards an increasingly 'conservative' political language which sought to preserve the Protestant succession and Revolution settlement.[46] Indeed, Gibson's circle could sound as Laudian as Heylyn himself, notably in Kennett's *Complete History*, where he advocated Laudian churchmanship, only criticising Laud for trying to introduce it too quickly.[47]

That geography and chorography could be drawn into this increasingly strident language of loyalism was apparent from Kennett's sermon on the accession day of Queen Anne. In *The Happiness of this Church and Nation* (1705), Kennett argued that in a 'survey [of] this immense Globe, we could not fix upon any Point of it more agreeable than this *Island*', thereby mirroring Camden's patriotic praise in *Britannia*. Further, Kennett added that British trade was selfless: 'not for Necessity, but for Choice; to oblige the World, rather than to serve themselves'.[48] This was exactly the argument Camden had used, saying Britain's location was ideal for trade, such that her commercial activities were 'for the general benefit of mankind' (*1695*, col. i). Finally, Kennett rhapsodised about the beauteous landscapes which liberty produced: 'Where Liberty sets her Foot; there Plenty, Ease, and Beauty chear up the Countenance of the very Earth ... But where Tyranny and hard Dominion reign; there any Garden of the World becomes in effect a Wilderness.'[49]

In this case, his source must have been the 1695 edition of *Britannia* rather than Camden's own editions, since descriptions of landscapes were an innovation in Gibson's edition, as he himself recognised in a note about Warwick (*1695*, col. 511, note h). It is clear that in drafting a sermon which eulogised the altered succession, Kennett had been drawn back to the edition of *Britannia* he himself had helped to construct in order to find geographical grounds for monarchical patriotism.

If *Britannia* could be drawn into the more strident and tense political language of loyalism in the early eighteenth century, it is unsurprising that this dynamic also worked in the other direction, the fraught loyalism of the 1700s being incorporated into the 1722 edition of *Britannia* by way of a more heavy-handed textual politicisation. By 1722, Gibson had become a key figure in church politics, having been elevated to the bishopric of Lincoln in 1716. In this position, he felt it necessary to respond to those who thought his scholarly labours 'less proper for the Character and Function of a *Divine*', arguing in his 'Preface' that 'no Diversion can be more innocent or laudable, than the History and Antiquities of our Native Country' (*1722*, [a.r]). In fact, this disclaimer of political intent was as disingenuous as Nicolson's in 1697 (see above). That this appeal to 'innocent pleasures' was a purely rhetorical move would already have been obvious to any reader who had looked at the 'Dedication' and became still more apparent when recourse was had to the textual notes of the 1722 edition.

The 1722 edition was dedicated to George I, and was a sustained exercise in the defence of the Hanoverian succession, attempting to link this to *Britannia* itself. Deploying the same providential language as the sermons Gibson and his circle preached in the fraught dynastic situation of the 1710s, *Britannia* was dedicated to George 'in a true sense of the signal Deliverance they [the British kingdoms] receiv'd, and of the manifold Blessings deriv'd to them, under a Succession of PROTESTANT Princes in Your Illustrious House' (*1722*, *ar). The 'Dedication' was one which all would agree with, bar those 'misled ... to serve those unworthy and unnatural Ends' of Jacobitism (*1722*, *av). The Protestant succession, of course, had first been secured by the events of 1688/89, and the 1722 edition of *Britannia* made explicit its support for these events and their being read providentially in a way the 1695 edition had not. At Torbay, Gibson added a note that:

> Of late years, it [Torbay] has been memorable for the landing of the Prince of Orange, afterwards King William the third, for the deliverance of these Kingdoms, when they were upon the point to be destroyed [*sic*] and swallowed-up by Popery and Arbitrary Power. (*1722*, col. 37; cf. *1695*, col. 39)

All this is in line with Gibson's hyperbolic praise of the Hanoverian succession and his sense of the continued threat of Catholicism in the wake of the 1715 Jacobite rebellion, as manifested in his sermons in the 1710s.

Where the praise of Hanover might seem gratuitous, Gibson tried to link it in more seamlessly to the texture of *Britannia* by his revised treatment of the Saxon material around which the text was structured. As previously stated, from its first publication in 1586, *Britannia* had been structured around a county framework, those counties then being grouped according to the kingdoms of the Saxon heptarchy. In 1722, Gibson attempted to give the Saxon theme a Hanoverian aspect and to tie the English to Hanover by the sanction of ancient precedent which he valued as a chorographer. Where George had only ascended to the throne eight years previously, Gibson gave this the sanction of history as follows:

> the ensuing Work points out a Relation between your Majesty and these Kingdoms, of a far more Ancient Date. Not only our Histories, but our Language, our Laws, our Customs, our Names of Persons and Names of Places, do all abundantly testify, that the greatest part of your Majesty's Subjects here, are of SAXON Original. And if we enquire from whence our Saxon Ancestors came, we shall find, that it was from your Majesty's Dominions in *Germany*. (*1722*, *av-*br)

As such, loyalty to Hanover was in fact loyalty to a more ancient claimant to the throne than the Stuart line. The common law/chorographical mind, then, was justified in supporting the altered Protestant succession. Eighteenth-century writers continually had recourse to notions of England's primitive Saxon liberties, but Gibson used this idea here to a far more specific dynastic end.[50] Once Saxon studies had a dynastic message, so Anglo-Saxon could become a politicised terrain, as it had been in the debate between Atterbury and Nicolson over the 'witena' which the 1695 *Britannia* had sparked. This politicisation occurs most notably in the 1722 edition's discussion of the etymology of the term 'Saxon' itself: Gibson objected to those who derived the term (as Camden himself did)[51] from 'the *Sacae* of Asia', preferring to derive it from 'the short swords call'd *Sachs*' which the Northern Germans carried, adding that 'the *Arms* of Saxony to this day ... *are two short swords a-cross*' (*1722*, col. clvi). In other words, if Britain derived its basic political geography from the Saxon era, as Camden had asserted and as Gibson *et al.* agreed, that geography linked Britain back to Hanover since the Saxon people and language were Germanic, as was the term itself. Ironically, as Camden had flattered the Stuarts, he had been paying a deeper implicit compliment to the dynasty which would supplant them.

The 'Preface' to the 1722 edition struck a second (false) note of political indifference, stating that it would include brief accounts of people who had served their country, which 'the Reader will find to be done with great *Justice* and *Impartiality*' (*1722*, [d2v]). In fact, if one looks at the individuals eulogised in the additional notes in 1722, they constitute a Whig pantheon. Some small fry were included, as at Reading, where a note praised William Cadogan, created Baron of Reading for his services under the Duke of Marlborough (*1722*, col. 170). In the case of Marlborough himself, Gibson's praise was hardly impartial: the note appended to the description of Woodstock gave a citation of his '*eminent and unparallel'd Services*' to the crown (*1722*, col. 298) scarcely less fulsome than Gibson's parallel of Marlborough and Joshua in his thanksgiving sermon occasioned by the defeat of the Jacobites in 1715.[52] Parallel praise in the realm of church politics was heaped upon Thomas Tenison, late Archbishop of Canterbury and Gibson's great mentor, for his anti-Catholicism under James II (*1722*, col. 391).[53] The most excessive praise, however, was unsurprisingly reserved for the royal family itself, and in particular the Prince of Wales. The 1722 edition differed from that of 1695 in that the notes were interleaved in the text and marked ({}), rather than being gathered at the end of each county. The 1722 mode of annotation allowed Gibson to both translate faithfully and then negate the potentially Jacobite resonance of Camden's praise of the Stuart family:

> the most potent Prince King James {the first;} his eldest son *Henry*, the darling and delight of these Kingdoms, was stil'd *Prince of Great Britain* ... that he might enjoy length of days, and all other

> Blessings; was the constant and hearty prayer of Great Britain: {But not more constant, nor more hearty, than are the Prayers of every true Briton at this day who has a just value and concern for his Religion and Liberty, That the like Blessings may be plentifully pour'd upon his Royal Highness, GEORGE Prince of Wales, together with his illustrious Consort, and their Royal Issue; upon whose Succession to the Throne (after his present Majesty, the best of Kings) not only the Happiness and Prosperity of these Nations, but the very Being of our Constitution, under God, entirely depends}. (*1722*, cols. ccxxxiv–v)[54]

Here Gibson prevented any possible transposition of the praise of James I onto the late James II or his son. He then added a long and unabashed eulogy to the Protestant Hanoverian dynasty, which was anything but the impartial account of characters his 'Preface' had promised. In fact, Gibson's method with respect to these pen portraits had been captured more accurately in a letter he had sent Ralph Thoresby in 1709 as he prepared the revised *Britannia*. Thoresby had suggested any revision should list those prominent in charitable activities by county. While Gibson declined as there was insufficient space, he thought such a project should be conducted in any case 'in justice to the Protestant cause (upbraided by the Papists with the want of charity)'. Further, he said *Britannia* would include at least 'the greatest and most considerable in each county'.[55] In the published version, it would appear charity was supplanted by more direct service to the Protestant cause in church and state, which Gibson saw as the greatest charitable service anyone could perform.

In the midst of the Convocation Controversy, the rage of party and the 1715 rebellion, Gibson made the political message of the 1722 *Britannia* far more transparent than that conveyed in 1695, by converting a diffuse loyalism into a blatant Hanoverian allegiance. With a Protestant succession secured, with Whig governments in power and with the Tories proscribed by George I, Gibson's Whiggish circle became unavoidably 'conservative', as they wished to conserve the political settlement achieved between 1689 and 1715. Equally, the threats to that settlement led them to an ever more unabashed monarchism, since they considered themselves to be preserving a providential dispensation to the British peoples, the importance of which made its defence by explicit interleaving in *Britannia* entirely justifiable in Gibson's eyes. The political language of the editions of *Britannia* described a trajectory which Whig thought followed more generally in the period: 'Sunderland's death in April 1722, and Walpole's emergence as Prime Minister, marked the culmination of Whig religious conservatism.'[56]

Britannia's (re)issue(s)

In his *Anecdotes of British Topography* (1768), Richard Gough described *England Illustrated*, a work printed in 1764 for Robert Dodsley, as 'nothing more than an abridgement of Camden, in a different method, all his errors adopted, and many new ones committed'.[57] In this *England Illustrated* was by no means alone, many hackwork geography books and chorographical studies plagiarising Gibson's editions of *Britannia* directly or copying books themselves drawn from this source. Yet this does not mean that the textual politics of Gibson's editions ramified down the eighteenth century, because it was far too delicate a plant to survive the plagiarists' methods of transplantation. Most plagiarisms took individual facts and descriptions of natural and antiquarian curiosities from *Britannia*, and were uninterested in the political messages which Gibson *et al.* had implanted by subtle use of the techniques of humanist textual editing.

Camden remained a great authority throughout the eighteenth century, but his errors were increasingly recognised. Above all, it was thought by antiquarians that Camden and his successors had been too reliant upon conjectural etymology, which was no substitute for fieldwork. This meant that quite contrary to Gibson's dichotomy between monkish scholasticism and Camden's Protestant antiquarianism, Camden could now himself be positioned as credulous and 'monkish' by Nathaniel Salmon's *New Survey of England*. The *New Survey*'s title page proclaimed it would treat 'the Defects of CAMDEN ... and the Errors of his Followers', adding: 'The two Causes of what I take the Liberty to call Error, are, The Indolence of the Age of Monkery, which made no Enquiries, but took all Things upon Trust; and the Fondness more discerning Men have shewed to Etymology'.[58]

As the reference to the errors of Camden's followers suggests, because of its reliance on textual evidence Gibson's circle itself could also be positioned on the side of 'Monkish' error, precisely as it had positioned Atterbury in the Convocation Controversy. The last great editor of Camden's *Britannia*, Richard Gough, was even more harsh than Salmon had been in characterising the state of chorography where no recourse was had to field archaeology:

> Incorrect pedigrees, futile etymologies, verbose disquisitions, crowds of epitaphs, lists of landholders, and such farrago, thrown together without method, unanimated by reflections, and delivered in the most uncouth and horrid style, make the bulk of our county histories.[59]

Gough's own edition of *Britannia*, published in three folio volumes in 1789, has certain similarities with Gibson's editions: dedicated to a Hanoverian king (George III in this case), Gough said 'Varro, Strabo, and Pausanias ... fall short in the comparison' with Camden, such that a correct edition of Camden became a patriotic duty – 'The love of my country and of honest fame are *my* incentives as they were *his*.'[60] Gough's edition retained a great interest in etymology as a key to topographical knowledge, but supplemented it with the evidence of archaeological digs. As such, it represents a liminal document, written as a humanistic chorography began yield precedence to field archaeology. In one sense, the transition had already been achieved by Gough's edition: on the eve of the French Revolution, Gough's *Britannia* did not engage in the humanistic textual politics which Gibson had practised in 1695 and 1722. The humanist methods of textual politicisation which yoked Camden and Gibson together were no longer appropriate, as the realms of politics and geographical scholarship began to be reconfigured.

7
Varieties of Orthodox Geography, 1700–50: Three Vignettes: Echard, Wells and Salmon

Laurence Echard

In some senses Laurence Echard was the successor to Edmund Bohun. Echard praises 'the Worthy Mr Bohun' in the 'Preface' to his *Most Compleat Compendium of Geography* (1691), a work which went through numerous editions in the first decades of the eighteenth century.[1] Moreover, it was in the *Compleat Compendium*, as we have seen, that Bohun published his 'Letter', vindicating himself from the charge of Jacobitism, but also lauding Echard. If we are looking for a genealogical continuation of Heylyn's line, then, we might look for it in Echard, given that both Bohun and he give it some textual foundation. Such a linkage might suggest that Echard was a Tory, and there is no doubt that he was an Anglican. Many contemporary critics viewed Echard as a Tory,[2] but recent commentators see him as a 'clerical whig'.[3] Both characterisations and the link with Bohun have in common the recognition that Echard was a stout defender of the church and state established by the Revolution settlement. Echard's, then, was *a* politics of orthodoxy and the establishment, although how to characterise it remains unclear and, as will become apparent from the two other vignettes presented in this chapter, his was not *the only* version of an orthodox geography on offer in the first half of the eighteenth century.

Geography had but a small part to play in the construction of Echard's orthodoxy, his main work being as an historian. The *Compleat Compendium* was compiled long before the historical works that made Echard's name and brought the critics down on him. It is a sparse collection of data about the nations of the world which, due to its minimal authorial presence, is designed in such a way as to make it uncontroversial. The *Compleat Compendium* does, however, contain an 'Appendix' which discloses an authorial voice, and helps to locate Echard's political project.

The most notable theme of the 'Appendix' is the utility of geography to a politico-theological argument. First, and invoking natural theology, geography refutes atheism:

> for the true observing of the vast variety of God's Works, His Infinite Power in Creating, and his Infinite Wisdom in disposing and ordering all the things of this Life ... [T]he right Consideration of these things is sufficient to baffle the crafty Wit of Atheists. (*CC.*, pp. 224–5)

But this argument from design is then supplemented with a rather more explicitly politicised argument:

> Another Advantage a *Divine* has from this Science, is by the right observing God's peculiar *Providence* in governing of all Nations and Kingdoms ... for to some People He gives hard and very severe Governments, where all are little less than Slaves ... But to others He gives mild and moderate Governments, where the People are very happy under such Protections ... From hence we may perceive the Punishments and Rewards of this Life, which are done according to God's Infinite Wisdom and good Pleasure. (*CC.*, p. 225)

Echard's view of the role of geography in the life of a clergyman, then, is not unlike Heylyn's framing argument in *Cosmographie*. But where for Heylyn the lesson of a geographical examination of providence is that all civilisations rise and fall, Echard is more confident that Europeans, blessed with Christianity, 'have been the greatest Sharers of God's special Blessings' (*CC.*, p. 226).

Echard had no doubt that history was a superior genre to geography, admitting in his *History of England* 'I have several times deviated and descended from the Dignity of an *Historian*, and voluntarily fallen into the lower Class of *Biographers, Annalists*, &c', which 'et cetera' included geographers.[4] But approaching his historical writings from the perspective established by the *Compleat Compendium*, it is possible to view them as exemplifications on the smaller canvas of the individual nation of the providence Echard had argued the geographer would find in the history and polity of all nations. Such a move is also countenanced by Echard, who reiterated the link between geography and history in Heylyn's terms: '*History* without *Geography* can never be well understood, or have its right Use, *but is as a Dead Carcass without either Life or Motion*' (*CC.*, p. 229).

Echard's writings on English history fit closely with the opinion expressed in the *Compleat Compendium* that the workings of providence can be discerned in the fortunes of a nation. His analysis in the *History of England* is divided into reigns, and at the close of each reign he gives a character of the monarch. These are notable for viewing monarchs as instruments of providence and as being judged on this earth by God. The character of Henry VIII combined both forms of providence:

> if his Crimes were still greater, it would be no just Aspersion upon the Reformation he effected; since Providence often brings about the noblest Designs by the most exceptionable Instruments. ... and the Hand of Heaven seem'd to lye heavy upon him and his Family; for notwithstanding all his extraordinary Methods to obtain Posterity, and his peculiar Establishments of the Succession, from six several Wives he was not able to produce one Grand-Child. (*HE.*, I 717)

Echard, then, writes 'not only [as] an English historian but also [as] an Anglican clergyman'.[5]

The Anglicanism Echard espoused, however, was a long way from that of Heylyn or Bohun. Indeed, Heylyn comes under Echard's scrutiny in his survey of the Civil War, and his character receives a very mixed reception: 'he never arrived at any higher Church-Preferment than the Sub-Deanery of *Westminster*; which seems most justly to be attributed to a violent and inflexible Spirit, which had prov'd highly prejudicial to that very Church he had so often strenuously and nobly defended' (*HE.*, III 94). Accusations of being a Tory notwithstanding, Echard's pantheon of great Anglicans was a world apart from the High Churchmen Heylyn singled out in *Cosmographie*. Echard's 'Heroes' (*HE.*, III 845) were the rationalistic Latitudinarians who sought a more tolerant Anglicanism capable of subsuming all Protestants, and explicitly distanced themselves from the Laudians.[6]

If the Latitudinarians were ideal churchmen, in Echard's opinion they rose to prominence as part of the providential blessing that was the Glorious Revolution. The third volume of Echard's *History* treats the events of 1688, and it is prefaced by a note which reverts to the language of the *Compleat Compendium*:

> those very Proceedings that caus'd the Ruin of the King and the *Popish* Cause, was so very near overthrowing that of the *Protestant* and *English* Establishment, that our Escape can be call'd no less than a mighty Deliverance; a *Deliverance* in which the Hand of Heaven appear'd eminently visible and conspicuous, not only here, but in other Parts of *Christendom*. (*HE.*, III A2r)

His narrative of the events of 1688 continues in this vein, suggesting, as Gibson was to in 1722, that the wind which blew William of Orange's forces to Torbay was an act of God (*HE.*, III 160). His portrayal of the constitutional outcome of 1688 is one appropriate both to Echard's Latitudinarianism, and to the Newtonian era in which he lived:

> It has been curiously observ'd, that the whole Machine, to the Glory of the Artificers, was so exquisitely contriv'd, and the Parts so duly adjusted, that when it was set together and put in Motion, it went on with little outward Assistance, in such Order and Regularity, that many of those who were carry'd by it, knew not that they were leaving their old Government, till they were quietly lodg'd under a new one; so that the Workmanship, tho' effected by visible Means, seem'd more Divine than Human.[7]

Echard's historical works, then, detailed the providential treatment of one nation's religion and polity, while the possibilities of a broader project treating the whole world in this manner had been signalled earlier in his *Compleat Compendium*. His *History of England* was condemned by many for its treatment of Dissenters. Interestingly, one condemnation linked Echard to Heylyn, suggesting he was too reliant on Heylyn's *Ecclesia Restaurata*: 'the Author our *Historian* made Use of, could not avoid writing Encomiums on their Patron, Archbishop *Laud*. *Heylin* wrote for Money as well as Malice'.[8] Yet it should be clear that however much Echard was textually ensconced in the Heylyn/Bohun geographical genealogy, and however much critics wanted to see him as a latter day Laudian, his political position was far from theirs. Where *Cosmographie* gave a cautionary tale of the fall of empires and the vanity of human wishes, even under an ideal Caroline–Laudian establishment, and where Bohun configured himself as Heylyn's successor, an embattled Anglican loyalist beset by the trials of a Popish sovereign he must obey, Echard is altogether more complacent. Heylyn and Bohun create providential narratives of trial and earthly hardship, where Echard sees the trials of the past as creating the ideal machine of church-state government in the present. While Echard's political theory seems less emphatic in its focus on loyalty and obedience than either Heylyn's or Bohun's, his political practice sits more unquestioningly and comfortably within the parameters of the church and state by law established than that of either of his predecessors.

Edward Wells

If Echard was a loyal Anglican Whig, Edward Wells was of the same genus, but of the Tory species. In many ways, the gap between the two men epitomises the stereotypes of Oxford and Cambridge universities in the period.[9] Cambridge had been the home of the Latitudinarianism and Newtonian physics which informed the language of Echard's description of the state machine established in 1688.[10] Echard wrote his *Compleat Compendium* in these surroundings, as a fellow of Christ's College, and his later work seems to reflect this intellectual milieu. Wells, by contrast, was educated at Westminster School and Christ Church College, Oxford. Both institutions were Tory and High Church heartlands and there can be no doubt that Wells was as influenced by this context as Echard had been by his.[11] I will focus initially on Wells's career as a pamphleteer in defence of High Church Anglicanism, before showing how this inflected his geographical writings.

Wells made his name by a series of pamphlets relating to church doctrine and politics, which sought to rule other denominations out of court and to legislate about intradenominational Anglican disputes. With respect to other denominations, Wells was uncompromising. A staunch defender of episcopalianism, Wells dismissed Presbyterianism as lacking scriptural basis or the sanction of the primitive Christian church:

> That Christ did leave the Government of his Church to his Apostles, is so evident from Scripture, as not to be denied by any sober Man. And that the Order of the Apostles was distinct from what we now call the Order of Presbyters or Priests, and the same with what we now call the Order of Bishops, is fairly to be gathered from the New Testament it self.[12]

Wells even goes so far as to say that Presbyterianism is the work of the devil, who, failing to prevent the purification of Christianity by the Reformation, 'made use of all his Cunning to blast the Fruits thereof, and he did it in no respect more, than in violating the Apostolical Form of Church-Government' (*Tracts*, p. 11). As for High Churchmen generally, then, ecclesiology was not merely adiaphorous, but essential to salvation.[13] Indeed, at several points Wells asserts that separation from the established church is a sin (*Tracts*, pp. 3 and 50). The attack on Roman Catholicism is less organised, which is unsurprising, given Wells's *Tracts* were penned against Dissenters, but as we saw with Heylyn it was a characteristic of High Churchmen that they tended to find as great a threat to come from Protestant Dissent as from Rome. Wells argues that Roman Catholics and Presbyterians are indistinguishable, both attacking the *via media* represented by Anglicanism. As well as claiming that the two groups had conspired to murder Charles I and destroy Laudian Anglicanism (*Tracts*, p. 192), Wells had no doubt:

> that *All the Adversaries of the Church of England* promote the *Same Common End*, viz. *Popery*, and that by many the *Same Common Artifices*; it ought therefore to be duly Considered, whether they may not *All* be very Properly and Truly comprehended in General under the *Common Name* of *Papists*, and so be subdistinguished into *Roman-Catholick Papists* on one hand, and on the other hand into *Fanatical-Papists*, otherwise called *Dissenters*. (*Tracts*, p. 47)

In viewing Anglicanism as the *via media* between extremes which were in fact indistinguishable, Wells followed a line taken by Anglicans of all shades.[14] But his writings also make it clear that he deployed the rhetoric of Anglican moderation for High Church purposes. This is most apparent in his equation of the High Church with the True Church, a standard Tory attitude in Queen Anne's reign:[15]

> he then that professing himself to be a *Member of the Church of England*, does accordingly keep up Conscientiously and Strictly to its Doctrines, Rites, and Ceremonies, the same does act herein *not* a whit *Higher* than his Duty requires, and therefore to act thus *High*, or to be in this Sense a *High* Churchman, is no other in reality than to be a *True* Churchman. (*Tracts*, p. 293)

For Wells, as for his High Church predecessor Heylyn, the Church of England is the independent and Catholic church in England (*Tracts*,

p. 6), rather than being merely the product of the Reformation. Being a Catholic church, it is seen by Wells to rightly take into account patristic traditions. This leads Wells to be antagonistic towards more rationalistic variants of Anglicanism. Thus he criticises the Cambridge Latitudinarians with whom Echard had aligned himself, 'not being of those *Latitudinarian*, alias *Comprehension*, alias *Moderation* Principles, which are Now adays become so Modish among some, to the almost Total Decay of True Christianity'.[16] This attack on the Latitudinarians comes in a pamphlet aimed at Samuel Clarke who, while an Anglican, had published a work on the Trinity which was widely viewed as disputing Christ's divinity. Wells's *Remarks* on Clarke attacked him in predictably High Church fashion as too reliant upon reason in his divinity and insufficiently attentive to the patristic evidence.[17] Just as Presbyterians and Roman Catholics could be lumped together as in error, so within Anglicanism the rationalism of Clarke and of the Latitudinarians were conflated and condemned from a High Church perspective prioritising tradition over (or more accurately as a form of) reason.

Given this context of theological and denominational argument in Wells's writing career, it would be surprising if his historical geographies of the New and Old Testament did not reveal a partisan purpose. The authorities Wells used in his historical geographies were eclectic, ranging from Jesuits to Calvinists,[18] but they were collected in the service of a project of far less ecumenical intent. This project is signalled by the dedication of the *Historical Geography of the New Testament* to Queen Anne, the idol of the Tory and High Church Faction. Wells praises Anne for having '*nothing more at Heart, than to preserve* (That True Branch of it [i.e. the universal church]) *the Church of* England'. Wells adds that 'in Conformity to Your Majesties Royal and Pious Example' he also sees his task as being 'to *Defend its Truly Apostolical Doctrine and Discipline against All Opposers Whatever*'.[19] Wells's geographical writings, then, are sign-posted by him as of a piece with his controversial writings. The influence of Wells's High Churchmanship upon his geographical writings can be seen in two ways: by the very project of writing geographies of the scriptures and the methodology deployed; and by occasional eruptions of polemical divinity into these texts.

The historical geographies of the Old and New Testament are premised upon the assumption that reason alone is not sufficient for the understanding of scripture. As such, Wells's entire geographical project rests on the same theological foundation as his dislike of

Latitudinarianism and the rationalistic religion of Clarke. Wells admits that the doctrinal side of scriptural exegesis is more important than the historical, but still considers the latter 'very commendable' (*NT*, A5r). In the texts themselves, the same mentality is displayed in Wells's willingness to rest arguments about places mentioned in the Bible on traditional attributions. Thus in his discussion of Sodom and Gomorrah, he confesses: 'As to their particular Situation, nothing certain can be (I think) determined concerning it; and therefore we must be content to acquiesce in the Situation commonly assigned them'.[20]

The High Churchmen were always more interested in church traditions and more open to traditionary reasoning due to their scepticism about the rationalistic reasoning espoused by Low Church divines. While it is true that Wells's texts 'show the tension between revealed knowledge (based on sacred texts) and experimental knowledge (based on mathematics, navigation and exploration)',[21] the extent to which he is still prepared to resort to tradition and etymological arguments locates his geographical enquiries in the ambit of his High Churchmanship.

Examining the texts themselves, Wells's polemical divinity is in evidence. Unlike Heylyn's *Cosmographie*, where Laudian interests were woven into the text subtly, however, with Wells the bulk of his two historical geographies comprised fairly anodyne hackwork, with long passages taken directly from his sources, but there was the occasional eruption of his own opinions which stands out from the text. Most notable is the three page defence of Trinitarianism in the *Historical Geography of the Old Testament*:

> It may be of good Use to observe here once for all, that by *God*, who is here and in other Places of Scripture, said to appear to the Patriarchs and other holy Men ... is to be understood, according to the general Doctrine of the Primitive Fathers of the Christian Church, *God the Son*. The Persons, whom I denote by the *Primitive Fathers of the Christian Church*, were such as lived in the very Age next to the Apostles ... Whence it appears, how derogatory to our Blessed Redeemer, how repugnant to the Doctrine of the Primitive Christian Church, and consequently, how much to be detested by all true Christians, is the Tenet of the *Socinians* or *Unitarians*, or whoever else deny Christ to be the *Very God of Very God*. (*OT*., I 362–4)

Two years before his attack on Samuel Clarke, Wells here articulates in a geography book the line he would take against Clarke, whose

Scripture Doctrine of the Trinity (1712) doubted precisely whether Christ was of the same status as God the Father. Indeed, as early as 1708 Wells had said the *Historical Geography of the New Testament* that Jesus was fully divine in his nature, '*Equal with God*, and *was* no other than *God*' (*NT*., p. 33).

Some of Wells's theological polemics were triggered more directly by the subject under discussion, and thus seem less incongruous, but they clearly moved beyond historical-geographical interests. A good example comes in his discussion of the Levites, the priestly tribe of the Israelites. Wells discusses the Levitical cities, which clearly were within the scope of his project: 'Among the *Levitical* Cities, there were ... thirteen assign'd to the Children of *Aaron*, the *Priests*, which are therefore by some Writers distinguished by the Name of the *Sacerdotal* or *Priests* Cities' (*OT*, II 254).

Wells was then drawn on to a defence of the primitive status of episcopalian church government of the sort he had given in his pamphlet attacks on the Presbyterians:

> For tho' *Aaron* and his Children, the *Priests*, were of the Tribe of *Levi*, and so were *Levites*; yet all that were *Levites* were not *Priests*; but the Priesthood appertain'd to the Children of *Aaron*. So that from the beginning of Gods giving Rules for the outward Order and Government of his Church here on Earth, there has been Three Orders Instituted among those that Minister about Holy Things; namely, the *High Priest*, the *Priest*, and the *Levite*, under the Legal Oeconomy, or in the *Jewish* Church; and answerable hereunto, the *Bishop*, the *Priest*, and the *Deacon*, under the Gospel Dispensation or in the Christian Church. (*OT*, II 254–5)

The Presbyterians, then, are refuted by historical geography: the mere existence of a distinction between the sacerdotal cities of Aaron's children and Levitical cities in general shows the antiquity of the distinctive functions of bishop, priest and deacon. Urban networks inscribe the geography of episcopalianism.

Having found ways in which to attack the Presbyterians in his geographical writings, it is unsurprising that Wells also found ways to attack the other form of Christian excess, which, as we have seen, he thought was interchangeable with Presbyterianism, namely Roman Catholicism. Wells's anti-Catholicism surfaces most explicitly at the close of his discussion of Nazareth. Here, in what he himself admits is an excursus (although 'not ... altogether impertinent'), Wells draws on the Protestant equation of Roman Catholicism with superstition:

> in how mean a Condition soever *Nazareth* may be at present, yet some Part of its Antient Buildings, I mean the Chamber wherein the Virgin *Mary* is said to be sitting, when the Angel brought her those joyful Tidings [i.e. the Annunciation] ... has had better Luck, even at the no less Expense than of a down-right Miracle, if we can believe the *Popish* Legends. For in these it is said, that this same Chamber, being after the Blessed Virgin's Departure, had in great Reverence by Christians, and remaining in *Nazareth* till the Holy Land was subdued by the *Turks* and *Saracens*, A.D. 1291, it was then most miraculously transported into *Sclavonia*. But that Country being unworthy of the Virgin's Presence, it was by the Angels carried over into *Italy*, and at length settled at *Loreto* then a Village in the Ecclesiastical State. (*NT*, pp. 21–2)

As might be expected given Wells's High Churchmanship and the methods he adopted in his historical geography texts, he was not attacking traditionary tales *per se*, but their Roman Catholic perversion ('popish legends'). Indeed, the *Historical Geography of the New Testament* in which this criticism of Roman Catholic superstition was made itself defended a story with no more than tradition to support it. The second part of the book related to the geography of the travels of St Paul, and Wells asserted Paul was 'the Great Apostle of the *Gentiles* in general, and the Proper Apostle of the *Britains* in particular, being (as Antiquity asserts) the First Planter of Christianity in these *British* Isles' (*NT.*, Part II, A3v–A4r). This was a tradition which Heylyn half a century previously had viewed as being less plausible than the belief that the primitive foundation of British Christianity was laid by Joseph of Arimathea. Whichever tradition was adhered to, however, its function was to demonstrate that the Church of England was both primitive and Catholic, with the supremacy of Rome being a later corruption of the Catholic and universal church.[22]

Within Wells's scheme, Roman Catholicism had enslaved people through superstitions such as that he related. As such, a distinction had to be drawn between tradition or antiquity and superstition or fable. Hence Wells's 'Geography of the Travels and Voyages of St Paul' could be based on tradition at the same time as his book as a whole lambasted superstition. This reinforces the earlier contention that the very undertaking of these historical geographies was a statement of

High Church intent. For Wells, to write a geography of the travels of St Paul was to select a subject which took on especial significance for a High Churchman, as St Paul gave the Church of England its primitive foundation and independence. In a discursive framework which relied on historical example to legitimate politico-theological action, the travels of St Paul sanctioned a pattern of church government. An historical geography which denied St Paul had travelled to England would threaten the apostolic status of an independent Church of England, and thus legitimate Roman Catholic attacks on it as modern and schismatic. As we have seen, Wells saw schism and sin as synonymous, such that the geography of the travels of St Paul was not merely a matter of scholarly interest, but rather of salvation.

Wells is representative of the Tory and High Church mentality in the ascendant in the reign of Queen Anne. He shares with Heylyn a High Churchmanship, but betrays no signs of feeling isolated. This is unsurprising given the power and patronage lavished on the Tories during Anne's reign. Working within the settlement of 1688, but appropriating Echard's smooth running machine for the Tory cause, Wells had every reason to be confident in the age of Bolingbroke, Harley and Swift. The Laudian era, the Restoration and the reign of Queen Anne were the high points of Anglican assertiveness, each seeing (progressively weaker) attempts to impose a confessional state on England. Wells therefore fits into a genealogy with Heylyn and Bohun, representatives of the two previous phases of militant Anglicanism, as a High Churchman whose geography books served such a vision of the church-state nexus. Wells's attempts to create a High Church geography were far less coherent than Heylyn's, which may in good measure reflect their respective intellectual abilities, but it may also reflect shifts in the nature of the Tory political language which became more apparent under the Hanoverians.

Thomas Salmon

Thomas Salmon's writing career began in the 1720s, as Wells and Echard laid down their pens. Salmon's scope as a writer has more in common with Echard, given that he wrote a number of historical tracts and never ventured into the realm of theology, but he gave his historical narratives a Tory twist which also aligns him with Wells. Unlike both men he had no formal education, and certainly the Oxbridge contexts which I have suggested informed the geographies of Wells and Echard were irrelevant to Salmon. Yet Salmon wrote what must be one of the most distinctive political and constitutional discussions in any geography book of the eighteenth century in his *Geographical and Historical Grammar* (1749). That discussion seems eclectic or even eccentric, and to see how he built up his palimpsest of views it is necessary to look at the historical and polemical works he produced before the *Grammar*.

Salmon produced three major historical works, 1722–40: *A Review of the History of England* (1722–24); *An Impartial Examination of Bishop Burnet's History of His Own Times* (1724); and *The History of ... Queen Anne* (1740). They show a high degree of consistency, each one taking the form of a Tory redaction of already-existent narratives which Salmon argued were misleadingly Whiggish. Interestingly, one of the historians he saw as too Whiggish was Laurence Echard, thereby agreeing with modern commentators and distancing himself from contemporaries who labelled Echard a Tory.[23] Salmon's strategy is clearly expressed in the *Review*:

> this Work ... is chiefly an Examination of such Histories and Annals of the three last Reigns [James II to Anne] as are already extant; and the principal Design of it is to prevent, if possible, partial Relations being convey'd down to Posterity as uncontested Truths; for our present Set of Historians seem all to write in one Strain, and to court one Interest.[24]

The last line may be a play on words, as the interest he claimed historians were courting was the Court, which was by the 1720s a Whig preserve under the Hanoverians.[25] Unlike the era of Queen Anne in which a Tory such as Wells could write confident of royal patronage, Salmon wrote as a Tory in an era of political proscription and this milieu is vital to the tone his works adopt, a tone which would eventually be transferred to his geographical writings.

In his redactions, Salmon consistently defended Tory administrations and policies, although in such a way as to make clear his hostility to Jacobitism.[26] This is perhaps most apparent in his *History of Queen Anne*, where Salmon's narrative was very positive about the Treaty of Utrecht (1713) brokered by the Tory government and argued that British foreign policy had to be centred on the sea, land war being favoured by the Whigs and Marlborough. His views on the latter issue give pointers to the ways in which he consistently criticised the Whigs:

> And thus the old Question was reviv'd, whether the Land or Sea Service ought to be most attended to: Not that the thing would bear

> a Question, but every one knows who commanded at land [i.e. Marlborough], what Alliances and Friendships he had contracted with the Treasury, and other Ministers.[27]

Salmon here suggests a Whig ministry would override the strategic interests of England in favour of nepotism, corruption and self-gain. Throughout the period 1720–42 Tory (and, indeed, opposition Whig) criticisms of a hegemonic Whig government would take forms such as this, suggesting that private (Whig) and foreign (Hanoverian) interests were being prioritised over public and national interests in a system driven by money and luxury rather than patriotic virtue.[28]

Like most Tories of the period, Salmon defended the Revolution Settlement as strongly as the Whigs,[29] but on a confusing medley of grounds. He denied that 1689 had been a contractarian settlement, but was equally dismissive of Bohun's argument that a just conquest had occurred.[30] Trying to characterise the constitutional framework established by 1689 more positively, Salmon's *Considerations on ... Naturalization* (1748) was equally miscellaneous. He referred, as Bohun had done, to the continental civil law tradition of Grotius and Pufendorf, but equally argued that from the time of Adam, fathers had possessed authority and that a contract had never been needed, appearing to allude to Filmer's *Patriarcha*.[31] As we have seen, this territory was trawled before by Bohun, but where Bohun rested supreme power in the king in a 'balanced absolutism', Salmon's image was of a balance of power more in accord with the ideal of checks and balances:

> If there is any fundamental or unalterable condition in our constitution, it is the distribution of the sovereign power between the crown and the two houses of parliament, and the poising of that authority so equally, that the one may not encroach upon much less subvert the rights and privileges of the other; for whatever tends to destroy this balance, and throw all the power into one scale, apparently tends towards the dissolution of this limited monarchy.[32]

Salmon's main concern, then, was not with political theory, but with acting as a monitor of the precarious constitutional balance. Where for the Whiggish Echard the constitution might seem to be a self-regulating machine, for Salmon it was a system reliant upon the public-spiritedness of its operatives, and as a Tory with no hope of preferment, he could play the patriot as guardian of the guardians.

In this role as monitor of virtue and corruption in public life to which so many Tory 'patriots' appointed themselves, Salmon lambasted the Whigs. As with many others, he chose to express this through historical writings.[33] This was most noticeable when he treated the exploitation of the British by their Dutch allies in the period when the Whigs refused to sign a peace with France to conclude the War of Spanish Succession:

> It will hardly be believ'd by Posterity, that our good Allies could have the Confidence to impose upon us at this Rate, much less, that our Ministry could suffer it for so many Years together: But if they reflect that this was done in an Age, when all publick Spirit was lost; when it was look'd on as the highest Folly to prefer the Welfare of our Country, to the making our private Fortunes, in an Age, when we were so far from being influenced by future Rewards ... [L]et Posterity consider this, and their Surprize will be less.[34]

This passage implicitly praised the Tories in the reign of Queen Anne for concluding the war at Utrecht in 1713, and aligned Salmon with Tories such as Swift and Arbuthnot, whose *History of John Bull* gave a similar account in allegory. Salmon clearly liked this passage, repeating it verbatim in his *History of Queen Anne*,[35] but its reference was not exclusively historical. For the language in which he condemned the Whigs of Anne's reign was readily applicable to the Whigs in his own era, and Salmon sign-posted this in the *History of Queen Anne*. First, he argued that the Treaty of Utrecht had 'banished for a Time' 'that covetous Stock-jobbing Spirit', a phrase strongly associated with the perceived corruption of Walpole's government. Secondly, in summarising Anne's character, he lamented that in the era after her death 'that little selfish Spirit which discovered itself soon after the Revolution is again revived'.[36] If the malfunctioning of the English political system was attributable to a corruption which had been checked by Anne, only to be revived subsequently, the criticisms Salmon made of the corrupt Whig stock-jobbing regime of Godolphin were implicitly allowed to be directly transferable to the age of Walpole.

For over 20 years, then, Salmon developed his persona as a Tory guardian of the virtues of the balanced constitution established by the Revolution Settlement. To do this, he envisaged the constitution as a fragile composition threatened by excess power in any element, this being fundamentally at odds with the vision that absolute power created balance articulated by Heylyn, Bohun and Wells. This allowed

Salmon to speak a language of virtue and corruption (rather than sin and traitorousness) not present in previous Tory political languages. As we shall see, it is this complex of ideas which fed into the very different image of English politics which is articulated in his *Geographical and Historical Grammar*.

At a cursory glance, the *Geographical and Historical Grammar* might appear to show no signs of any distinctive political language. The section on England starts with the conventional patriotic pæans to the climate, which avoids the continental European extremes of heat and cold, and follows this up by seeing the English themselves as similarly 'a good Medium between the Dutch and the French'.[37] The image of the constitution appears initially to follow the same conventional pattern, describing it as a balance of extremes: 'Every *British* Gentleman is sensible, that he lives in a Country where Life, Liberty, and Property are better secured than in any Kingdom in *Europe*' (*GHG*, p. 314). Yet at this point Salmon's own political position, detailed above, makes its impact, for his fear is that this constitution is not Echard's smoothly functioning machine, but an impossible balancing act which is always being pushed towards the extremes of tyranny by self-interested parties who wish to monopolise power. He starts this analysis by criticising the notion that power does (or can) come directly from the people:

> The Commons are said to represent the People, tho' they do not in Reality represent a fourth Part of them ... If there was any stress therefore to be laid on that Maxim, *That all just and legal Power is derived from the People* (from the Multitude) then there has been very few just or legal Governments in this or any other Nation. (*GHG*, pp. 314–5)

Salmon's point is that a full democracy is an impossibility, and he thus recapitulates the point he made about the events of 1688: that a contractarian understanding of the origins of political legitimacy is a delusion. But in making this point he exposes the nature of British 'representative' democracy in the period, when patronage and the control of constituencies by local elites was a vital part of political life and when the franchise was carefully limited by money or property requirements.[38] Salmon adds a further point about the limitations of the franchise, which has exercised modern historians more than contemporaries: 'the Ladies also may think it a Hardship, that they are neither allowed a Place in the Senate, or a Voice in the Choice of what is called the Representative of the Nation' (*GHG*, p. 315).

Given this image of the operation of the British constitution as a system in which democracy is severely but unavoidably limited, Salmon by extension is suggesting that the British balance of liberties relies on none of those elite groups which have access to power being allowed to monopolise that power. Unfortunately, the picture Salmon paints is of a political system being driven towards the excess concentration of power in the hands of the few. One potential source of such monopolisation is the monarchy: 'And, if the Crown should assume an absolute Dominion, how can this be remedied? For the Executive Power ... and the Command of the Forces by Sea and Land ... are vested solely in the Crown by Law' (*GHG*, p. 315).

Salmon adds that the crown has the power to suspend *habeas corpus* as well, such that the legal system cannot prevent the infringement of liberties, but as is clear from the conditional grammar of the quoted passage, he does not regard this as a perversion of the British constitution which is actually taking place, only one for which the constitution gives no redress. Far more serious for Salmon, because in the process of actually undermining British liberties, is the concentration of power in the hands of factions in church and state. Salmon paints a bleak picture (entirely within the standard Tory rhetoric) of the church in danger:

> Of the Religion professed in the *British* Isles it may be said as of the Constitution, that it was originally the best Institution in the World, but has been so metamorphosed and deformed, the Professors broken into so many Sects and Parties that bear an implacable Enmity to each other, that the Spirit of Christianity is in a Manner lost; Vice and Profaneness reign triumphant; the sacred Truths of Christianity are questioned and disputed; and a Man that is not an Infidel is scarce allowed to have common Sense among those that look upon themselves to be the polite World. (*GHG*, p. 318)

For Salmon, the established church should be given far more unequivocal support. That he believes the Whigs, who had been in power from 1715, had failed in this is testified by his barbed comment equating the state with the irreligious:

> An Episcopal Church is said to be established in *England*, but so weakly established that every one is at Liberty to dissent from it, and strike out what Religion he pleases; he may declare himself an Infidel with Impunity, and these are esteemed by some the best Friends to the State. (*GHG*, p. 318)

This defence of the established church shows clearly that Salmon's political position was not simply that of a disgruntled opposition Whig out of power, but of a Tory, given that opposition Whigs betrayed an anticlericalism which always left them at some distance from Tories.[39]

Discussing the present state of British government, Salmon remained circumspect, speaking in vague but easily decipherable terms about the dangers to liberty posed by the 'friends to the state'. The equation of the problems of corruption with the Whig supremacy was made clearer in his long historical section on 'Revolutions and Memorable Events'. He continually refers to seditious books published by Whigs which undermined the church (as Defoe's *Shortest Way with Dissenters*, *GHG*, pp. 319–20), and adds a passage from his personal *bête noire*, the Whig bishop and historian Gilbert Burnet, which was clearly designed to be read in the context of present politics as well as those of the 1690s: 'Bishop *Burnet* observes, that *few were preferred at this Time but* Whigs, *except they purchased their Places, for the* Whigs *he says set every Thing to Sale*' (*GHG*, p. 310). Finally, and in what is the most notable attack on the Whigs, Salmon maintains that it is the Whigs rather than the Tories who created the problem of Jacobitism. The experience of complete proscription from patronage and power under George I destabilised relations within the oligarchy controlling the quasi-democratic system of British government:

> His Majesty [George I] being persuaded, that the late Queen [i.e. Anne], as well as her last Ministry [Harley's Tories], were engaged in a Design to bring in the Pretender, though this is generally supposed to have been an Artifice of the *Whigs* to get into Power, and keep their Enemies out of the Administration. Had his Majesty been pleased to have encouraged all his Subjects equally at his Accession, the Pretender's Name had never been mentioned. (*GHG*, pp. 332–3)

While Salmon wrote this well after the Jacobite rising of 1745, the Jacobite threat was still credible, riots and demonstrations of loyalty to the Stuarts continuing to perturb the government.[40] As such, Salmon's analysis in an era when the Tories remained in the wilderness also implied that the Jacobite threat would remain intact for as long as the Whigs monopolised power and marginalised Tory families with a legitimate claim to participate in the distribution of patronage and rewards.

Salmon's *Geographical and Historical Grammar*, then, continues the Tory opposition to the Whig monopolisation of power which his historical writings had articulated. Salmon gives an abridged version of his argument that the British constitution only functions smoothly when disinterested or at least balanced patronage is dispensed. Given the limited size of the ruling group, to exclude one element of that group is to imperil the constitutional balance. The signs of decay Salmon finds in the modish contempt of the established church, in the proliferation of seditious pamphlets and in the continued threat of Jacobitism are all explained by just such an imbalance in the distribution of power created by the Whigs at the accession of George I. Within this piece of hackwork, then, Salmon managed to develop an appropriately simplified version of the Tory deployment of patriot rhetoric.

By the late 1740s when Salmon wrote his *Geographical and Historical Grammar*, the patriot rhetoric of virtue and corruption was already losing much of its popularity. After Walpole was finally ousted in 1742, the succeeding administrations which had deployed patriot language when in opposition signally failed to change anything about the operation of the patronage system, which led to a widespread disillusionment with, and drift from, that language.[41] In this context, it is worth noting that posthumous revisions of Salmon's enduringly popular geography book toned down the political position and language he had adopted, as it was so clearly the product of an earlier era of political argumentation which no longer seemed relevant.[42] The thirteenth edition of 1785, for example, edited the political stance of the book into anodyne national triumphalism by retaining Salmon's praise of the constitutional structure while softening his fears over the precariousness of that structure. The established church was said to be episcopal with no threats to it being discussed, and no sense of Whig domination was given by the discussion of parliament.[43] In part this must reflect the readmission of Tories into government with the accession of George III in 1760. It also shows the opposite process of editing from that which Bohun had suffered in 1689. Where editorial practices had exposed Bohun's *Geographical Dictionary* to the charge of Jacobitism, they toned down the political partisanship of Salmon's *Geographical and Historical Grammar*. Both examples show that geography books, so many of which ran through numerous editions, did have their political language and stance changed gradually over time, although they appear to be (and often were) little more than reprints with the title page changed to boost sales. Given the shifting tides of political context and language, to leave a book as it was laid it open to misreading in the light of subsequent political languages. The alterations to Salmon's book took it out of the context of Walpole's premiership in the 1720s–40s, but did not give it a clear political positioning in the 1780s. The language of the marginalised Tory was itself marginalised textually in an era when, with the Tories back in power, it could no longer serve any function.

8
The Denominational Politics of Travel-Writing: the Case of Tory Anglicans in the 1770s

Travel writing, politics and landscape

We have already seen in Chapter 2 that geographers distinguished their practice textually from travel writing, seeing the latter as raw material for geography. Travel writers agreed with this textual division of labour. Thus Daniel Defoe wrote in his *Tour* upon reaching Cambridgeshire: 'As my Business is not to lay out the Geographical Situation of Places, I say nothing of the Buttings and Boundings of this County'.[1] For Defoe, as for geographers, the traveller's art was to respond to places visited, where the geographer was concerned with mathematically-determined location and with the collation of information provided by travellers. But in his disclaimer, as in the geographers' careful delineation of the spheres of geography and travel, lay the recognition that the two forms of writing were closely related, travel acting as a form of geographical knowledge, even if it was not geography *sensu stricto*. One way in which geography and travel writing were related was that they both encoded political positions. As we shall discover, true to its different generic mode of operation, travel writing offered different sites for politicisation from those we have analysed for geography books, but unsurprisingly the key issues were the same.

There can be no doubt that travel as an activity and the travel account as a genre were seen as within the ambit of political discourse in the eighteenth century. This can be gleaned by investigating any of the numerous advice books which aimed to instruct would-be travellers in what to observe and how to record it in order to gain maximal improvement from their trip. Advice books for travellers were in no sense a new phenomenon by the eighteenth century, the English-language having seen examples for at least two centuries.[2] A popular example was Josiah Tucker's *Instructions for Travellers* (1758). Tucker was to become a prominent political commentator in the 1770s in the unavoidably geopolitical crisis of the American War of Independence,[3] but a decade earlier his *Instructions* recognised at a more general level that geography, travel and politics were interwoven. For Tucker, the model traveller 'must dedicate his principal Studies towards tracing such secret, though powerful Effects and Consequences, as are produced by the various Systems of Religion, Government, and Commerce in the World'.[4] Travel, then, and travel writing were primarily politicised activities. In discussing how travellers should discern the state of religion and government in the areas through which they travelled, Tucker laid particular emphasis upon the appearance of the landscape as a reliable gauge:

> during his Travels he should constantly bear in mind the grand Maxim, That the Face of every Country through which he passes, the Looks, Numbers, and Behaviour of the People, their general Cloathing, Food, and Dwelling, their Attainments in Agriculture, Manufactures, Arts and Sciences, are the Effects and Consequences of some certain Causes; which Causes he was particularly sent out to investigate and discover. ... [C]ertain it is, that every considerable Effect must be ascribed, and may be traced up to, one or more of these Causes ... Natural, – Artificial, – Political, – and Religious.[5]

Using landscape as an indicator of political health became something of a commonplace in the eighteenth century, being repeated by many travellers such as Tobias Smollett.[6] But where Smollett made his comments in France and Italy, Tucker's *Instructions*, like most British advice books for travellers, saw European journeys as something to which young travellers only graduated having first grown familiar with their own nation: 'Now Travelling into foreign Countries for the Sake of Improvement, necessarily pre-supposes, that you are no Stranger to the Religion, Constitution, and Nature of your own. ... Therefore let a young Gentleman begin with the Tour of his own Country'.[7] The reasons for this emphasis on the virtues of domestic travel were twofold, both being political. First, domestic travel would engender the same rational pride in the constitution and church government of Britain as it had in Camden's case. Consequent to this, and secondly, such patriotic pride was essential to prevent young minds being seduced by the forms of government which they would encounter on the continent, and more particularly by the liturgical 'pomp' of Roman Catholicism. As Tucker put it, in terms entirely commonplace from the Reformation on,

> We must first begin with Religion, not only because it is the most Important in its Nature, but because if a Traveller is not well grounded in the Principles of it before he sets out, he will run the Risk either of having none at all during his whole Life, or of being made a Convert to a very bad one, I mean the Popish.[8]

For Tucker, then, and for a host of similar authors of advice literature for British travellers, travel and travel writing were political activities, 'politics' being understood to include constitutional, governmental, theological and ecclesiological issues. The landscape, or 'face of the country', was a key gauge by which the intelligent traveller would adjudicate on political questions, starting with the British Isles themselves. Accordant with this vision of how travel and travel writing were politicised, in this chapter I will analyse the textual politicisation of travel books mainly about the British Isles, focusing on three writers in the 1770s and paying close attention to religious politics. This should redress an imbalance in recent scholarship, which has tended to focus on governmental rather than religious politics, and on the politics of British travel in a European rather than a domestic context.[9] I will particularly trace the ways in which theological, ecclesiological and liturgical issues inflected the presentation of landscape in various travel writings: as landscape was viewed by contemporaries as a key political indicator, it provides a revealing site for comparison between travellers.

Thomas Pennant: a travelling Tory Anglican Scotophile

Thomas Pennant described himself as 'a moderate Tory', going on to explain what he meant by this:

> I admired the virtues of Lord *Russel*, and pitied his fall. I should have been a Whig at the Revolution. There have been periods since, in which I should have been, what I now am, a moderate Tory; a supporter, as far as my little influence extends, of a well-poised balance between the crown and people.[10]

Writing after the French Revolution, Pennant had no doubt that the greatest threat to the constitutional balance of 1688, which he saw as the embodiment of political wisdom, came from populist Jacobinism. In this connection, Pennant threw himself into anti-Jacobin politics in his native Wales, founding a Loyal Association in 1792, of which he was appointed chairman.[11] As Pennant's comment on his 'moderate Toryism' implies, this was not the first time he had opposed the forces of what he saw as anarchic populism: believing that the American secession was 'the most deplorable event in all the annals of *Great Britain*', he had tried to quell the unrest of Wilkes's 'pretended patriots' in the 1770s and written his *American Annals*, accusing those who had led the British forces in America of a culpable incompetence which had led to the victory of a deluded mob.[12] As Pennant's references to Great Britain make clear, he did not actually defend the constitutional settlement of 1688, but rather that which also included the 1707 Union with Scotland. Indeed, Pennant was hostile towards those who threatened to disturb the Union, arguing in his *Literary Life* that his two published tours of Scotland had 'labored earnestly to conciliate the affections of two nations, so wickedly and studiously set at variance by evil-designing people'.[13] Pennant's moderate Toryism, then, was one which continued the line we have already seen Edmund Gibson and the contributors to *Britannia* establishing, being conservative of the constitution in church and state established by the events of 1688 and 1707. In Pennant's lifetime this necessitated an opposition to Jacobitism, American secession, Wilkesite patriotism and finally Jacobinism.

Pennant made his name as a travel writer. The question arises as to how his political position was transferred into this genre. We can answer this at two levels: first, and structurally, by looking briefly at the textual sites which travel writing presented to Pennant (and others) for the expression of a political ideology; and secondly by looking at how he actually exploited those sites in his *œuvre*.

Pennant used the places he visited to encode his political position within travel narratives. The 'Preface' to Pennant's posthumous *Journey from London to the Isle of Wight* (1801) credited him with being 'the first who enlivened the dryness of topographical research with historical and biographical anecdote',[14] and it was these digressive anecdotes which gave Pennant scope for political expression. As an example, take Pennant's comments on the church at Mallwyd in his *Tour in Wales*: 'Reach *Mallwyd*, remarkable for the situation of the altar, in the middle of the church; which Doctor *Davies*, author of the dictionary, then incumbent, in defiance of the orders of archbishop *Laud*, removed again from its imaginary superstitious site at the east end.'[15]

Here Pennant articulated his hostility to the Laudian regime of which Heylyn had been an important proponent, also expressing a certain patriotic pride in Welsh 'tenacity of ... liberty', as embodied in John Davies.[16] The church fabric at Mallwyd, then, gave Pennant a means by which to express his ecclesiological convictions, opposing what he saw as the tyranny of Laud. In visiting a site of historical interest Pennant could be led on to political reflections, especially given the strong political import of history which we have already noted in connection with geographical writing. In other words, and generalising from this case, visiting an historical site provided the travel writer with an opportunity for historicised political interventions in present day debates akin to that which sections concerning national history provided for the writer of geographical gazetteers.

One particularly important subspecies of historical site should be mentioned, as Pennant repeatedly used it to express himself politically: the portrait. Pennant visited numerous country houses, and was often led from pictures of important actors in British history to evaluative judgements about the characters in question. As a companion piece to his comments on Mallwyd, we can take Pennant's reflections at Gorhambury on a portrait of George Abbot, Archbishop of Canterbury under James I, whose Calvinist beliefs we have already come across in Chapter 3, as expressed in his geographical work, the *Briefe Description*:

> the Calvinistical principles with which he was strongly imbued ... [were] probably too favorable to the tenets which, under him, acquired strength, in the following reign, to subvert both church and state, with the assistance of the contrary conduct of the indiscreet and furious Laud. How difficult is the virtue of moderation![17]

Pennant here espouses a moderate Anglican *via media* of the variety which 1688 put in place, toeing a line between Calvinism and Laudian Arminianism, while imputing to the extremism of both positions the environment in which the Civil War could develop. Abbot's portrait, then, allowed for an historical reflection which aligned Pennant with loyalism and moderation, implying that these virtues were codified in the period after the Civil War. Once more, a generic parallel can be made: portraiture allowed for political interventions akin to the character sketches in historical sections of geography books.

Having analysed the textual sites at which Pennant could politicise his travel writings, we can now look at the extent to which he embraced these opportunities. In matters of political governance, Pennant supported the balance of power between state and people which he believed had been established in 1688. On the one hand, this was expressed in Pennant's hostility to populism. Like so many landowners, Pennant feared the lawlessness of forest areas, commenting at Threapwood that

> it is to be hoped, that [the] legislature will take an opportunity of rendering the magisterial power as valid here as in other places; especially when it is to be considered, that there are to the amount of two hundred and sixty-seven inhabitants, who want instruction in the doctrine of universal submission to law.[18]

For Pennant, acquiescence to the law should not vary geographically. A parallel case, but far more serious to Pennant as we have already seen from his *Literary Life*, was the geopolitically-enabled want of submission to the law which the American War of Independence represented. Here Pennant, in an extraordinary passage in his Isle of Wight *Journey*, adopted a providential and conspiratorial language worthy of Foxe's *Book of Martyrs*. Visiting the armoury at Purfleet, Pennant discussed Benjamin Franklin's advice concerning the protection of the gunpowder store from potentially lethal lightning strikes, being led on to a politicised parallel:

> The very same philosopher, who, living under the protection of our mild government, was secretly playing the incendiary, and too successfully inflaming the minds of our fellow subjects in *America*, till the great explosion happened, which for ever disunited us from our once happy colonists.

Pennant went on to point out that Franklin's system did not work, the Purfleet store being struck by lightning in 1777, and here he cast Franklin in the role of Guy Fawkes: 'the evil genius of the wily philosopher stood victorious, and our capital narrowly escaped subversion'.[19] Pennant, then, in his travel works as in his public life, opposed what he saw as false patriotism, which misled the people about the quality of their governance and encouraged American schism.

But Pennant was scrupulous in his adherence to a balance between the people and the state in his travel writing, taking a number of opportunities to lambast those he saw as threatening the constitutional balance from the other, absolutist, side. This led Pennant to be hostile towards the reign of Elizabeth I, who thought parliaments 'the mere instruments of giving sanction to her will ... Her prerogative was the rule of government ... and this was the GOLDEN REIGN OF ELIZABETH!'[20] Predictably, Pennant criticised the 'absolutist' monarchs of the Restoration, whom Ogilby's geographical iconography had supported, 'the royal rakish' Charles II and James II, who 'proved false to his high trust' and was replaced by 'the sullen mistreated hero William'.[21] Pennant, then, true to his self-declared moderate Toryism, used his travel writings to defend the balance of constitutional powers which he viewed as enshrined in the Revolution settlement.

If Pennant at a specific level defended the outcome of 1688/89, he also deployed the more general political language of civic humanism in his travel writings, lamenting the threat to *virtu* posed by the extension of luxury and commerce. Where Pennant found pockets of yeoman virtue, he saluted it: he found the Eglwysey valley 'chiefly inhabited (happily) by an independent race of warm and wealthy yeomanry, undevoured as yet by the great men of the country'.[22] But this was the exception to a rather more bleak rule, which Pennant, true to the method of Tucker's *Instructions*, saw embodied in the landscape of Northumberland:

> A humour fatal to the commonwealth prevails over many parts of the north, of flinging numbers of small tenements into a large one ... [People are] forced to exchange the wholesome, the vigorous, the innocent lives of the rural oeconomists, for the sickly short-lived employs of manufacturers in *Birmingham*, and other great towns, where disease, and often corrupted morals, cause double the consumption of people ... And all these improvements result from the unprincipled and iniquitous notion of making the buyer of the produce pay ... to enable the farmer to make a princely fortune, and to live with a luxury the shame of the times. They have lost the respectable character of the old English yeomanry.[23]

It was in this context of a decay of civic virtue grounded in the land that Pennant looked to Scotland as potentially redemptive of British moral, martial and political life. Pennant alleged that he started the vogue for Scottish tours,[24] and he certainly gave North Britain a key role in defending the achievements of 1688. He saw the final defeat of the Jacobites in 1745 and the destruction of Highland culture as extreme but essential measures, which finally extirpated feudalism from the British Isles. The townscape of Perth was taken as a good indicator of the effects of 1745: 'this town, as with all *Scotland,* dates its prosperity from the year 1745; the government of this part of *Great Britain* having never been settled till a little after that time. The rebellion was a disorder violent in its operation, but salutary in its effects'.[25]

As this suggests by implication, the rural landscape was also seen by Pennant as a fair gauge to the improvements which enlightened and constitutional government in the mould of 1688 had brought to Scotland:

> inattention to agriculture continued till lately; but on reaching *Eytoun* [i.e. Ayton], the scene was greatly altered; the wretched cottages, or rather hovels of the country, were vanishing ... the lands are inclosing, and yield very good barley, oats, and clover; the banks are planting: I speak in the present tense; for there is still a mixture of the old negligence left amidst the recent improvements, which look like the works of a new colony in a wretched impoverished country.[26]

The Scottish landscape, then, bespoke a country in political transition, but this did not simply mean negatively that Scotland was 'catching up' with England, but positively that it had yet to experience the decay in civic valour which commerce and luxury had engendered among the English yeomanry:

> The noxious part of the feudal reign is abolished; the delegated rod of power is now no more. But let not the good part be lost with the bad: the tender relation that *patriarchal* government experiences, should still be retained: and the mutual inclination to beneficence preserved.[27]

It was this ancient Scottish *virtu* which could redeem modern Britain. This was expressed in Pennant's rhapsodic Ossianic 'vision', with which he closed the first volume of his second, 1772, Scottish *Tour.* In this vision, Pennant fabricated a dream in which a chieftain from the age of Ossian (James Macpherson's supposedly pre-Christian Highland epic poet) spoke to him of the role the Scots could play in the era of the traumatic American War of Independence:

> they [the Scots] will become willing substitutes for their Southern brethren. Occupied in the soft arts of peace, THOSE should extend

> your manufactures; and THESE would defend your commerce. ... Have not thousands in the late war proved their sincerity? Have not thousands expiated with their blood the folly of rebellion, and the crimes of their parents?[28]

Pennant's travel writings, then, expressed his perpetual vigilance in defence of the constitutional balance created in 1688 and 1707. Opposing both populism and absolutism by various textual interventions and historical digressions, Pennant's response to a century of commercial change after the 'Glorious Revolution' was to envisage a division of labour between England and Scotland, which would cement the union between the two, creating a political unit with both commercial vitality and martial valour.

The constitutional balance which Pennant upheld in the political sphere was one he ultimately justified in theological terms:

> Feudal governments [*sic*] like that of unmixed monarchy has its conveniences and its blessings. The last rarely occur from the imperfection of human nature: One BEING only can lay claim to that; therefore it is the business of every honest man to resist the very appearance of undivided power in a prince, or the shadow of independency in a subject.[29]

Absolute power was the preserve of God, and to lay claim to it as a mortal of limited understanding was impious hubris which would lead to political catastrophe. This raises an important question: What theological outlook did Pennant espouse in his travels? Pennant described himself as 'a sincere well-wisher to the pure form of worship of the church of *England*', a description with which his son's biographical memoir concurred.[30] Pennant's definition of a 'pure form', as we shall see, was a fairly low-church one, unusually sympathetic to Presbyterianism for an Anglican communicant.

We have already seen Pennant's use of several sites to attack the Laudian vision of the Church of England, and Pennant renewed his attack on High Church Anglicanism elsewhere, as when he reached the parish of Selatyn, where 'that high-church meteor, that party-tool, Dr *Sacheverel* [*sic*] was, in 1709, presented to this living'.[31] Given this, it is unsurprising that in his early tour around Europe Pennant was uncompromising in his criticism of Roman Catholicism.[32] But as in his comments on political government, so in ecclesiastical polity, Pennant attacked all extremes, likewise denouncing the Calvinist Covenanters in Scotland as 'false prophets' and being repulsed by John Knox's 'fanatic fury'.[33] In the classic vein of Anglican *via media* argumentation, Pennant saw extreme Puritanism and Catholicism as essentially the same, both being politically motivated movements with little or no true spirituality. Thus he compared French and Scottish religious extremism:

> where rebellion is the object, religion *mutatis mutandis* makes little difference. The Preachers of the popish league, and those of the puritanical were the common incendiaries of each people; the pulpits of the Popish religious and of the disciples of Knox resounded equally with sedition, and the consequences equally destructive to their respective countries.[34]

Pennant's attack on the Covenanters and Knox was not the lead-in to a standard Anglican critique of Presbyterianism of the variety we have already witnessed in Edward Wells. On the contrary, Pennant's travel writings show that he was highly favourable to the modern Scottish kirk. Discussing belief in witchcraft, he denied that the Scottish were a more superstitious (and therefore less religiously educated) people than the English, given 'that ridiculous imposture in the capital [i.e. London] itself, in 1762, of the *Cock-Lane* ghost, which found credit with all ranks of people'.[35] But Pennant went further than this, arguing that in pastoral, political and theological terms, the Scottish church was superior to the Anglican. At the pastoral level, Pennant praised the modern, moderate Presbyterian minister:

> The clergy of *Scotland*, the most decent and consistent in their conduct of any set of men I ever met with of their order, are at present much changed from the furious, illiterate, and enthusiastic Teachers of old times ... They never sink their characters by midnight brawls, by mixing with the gaming world, either in cards, cocking, or horse-races, but preserve with a narrow income, a dignity too often lost among their brethren south of the *Tweed*.[36]

At the political level, Pennant was drawn to Presbyterian church government as it seemed to be the equivalent of the political balance he believed had been enacted by the events of 1688: 'upon the whole it appears that in the judicatories of the church of *Scotland*, there is an equal representation of the Laity as of the Clergy, which is a great security to the Laity against the usurpations of the Clergy'.[37]

Episcopalianism, by contrast, gave too much power to the clergy such that they, paradoxically, were in danger of assuming a profane, God-like absolute power which they lacked the faculties as mortals to merit. Akin to Edmund Gibson, then, Pennant by implication supported the reining in of the Anglican Convocation. Structurally, Presbyterianism was superior as it prevented such hubris, enshrining church–political checks and balances. Finally, at the theological level, Pennant gave gentle voice to the notion that Presbyterianism was the primitive model of church government, rather than the Cyprianic episcopal vision which Heylyn and later generations of Anglicans had espoused. Visiting Iona, Pennant was led into this pro-Presbyterian theological discussion:

> *Bede* speaks of the singular pre-eminence, and says that the island always had for a governor an Abbot-Presbyter, whose power (by a very uncommon rule) not only every province, but even the bishops themselves, obeyed. From this account, the enemies to episcopacy have inferred, that the rank of bishop was a novelty, introduced into the church in corrupt times; and the authority they assumed was an errant usurpation ... In answer to this, archbishop *Ussher* advances, that the power of the abbot of *Jona* [*sic*] was only local; and extended only to the bishop who resided there ... But notwithstanding this, the venerable *Bede* seems to be a stronger authority.[38]

If politically Pennant saw the Scottish as a potential source of civic and martial virtue to counterbalance England, it is likewise apparent that he was well predisposed towards the Scottish church, seeing it as more pure in organisational and pastoral terms than its English counterpart. Again, then, Scotland provided a site which had resisted corruption, and which could provide moral fibre in the post-1707 context. For a Tory Anglican, Pennant was unusually Scotophiliac.

Varieties of Tory Anglican traveller

Thomas Pennant was only one variety of self-declared Tory Anglican traveller, and one who proved unusually positive in his response to Scotland. I now turn to two further travellers of the 1770s who published accounts of their journeys, both of whom were Tory Anglicans, but whose visions of Anglicanism being different from Pennant's led them to adopt very different views of the landscapes through which they progressed, notably with respect to Scotland.

John Wesley: Tory Evangelical Anglicanism in Scotland

After the death of his brother, John Wesley defended Charles and himself from the charge of Jacobitism, whilst spelling out his own political position:

> Most of those who gave him [i.e. Charles] this title [of Jacobite] did not distinguish between a Jacobite and a Tory; whereby I mean 'one that believes God, not the people, to be the origin of all civil power'. In this sense, he was a Tory; so was my father; so am I. But I am no more a Jacobite than I am a Turk; neither was my brother.[39]

As is clear, Wesley's definition of his political position was unavoidably religious, allying Toryism with some form of *jure divino* theory. But where this might appear to link Wesley closely with Pennant, who also espoused a theologically grounded Tory politics, Wesley in fact stood at some considerable distance from Pennant. Discussing Pennant's Scottish tours, Wesley commented that he was:

> a lively as well as judicious writer. Judicious, I mean, in most respects; but I cannot give up to all the Deists in Great Britain the existence of witchcraft till I give up the credit of all history, sacred and prophane.[40]

Wesley, then, was aligning himself with a position – the belief in witchcraft and ghosts – we have already seen Pennant dismiss as superstitious, and thereby distancing himself from Pennant's low-church rationalistic Anglicanism.[41] While both Pennant and Wesley were professing Anglicans, Wesley was more doctrinally so, arguing that the Church of England was 'nearer to the Scriptural plan than any other in Europe', which, as we have seen, is not a view Pennant held.[42]

Wesley's response to Pennant as a traveller was part of a more general pattern he exhibited, evaluating travellers according to the impact their works would have on faith in the Christian message.[43] Thus Wesley attacked both the Abbé Raynal and Lady Mary Wortley Montagu for their positive comments about other, non-Christian cultures.[44] This evaluative framework became most apparent at the end of Wesley's life, as he became increasingly incensed by the praise heaped upon the newly-discovered South-Sea Island cultures for their apparently prelapsarian purity:

> Meeting with a celebrated book, a volume of Captain Cook's *Voyages*, I sat down to read it with huge expectation. But how was I

> disappointed! I observed (1) things absolutely incredible: 'A nation without any curiosity'; and, what is stranger still (I fear related with no good design), 'without any sense of shame! ...' Hume or Voltaire might believe this, but I cannot. (*Journal*, VI 7)[45]

For Wesley, then, the geographical knowledge produced by travellers was to be assessed primarily by the extent to which it chimed with Christian doctrine.

Given Wesley's religious assessment of travel writing, it is unsurprising that he placed his own incessant travelling within a religious context. In his sermon, 'The General Spread of the Gospel', Wesley argued that at the Second Coming of Christ, 'the loving knowledge of God ... shall cover the earth'. Within this framework, Wesley believed that the commencement of this evangelical process had been witnessed in Oxford, with the conversions to faith in the Holy Club:

> From Oxford, where it first appeared, the little leaven spread wider and wider. ... It afterwards spread to every part of the land, and little by little one became a thousand. It then spread into north Britain and Ireland, and, a few years after, into New York, Pennsylvania, and many other provinces in America.

In other words, an evangelical purity was diffusing geographically from its Oxonian beginnings, and Wesley was the key facilitator by virtue of his travels, which meant he had seen more conversions 'within fifty years last past than anyone in England or Europe'.[46] Further to this, Wesley pictured travel in his sermons as the activity of Fallen mankind, commenting that before the Fall 'man being ... (as he will be again after the resurrection) equal to angels, was able to convey himself at his pleasure to any given distance'.[47] In his sermons, then, Wesley saw his own travels as prefiguring the Second Coming of Christ, and being the archetypal labour of Fallen Man in search of redemption.

It follows from Wesley's theological reading of the activity of travelling that his account of his own peregrinations, the *Journal*, should fit within a squarely Evangelical framework, especially as the published version 'was essentially a vehicle for apologetic and propaganda, a selective and slanted account'.[48] We can see this by focusing on Wesley's depiction of nature and landscape in the *Journal*, which rendered them a gauge for his own Evangelical mission, giving a Methodist turn to the reading of the landscape which Tucker's *Instructions* had recommended to travellers.

One particularly noticeable feature of the *Journal* is its constant appeal to providence in interpreting both everyday and extreme natural events. This is most apparent when Wesley records sudden changes of weather conditions as he is about to preach, as at Yarm:

> Just as I began preaching ... the rain began; but it stopped in two or three minutes, I am persuaded, in answer to the prayer of faith. Incidents of the same kind I have seen abundance of times ... and they are nothing strange to them who seriously believe 'the very hairs of your head are numbered' (*Journal*, V 17)[49]

This constant recourse to providence as an explanatory framework when describing natural events as a travel writer was entirely consonant with Wesley's belief in God's continued intervention in daily life, which was backed up by an affirmation of the limits of the natural knowledge attainable by finite beings.[50]

If Wesley's recourse to providential explanation, which Pennant would have regarded as superstitious, implicitly spiritualised the operation of the natural world in his travel writing, other techniques he used to that end were both more self-conscious and more literary. Two major techniques are readily apparent. First, with respect to gardens, Wesley repeated invoked the vanity-of-human-wishes trope in his response. For whilst at the end of his life Wesley opined that there was 'no more sin in keeping a flower garden than in smelling a rose',[51] most examples of actual gardens he came across in his travels were described as errant displays of worldliness. As one example among many:

> We spent some hours at Wrest, a seat of the late Duke of Kent, who was forty years laying out and improving the gardens, which I cannot but prefer even before Lord Cobham's. But how little did the place answer its name! How little rest did its miserable master enjoy! Thou, O God, hast made our heart for Thyself, and it cannot rest till it resteth in Thee. (*Journal*, IV 95–6)

Wesley uses wordplay about 'rest' to make his point, which imports a homiletic tone into his travel writing. Wesley was more austere in his response to Stourhead in Wiltshire which, comparable with his response to travel writing, he evaluated according to its consequences for Christian belief:

> Others were delighted with the temples, but I was not ... because I cannot admire the images of devils [i.e. the statuary], and we know the gods of the heathens are but devils. (*Journal*, VI 128)

Indeed, Wesley only approved of a garden in the theologically congenial surroundings of Trevecca, which was created by the Welsh evangelist, Howel Harris: 'the gardens, fish-ponds, and the mount adjoining make the place a little paradise. He [Harris] thanks God for these things, and looks through them' (*Journal*, V 25).

Wesley's second more 'literary' spiritualisation of the landscape in the *Journal* was his interweaving of literal and metaphorical usages of topographical terminology.[52] An example of this came as Wesley made one of his early forays into outdoor preaching:

> I preached near the house we had a few days before began to build for a school, in the middle of Kingswood, under a little sycamore-tree, during a violent storm of rain, on these words, 'As the rain cometh down from heaven, and returneth not thither, but watereth the earth and maketh it bring forth and bud: ... so shall My word be that goeth forth out of My mouth'. (*Journal*, II 228–9)

Wesley skilfully uses his situation and the weather to visualise a Biblical simile, thus rendering the imagery concrete and the landscape spiritual. While he returned to this technique in a number of contexts, Wesley's favourite form was a parallel between agricultural and spiritual improvement:

> From the top of the next enormous mountain we had a view of Weardale. It is a lovely prospect. The green, gently-rising meadows and fields on both sides of the little river, clear as crystal, were sprinkled over with innumerable little houses, three in four of which (if not nine in ten) are sprung up since the Methodists came hither. Since that time, the beasts are turned into men, and the wilderness into a fruitful field. (*Journal*, V 464)

Just as with Pennant, and in accordance with Tucker's *Instructions*, the landscape is taken as a gauge of religio-political health, but with two important differences from Pennant. First, the health gauged is evangelical, not political and ecclesiological. Secondly, the way in which landscape is assessed makes it hover between being a literal gauge and a metaphorical symbol.

It should not be imagined, however, that Wesley's evangelical reading of the landscape was devoid of political import, as can be seen by examining his opinions concerning Scotland in the *Journal*. Again, Wesley liked to play on the parallel between the spiritual and agricultural improvement of the landscape (see *Journal*, VI 500–1). But furthermore, and like Pennant, Wesley argued that the spiritual improvement of Scotland surpassed that of England, highlighting the religiosity of the people rather than their Presbyterianism,[53] true to his overarching evangelical criterion of assessment: 'Oh what a difference there is between South and North Britain! Every one here [in Scotland] at least loves to hear the word of God; and never takes it into his head to speak one uncivil word to any for endeavouring to save their souls' (*Journal*, V 75).

This, of course, was an implicit critique of the Anglican church of which Wesley was a minister, and it was precisely such comments which, given the interconnection of church and state, led some to see Methodism as a destabilising political force (and to make accusations such as that with which this discussion opened of Wesleyan Jacobitism), despite Wesley's own overt loyalism. Doubts over Wesley's orthodox Anglicanism could only be exaggerated by his willingness to preach in a Presbyterian kirk (see *Journal*, IV 63) and his attendance of Church of Scotland services (*Journal*, VI 105), even if he preferred the Anglican service and church interiors (*Journal*, V 459 and IV 62), and expressed his contempt for John Knox's fanaticism (*Journal*, V 366 and VI 506). Wesley's expressed antagonism to Scottish Episcopalianism (*Journal*, V 248 and V 461–2) could also raise doubts over his religious and monarchical Toryism.

To summarise, Wesley's approach as a Tory Anglican travel writer was driven by an Evangelical fascination with questions of eschatology, which placed him at some distance from Pennant, who made more ecclesiological and liturgical forays into religious politics. Wesley's travels, as his sermons suggested, never escape the providential framework of the unfolding of God's plan through the course of human history, his own work as an evangelical traveller being cast as at the hinge point in the spread of the gospel which would prefigure the redemption of mankind.

Samuel Johnson: Tory High Church Anglican Scotophobia

Famously, Samuel Johnson defined 'Tory' in his *Dictionary* of 1755 as 'One who adheres to the antient constitution of the state, and the apostolical hierarchy of the church of England, opposed to a whig', while defining 'Whig' as 'the name of a faction'.[54] By these partisan definitions, there can be no doubt that Johnson was declaring himself to be a Tory of sorts, an impression reinforced by other declarations, such as Whiggery being 'a negation of all principle' and his famous comment that the first whig was the devil.[55] Despite their numerous disagreements, Johnson's three main contemporary biographers all accepted that Johnson was a Tory,[56] and modern disagreements about Johnson's political stance have contested his Jacobitism, not his Toryism.[57] Johnson was also a dedicated Anglican: 'a most unshaken church of England man' in Piozzi's words, who wrote at least twenty eight sermons, in which 'there are several peculiarly Anglican marks'.[58]

Johnson in some respects admired both of the Tory Anglican travellers discussed so far. In Wesley's case, Johnson's admiration was connected with his evangelical travelling labours, Johnson saying 'he could scarcely doubt the sincerity of that man, who travelled nine hundred miles in a month, and preached twelve times a week' (*Life*, II 123). Furthermore, Wesley's and Johnson's political positions came so close by the 1770s as to intersect, Wesley being converted by Johnson's political pamphlet *Taxation no Tyranny* (1775) to a hostility towards American independence, plagiarising Johnson in his own *Calm Address to the American Colonies*.[59] In the case of Pennant, Johnson expressed his appreciation of his abilities as a travel writer: 'he's the best traveller I ever read' (*Life*, III 274). Yet Johnson's nexus of Toryism, Anglicanism and travel was different from either Pennant's or Wesley's. He was, like other critics of Methodism, fearful of the possible political consequences of Wesleyan Evangelism: 'Speaking of the *inward light*, to which some methodists pretended, he said, it was a principle utterly incompatible with social or civil security' (*Life*, II 126).

For Johnson, personal revelation of the sort claimed by Methodists undermined the hierarchical basis of Toryism and Anglicanism, as given in his *Dictionary* definition, such that Wesley's spiritualising of the landscape was a menace. Pennant fell under the same opprobrium, for if Whiggism was the negation of all principles, he saw Pennant as 'a *Whig*; a *sad dog*' (*Life*, III 274; see also *Life*, V 157). Pennant agreed in placing his moderate Toryism, constructed around the principles of 1688, at some distance from Johnson, whose Tory piety he saw as 'mingled with ... numerous weaknesses and prejudices', not least of which must have been Johnson's willingness to investigate the case of the Cock-Lane ghost (*Life*, I 406–7 and note), which (as we have seen) Pennant thought unworthy of refutation by rational investigation.[60] For Pennant, then, both Johnson's and Wesley's Tory Anglicanism were superstitious. How did Johnson's different and third incarnation of a Tory Anglican manifest itself in his travel book, the *Journey to the Western Islands of Scotland* (1775),[61] and what was the intellectual basis of this variant?

Looking first at Johnson's difference from Wesley, his scepticism about inward light is reflected in the total absence of providential argumentation in his *Journey*. The closest Johnson comes to invoking providence is at Elgin cathedral, where the stripping of the roof leads provoked him to comment, that 'I hope every reader will rejoice that this cargo of sacrilege was lost at sea' (*Journey*, p. 18), but he does not actually ascribe this to God's vengeance. Johnson never spiritualises 'Nature' in the *Journey*, but discusses it as an assemblage of climate, rocks, soil, vegetation and minerals. In this he was part of 'the nominalistic impulse', which was reluctant to ascribe active powers to nature.[62] That Johnson held this view is most clearly revealed in the 1773 edition of his *Dictionary*. Here, under the key term 'nature', Johnson added a long epitome of Robert Boyle's *Free Inquiry into the Vulgarly Receiv'd Notion of Nature* (1685), which attempted to spell out the acceptable and unacceptable uses of the term, these being determined largely by their theological implications. Thus in Johnson's epitome:

> '*Nature* sometimes means the Authour of Nature, or *natura naturans* ... For *nature* in this sense may be used the word *creator*.'
>
> '*Nature* is sometimes indeed commonly taken for a kind of semideity. In this sense it is best not to use it at all.'

Boyle's conception of nature appealed to Johnson because it coopted the empirical approach into a framework where it could not lead to idolatry, or the vain ascription of natural operations such as rain ceasing or starting to God's particular providence. The argument for seeing special providence in the operation of nature was one Johnson undermined at Slains Castle: 'storms, whether wished or not, will sometimes happen' (*Journey*, p. 13). At Loch Ness, Johnson theorised that aside in his scepticism about the claim that the Loch does not freeze, which he saw as a putative exception 'from the course of nature' (*Journey*, p. 23). The burden of proof lay with those who would assert such an exception to the regularities observed in the operation of nature. Reasoning from first principles, Johnson believed 'its profundity ... can have little part in this exemption' from the course of nature, but if the Loch indeed did not freeze, Johnson wished to invoke other natural processes to explain this.[63]

Equally, Johnson did not try to invoke the laws of nature as a proof of God's beneficence in the *Journey*, the so-called design argument, which was so popular in eighteenth-century Anglican apologetics, particularly of the rationalist, low-church ilk. Johnson's distance from the design argument was noted by Piozzi:

> Rousseau is not like Johnson when he thinks a mute & sublime Admiration of his works the best Worship of the Creator, altho' that Admiration should excite no Act of any sort, but end wholly in itself – Johnson thought that God Almighty sent us here *to do* something – not merely to *stare about.*[64]

Subsequent travellers in Scotland, approaching travel writing from different theological positions from Johnson's, noted the absence of any design argument in the *Journey*. Thus the picturesque tourist William Gilpin, whose aesthetics were an attenuated form of the design argument consonant with his own low-church Anglican spirituality,[65] complained that:

> Dr Johnson says, the Scotch mountain has the appearance of matter *incapable of form, or usefulness*. As for it's [*sic*] *usefulness*, it may for any thing he can know, have as much use in the system of nature'.[66]

Gilpin and Johnson had different notions of utility, Johnson's being an immediate ability to aid human survival, Gilpin's a more cosmological contribution to the overall divine plan. But it should not be thought that, in refusing to spiritualise nature, Johnson was operating in secular terms: on the contrary, he believed theology was a system of propositions, the merit of which could not be gleaned from the landscape, and that faith was precisely supposed to be based on 'evidences and sanctions ... not irresistible, because it was intended to induce, not compel'.[67] To view all operations of nature as providential interventions in one's favour, or as law-governed knock-down proofs of God, was to cease to have to make the leap of faith demanded from all Christians, and replace it with a false enthusiasm or a false rationalism in religion. Johnson, then, was constructing an Anglican *via media* accordant with that of Edward Wells.

Turning to Johnson's relationship with Pennant as a Tory Anglican traveller, it would appear that Johnson's rejection of providential readings and his refusal to invoke design arguments would align him with Pennant. Both, in short, were concerned as travel writers with the visible fabric of the church, and both used such sites to comment on ecclesiological and liturgical issues. Looking at the religio-political comments in Johnson's *Journey* and those made by Pennant (and, in fact, Wesley) in Scotland, there are considerable areas of agreement. Johnson shared with the other two a revulsion at 'the tumultuous violence of Knox' (*Journey*, p. 17), and expressed regret as they had at the destruction of old churches and monasteries, most famously and movingly on Iona:

> We were now treading that illustrious Island, which was once the luminary of *Caledonian* regions, whence savage clans and roving barbarians derived the benefits of knowledge, and the blessings of religion. ... That man is little to be envied ... whose piety would not grow warmer among the ruins of *Iona*! (*Journey*, pp. 123–4)[68]

But the areas of agreement between the three travellers should not mask their overall divergence. Johnson's tone regarding the state of religion and the established church in Scotland is very different from Wesley's admiration and completely divorced from Pennant's eulogy, because of his hostility to Presbyterianism. In conversation with Boswell, Johnson gave a preference to Roman Catholicism over Presbyterianism (as had Heylyn and Wells), as 'the Presbyterians have no church, no apostolical ordination' (*Life*, II 103). This hostility was encoded in Johnson's *Journey* as surely as Pennant's sympathy had been in his travel writing. Thus, in a general assessment of the Scottish church, Johnson's positioning of Anglicanism as a superior confession was apparent if implicit:

> The ancient rigour of puritanism is now very much relaxed, though all are not yet equally enlightened. I sometimes met with prejudices sufficiently malignant, but they were prejudices of ignorance. ... Reason and truth will prevail at last. The most learned of the Scottish Doctors would now gladly admit a form of prayer, if the people would endure it. (*Journey*, p. 87)

Further, and unlike Wesley, Johnson refused to attend a Church of Scotland service, saying 'I will not give a sanction, by my presence, to a Presbyterian assembly' (*Life*, V 121; see also *Life*, V 384). He praised instead the Episcopal churches in Scotland, which used the Book of Common Prayer service (*Journey*, pp. 12–13).

Johnson, then, used his textual interventions as a travel writer to make the reverse point to that of the low-church Pennant: the Church of Scotland was inferior to the Anglican Church in organisation and modes of worship, and was only praised to the extent that it was showing signs of introducing the liturgical forms of episcopal Anglicanism. As such, Johnson's attacks on Knox and his regret at the destruction of cathedrals and monasteries by the Reformation had a more contemporary resonance than similar sentiments had in the writings of Pennant and Wesley, because he saw moderate Presbyterianism as the lineal successor of such actions, not as a decisive break with them. Where for Pennant Presbyterianism checked the potential tyranny of an episcopal body like the Anglican Convocation, for Johnson it opened the way to individualistic anarchy, just as the Wesleyan doctrine of inward light did: it was 'dangerous to be without' an apostolic ordination, and to have no public mode of worship left it to individual judgement 'whether they will join' (*Life*, II 103 and 104). Without firm organisation and hierarchy, Johnson's Tory Anglicanism saw the spectre of a reversion to iconoclastic excess as always waiting in the wings.

As Pennant and Wesley saw spiritual health in Scotland mirrored in a vibrant landscape, so Johnson's infamous negative comments about the Scottish landscape seem to be of a piece with his portrayal of the nation's religious culture. Johnson saw the Lowland landscape as 'extended in uniform nakedness', denuded because 'many centuries must have passed in waste without the least thought of future supply'. Like Pennant, Johnson explicitly ascribes this barrenness to border conflicts with England, but a denominational resonance is inescapable in his suggestion that the 'improvidence ... probably began in times of tumult,' sandwiched as it is between his comments on the destruction of St Andrews cathedral by 'the violence of Knox' and his account of the ruins of Arbroath monastery (*Journey*, pp. 6–7), which equate the Scottish Reformation with tumult.

Using the other two positions of Wesley and Pennant as Tory Anglican travellers to triangulate with Johnson's, we can move from asking how Johnson's view was distinctive to explaining its intellectual basis.[69] In contrast with Wesley, Johnson refused to invoke God's hand in his presentation of the natural world. Johnson's variant of Tory Anglicanism, then, focused on church politics, as did Pennant's, both using sites visited to encode their stance. But Johnson's was a staunch episcopal viewpoint emanating from seventeenth-century denominational politics, where Pennant adopted a more ecumenical line inspired by the events of 1688 and 1707. As Johnson's contemporary biographer, Sir John Hawkins, remarked:

> I have heard him assert, that, since the death of Queen Anne, it had been the policy of the administration to promote to ecclesiastical dignities none but worthless and undeserving men: nor would he then exclude from this bigoted censure those illustrious divines, Wake, Gibson, Sherlock, Butler, Herring, Pearce, and least of all Hoadly; in competition with whom he would set Hickes, Brett, Leslie, and others of the nonjurors.[70]

In short, the intellectual basis which distinguished Johnson's Tory Anglican travel writing from Pennant's and Wesley's was theological, and rested in his High-Churchmanship. This suggestion is strengthened by the substantial concurrence we have noted between Johnson's denominational politicisation of travel and the High Church geographies of Heylyn and Wells. For Johnson, faith was textually grounded, could not be gleaned from the appearance of the landscape, and could only be defended by a primitive, episcopal system of church government with a defined liturgy. In the joint concern for the fabric of the visible church and the diminution of the role of the proof of God from nature vis-à-vis other routes to faith (whether in the form of providential interventions in nature or the law-like design argument), both of which were reflected in the characteristic presences and absences in Johnson's travel writing, Johnson's response to religion in his travel writing partook of two of the characteristics which distinguished eighteenth-century High Churchmanship from Evangelical and Low-Church Anglicanism.[71] It was this, then, which led Johnson to develop a third variant of Tory Anglican travel writing.

Denominationally debating Tory Anglican travel writing

Thus far, I have deliberately restricted myself to a narrow remit, ringing the changes on varieties of Tory Anglican to highlight how theological and ecclesiological differences could feed into different stances as a travel writer. This analysis, of course, could easily be widened to take in other political positions and other denominations. While there is not space to engage in an extended demonstration of this here, I will at least sketch how such a religious politics of travel writing operated in the eighteenth century by analysing two responses to Tory Anglican travel writing, also thereby proving that contemporary readers were alive to the ideological resonances of travel in the realms of both church and state.

Responding to Wesley's travels: Graves's *Spiritual Quixote*

There were numerous literary responses to Wesley and Whitefield's Evangelical revival,[72] but perhaps the most able which centred on questions of religion and travel was Richard Graves's *Spiritual Quixote* (1773). This satire, modelled as the title suggests on Cervantes, had a serious political purpose: to laugh Methodism out of court. This was a personal issue for Graves, since he was an Anglican clergyman and his brother, Charles, had converted to Methodism while at Oxford and eventually been declared mad.[73] Graves saw Methodism as the lineal descendent of the Puritanism which had engulfed England in Civil War in the previous century, and his titular quixote, Geoffry Wildgoose, was said to be versed in the writings of

> some Presbyterian, some Independent, some Anabaptist, some Fifth-Monarchy men ... all differing somewhat in their principles, but all agreeing in their inveteracy against the Church of England.[74]

For Graves, as for Johnson, Methodism's leaders had good and pious intentions, but their movement 'must end at last in utter confusion' (*Quixote*, p. 32), threatening to destroy the political stability of England. This was signalled by the fact that the Wildgoose's Sancho, Jeremiah Tugwell, was described as a Jacobite (*Quixote*, p. 24).

The novel skilfully parodies the language of Wesley and Whitefield's travel journals, thereby undermining the sanctification of their labours and the spiritualisation of nature and landscape which they tried to effect. First, Graves imitated Wesley and Whitefield on the providential changes in weather conditions which accompanied their preaching: as Wildgoose begins to preach, a storm abates, 'dispelled by a single ejaculation', which he interprets, using language lifted from Whitefield's actual *Journals*, as 'the course of nature ... altered in favour of his harangue' (*Quixote*, p. 234). But then the storm resumes, which Wildgoose (as Whitefield) again interprets as providence. Graves's point, as Johnson's, was that the course of nature could not be read as proof of God's particular providence, except by an ignorant fanatic. Graves also picked up on Wesley's incessant moralising of gardens, and combined praise of his friend William Shenstone's garden, The Leasowes, with a parody of Wesley. Wildgoose visits Shenstone, who, as a good Anglican, tries 'combatting his friend's enthusiastic notions'.

But Wildgoose responds by destroying Shenstone's waterworks and pulling down a statue. Here Wildgoose responds to 'pagan' statuary as Wesley was to three years later at Stourhead (see above), arguing 'that Mr Shenstone quite idolized or worshipped those lifeless objects'. This section also ably parodied Wesley's habit of literalising Biblical topographical imagery: Wildgoose writes to Shenstone that in his garden he has 'forsaken the fountains of the living Lord; and hewn you out cisterns, broken cisterns, that will hold no water'. In response Wildgoose 'had forced open his sluices, and emptied his reservoirs, so that, in a literal sense, his *cisterns could hold no water*'. Where Wesley hoped such imagery would spiritualise nature and literalise the Biblical message, Shenstone 'could not forbear laughing at his old friend's frantic proceedings', the response Graves hoped to encourage in all Anglicans (*Quixote*, pp. 329–32).

Graves's general strategy with respect to the Methodist rhetoric of evangelising travel and travel writing was made apparent in Book II, Chapter III, where Tugwell raised objections to Wildgoose's desire to model his itinerant preaching on travel in Biblical times. Tugwell deflated Wildgoose's quixotic pretensions. Thus, the two discussed whether they should take food and money on their journey, Wildgoose answering in the negative:

> 'Was not Elijah fed as well by ravens, dost thou think, as he would have been from the king's table? And did not he sleep as sweetly under a juniper-tree, as he would have done upon a bed of state?'
>
> 'Yes, yes, (says Jerry) that is very true; but then that was in the Holy Land; where belike it is very much warmer than it is in this country. And then mayhap there were no inns nor ale-houses in those days; nor any thing to be got for love or money, in the wilderness where Elijah travelled: so he was obliged to trust to Providence, as a body may say. — However, folks were not so hard-hearted in those days'. (*Quixote*, p. 36)

By taking the Methodist elision of Biblical and modern circumstances and imagery to a literal extreme by his ignorance, Tugwell makes the whole rhetorical structure risible. Graves portrayed Tugwell as ignorant of 'the niceties of chronology and geography' (*Quixote*, p. 23) and Wildgoose as no longer interested in useful travel information – 'the trade or manufactures ... or what curiosities ... were worthy of a stranger's notice' (*Quixote*, p. 69) – since his conversion to

Methodism. As reconstrued in the *Spiritual Quixote*, then, Methodism was an unintelligent response to geography and travel, a series of rhetorical moves of no devotional worth, and a potentially destabilising force in the British church–state polity.

Responding to Johnson's travels: McNicol's *Remarks*

Just as Graves linked Wesleyan Methodism to Jacobite threats to the English throne, so Johnson's *Journey* was seen by some contemporaries as a covert pilgrimage to the ground where the Stuarts had been defeated. Thus the reviewer of the *Journey* in the *Gentleman's Magazine*, noting that 'our author must ... have passed near, if not over, the field of Culloden', added 'we cannot help wondering at his studiously avoiding to mention that decisive battle ... where the fate of three kingdoms was really determined'. In a later instalment, the same reviewer noted the 'striking peculiarities' in the *Journey*, foremost amongst them its author's refusal to use 'the invidious word *rebellion*' for the events of 1745, which would have amounted to an acceptance of the legitimacy of the Hanoverian regime.[75]

Yet the most vituperative contemporary criticism Johnson's *Journey* encountered was for its interventions in the realm of denominational rather than dynastic politics. We have already seen (albeit briefly) that Johnson's High Church variant of Tory Anglican travel writing could elicit responses because of its theological positioning in the case of William Gilpin's low-church Anglican picturesque. The longest response to Johnson's *Journey*, however, came in the form of Donald McNicol's *Remarks* (1779), a Presbyterian and patriotic rejoinder to Johnson's perceived Scotophobia.

McNicol's main and repeated theological point was that Johnson's travel writing pursued a crypto-Catholic agenda. As we have seen, from a Presbyterian perspective this might have some validity, Johnson having preferred Catholicism over Presbyterianism in conversation. McNicol picked up on the Johnsonian lament on Iona (see above), and interpreted it in a Catholic light: 'The *Scots*, at least, do not carry their veneration for such relics so far as the Doctor did in the island of *Jona* ... a circumstance which is no bad index to his religious *creed*.'[76]

McNicol was, of course, implying that any veneration of monastic ruins was an admission of Catholicism.[77] He repeated his allegation at a number of points, as when responding to Johnson's comments on the murder of Cardinal Beaton (*Journey*, p. 3; *Remarks*, pp. 23–4) and on the Catholic island of Eigg (*Journey*, pp. 105–6; *Remarks*, pp. 188–90). Furthermore, McNicol had no doubt that Johnson's criticisms of the

Scottish landscape were in part religiously driven. Refuting Johnson on a point of fact, he opined:

> Not far from this town [Forres], in his [i.e. Johnson's] way to Nairn, he had an opportunity of seeing the castle of *Tarnaway* [i.e. Darnaway], an ancient and noble seat of the Earls of Murray. Here he would have found, what he pretends so often to have looked for in vain, parks, plantations, and natural woods in abundance; which, with other beauties of nature and art, might sufficiently compensate for the trouble of a short peep as he went along; it would not have taken him much out of his way, and he would have made a shift to visit a *popish* church, or even the ruins of one, at a greater distance. (*Remarks*, p. 54)

Where Johnson had placed Presbyterianism as the successor and inheritor of Knox's fanaticism, McNicol as a Presbyterian minister responded by reorientating the evidence to construe Johnson as the credulous one, threatening the stability of British church and state politics. Where Pennant (as we have seen) discussed Johnson's religious 'prejudices' and the superstition of the Cock-Lane ghost separately, McNicol linked the two, suggesting Johnson 'sat up whole nights ... repeating *paternosters*' in fear of the ghost (*Remarks*, p. 198). Unsurprisingly, McNicol preferred the more positive account of Scotland provided by Pennant:

> That gentleman found abundance of woods, and even *trees*, in different parts of the country ... But *he* travelled with his judgment *unbiassed*, and his eyes *open*. (*Remarks*, pp. 19–20)

Pennant's approval of the Scottish landscape, as we have seen, viewed it as a gauge of the vibrant civic and religious health of the nation, and McNicol followed him, while couching his patriotism in this language of landscape.

Yet McNicol was himself open to criticism from a position more sympathetic to Johnson's Anglicanism. Thus the *Critical Review*'s discussion of McNicol's *Remarks* acknowledged his learning and even accepted his Scottish patriotism, but then concluded damningly:

> we cannot regard with any degree of indulgence, the illiberal strain of personal abuse, of petulant invective, and of groundless insinuation into which he has frequently descended. We may add, that

> the Remarks would ... have carried greater force, as well as dignity, had the author discovered fewer symptoms of unaccommodating prejudice, and intemperate zeal, where the subject concerns religion.[78]

The *Critical Review*'s point was that McNicol had failed to adhere to the models of polite discourse championed in the eighteenth century in his engagement with the denominational politics of Johnson's *Journey*. The 'insinuations' referred to may well have included those of Johnson's 'popery'. Clearly McNicol's ardour was seen by the reviewer as fatally weakening his case, as it aligned him not with the moderate, modern Presbyterians whom Pennant eulogised, but with the bigoted fury of the Knoxian Reformation which Johnson, Wesley and Pennant had condemned. The *Critical Review*'s comments form a fitting final testimony to the centrality of denominational politics to eighteenth-century travel writing, McNicol's Presbyterian rejoinder to Johnson's High Church Anglican *Journey* itself facing reproof for failing to adhere to models of polite religious dispute, tipping over instead into a dangerous Knoxian fanaticism. Amidst the welter of claims and counter-claims travel writers made, denominational politics clearly bulked large.

Conclusion

Looking at the travel accounts of the 1770s, it can be seen that the genre provided sites in which political and religious languages could be encoded. Contemporaries were clearly sensitive to this possibility as readers, responding to it both seriously and in a satirical vein. In this respect, travel as a form of geographical knowledge was akin in its political functioning to geography books *sensu stricto*. On a European scale, Black has shown that ideological prejudices continued to be vital to later eighteenth-century British travel writing.[79] This conclusion also holds good on the domestic front, where the sorts of denominational debates which fed into the political argumentation concerning the American War of Independence and the controversy over subscription to the thirty-nine articles in the 1760s and 1770s found a congenial textual home in the travel account.[80]

9
The Scottish Enlightenment and British Geography (I): Guthrie and Pinkerton, c.1770–1802

Guthrie's *Geographical Grammar*: the impact of the Scottish Enlightenment[1]

William Guthrie's political career started in the same political milieu in which Thomas Salmon's reached its fruition. While Guthrie had been a hack writer for the *Gentleman's Magazine* in the late 1730s with Samuel Johnson, he achieved notoriety writing in the *Old England Journal*, which was the most prominent patriot periodical in the early 1740s, opposing first Walpole and then the succeeding governments. A collection of essays from that journal (some of which were by Guthrie) was reprinted as *A Collection of Letters Publish'd in Old England* (1743), cashing in on the *succès de scandal* caused by the imprisonment of the printer. Guthrie was soon silenced by a pension from the government.[2]

But the political context in which Guthrie made his contribution to the British geographical tradition 30 years later was a rather different one. Despite his pension, Guthrie remained reliant on writing to sustain his livelihood, and turned in the 1760s to historical works. Here he was part of a Scottish movement which sought to re-examine English, Scottish and British history in a manner that was alleged to overcome the partisan Whig and Tory bias of the histories which had been produced by English writers.[3] This historiographical programme was one of the key elements of the Scottish Enlightenment. Given the traditional interconnections of geography and history, this unavoidably brought geography within the ambit of the Scottish Enlightenment, by raising questions concerning the contexts and arguments within which geographical evidence could be deployed. Geographical information began to be used by Adam Ferguson, John Millar and others in the construction of general models of societal development. Heylyn had drawn such a model in *Cosmographie*, seeing geography as enforcing a message of human and societal transience which was conducive to piety. The Scottish conjectural models are more secular and confident of social progress. In neither case, however, was such a model devoid of a political message, for in both cases the generalising tone also supported a more specific theological politicisation. Despite the Scottish Enlightenment interest in history and in the philosophical use of geographical information, however, it was left to William Guthrie to introduce such modes of thinking into geography *sensu stricto*. It should be emphasised that Guthrie worked at some distance, intellectual and geographical, from this movement. Where Guthrie lived in London, the key figures in the Scottish Enlightenment not only stayed in Scotland, but also never had to write to earn a living as he did.[4] But Guthrie did introduce some of the tone of Scottish Whig historiography into his historical writings, and thereafter into his *Geographical Grammar* , a trajectory we can now review.

We can discern Guthrie's historical approach from his *History of Scotland* (1767–68). He was insistent that the Scottish understanding of constitutional liberty was actually older and more secure than the English. By 1285 the Scots were 'a far more civilised people than either the English or the French', a view confirmed by the Declaration of Arbroath (1320):

> It was remarkable equally for the spirit of civil and religious independency, as to every point relating to their national liberties ... the nobility of Scotland, at that time, thought that their laws were superior to their king; that he was no more than the guardian of his people's freedom, and dismissible from his charge and sovereignty the moment he attempted to subvert it.[5]

Scottish Whig historians wished to see their constitution as embodying the principles of the Revolution Settlement of 1689 long before that time. They wished to rewrite Scottish history as the triumph of liberty rather than as a series of bloody usurpations as part of a more general programme to bring 'politeness' to Scottish political and social life.[6] Clearly Guthrie worked within this framework here, and he made the point more explicitly in discussing the events of 1688/9. He dismissed the English argument that James II's heir was illegitimate as an unnecessary lie:

> The English Whigs, in direct contradiction to the strongest evidence, and indeed to their own principles, have disputed the reality of that birth, chiefly upon the frivolous grounds advanced by bishop Burnet, the most credulous writer in the world. ... The principles upon which the Revolution was founded have no regard to such suppositions; for they must have taken place independent of all consideration of the birth, or of any right derived to the offspring of the king and queen.[7]

Guthrie here moved the debate about the Glorious Revolution away from the dynastic terms in which Bohun and others had couched it. He preferred to see the grounds on which a monarch could be deposed in terms of a violation of natural or legal rights, the same contractarian language he had used to describe the Scottish constitution as defined in the Declaration of Arbroath. Given their long history of understanding the true nature of liberty, it is not surprising that Guthrie believed the Scots had theorised the events of the Glorious Revolution better than the English, a point he returned to in the *Geographical Grammar* (see below).

The other point to note about Guthrie's *History* is its position concerning Scottish Presbyterianism. With the exception of Pennant's tours, the geographical works discussed so far took Presbyterianism as a form of fanaticism which destroyed Charles I and was a permanent threat to the Church of England's recovery of the primitive pattern of Christianity. Guthrie, whose own background was Episcopalian, agrees that the Scottish Reformation displayed excesses of iconoclasm and that the Covenanters, 'ignorant, bloody, vindictive, madmen and enthusiasts' were 'accessory to the tragedy' of the murder of Charles I.[8] Moreover, the treatment of the Episcopalian clergymen who were expelled at the Revolution Settlement is taken as a sign of how recently the Presbyterians remained as ruthless as their Reformation forebears.[9] But for Guthrie the modern Presbyterians are of a different ilk, being more rational and tolerant. Despite his Episcopalian background, then, Guthrie's narrative links him to the moderate Presbyterian view of Scotland, something also evident in his highly positive assessment of the Union of 1707.[10] As recent work has shown, this was the dominant tone of the Scottish Enlightenment, rather than the much-publicised agnosticism of David Hume. The Scottish Enlightenment differed from the French in that its leaders were prominent members of the church, judiciary and other powerful institutions. Indeed, Enlightened politeness was seen as embodied in the moderate Church of Scotland rather than being its antithesis.

Guthrie's *Geographical Grammar* was published in 1770, the year of his death. The main achievement of that work was to transfer the sort of moderate Presbyterian analysis developed in his *History of Scotland* to the genre of the geographical compendium. The result is a radical disjunction in political viewpoint between Guthrie's geography and those discussed so far.

If we start by focusing on Guthrie's description of England, his understanding of the church–state nexus is very different from that of his predecessors. In politics, he casts the English story as one of progress towards the securing of liberties for the people. Unsurprisingly such a narrative sees the attempt to impose the 'Norman Yoke' on the English as a failure, and Magna Carta as an important constitutional moment.[11] His conclusion is that:

> the English constitution has continued in its full vigour for above 500 years. It must, at the same time, be admitted, that it has received, during that time, many amendments, and some interruptions, but the principles are the same. (*Grammar 3*, p. 234)

Here we can see Guthrie adopting the disinterested pose of a Scottish historian, both asserting the antiquity of the English constitution as Whig historians had tended to and accepting the interruptions in this 'ancient constitution' caused by Conquest and the like in the manner of Tory historians.[12] This impression is reinforced by the fact that Guthrie both sees 1688 as restoring the true balance of liberties to the English, and, as in his Scottish history, as an event more fully understood by the Scottish people than the English:

> At the Revolution, the Scots gave a fresh instance how much better they understood the principles of liberty than the English did, by omitting all pedantic debates about abdication, and the like terms, and voting king James at once to have forfeited his crown. (*Grammar 2*, p. 89)

In this, the *Geographical Grammar* also recreated the narrative of English history and the authorial position Guthrie's *History of England* (1744–51) had adopted.[13]

The story did not end at the recreation of the ideal constitutional balance in 1688. England is depicted in the *Geographical Grammar*, as in Pennant's contemporaneous travels, as in danger of taking liberty and commerce to an excess, thereby degenerating into luxury. Thus the present state of England is not a happy one: 'Many, especially foreigners, have attributed that remarkable self-dissatisfaction of the English, which too often proceeds to acts of suicide, to their air and climate; but however these may operate, the evil probably lies in the people's manner of living, which is more gross and luxurious, than that of any other nation' (*Grammar 3*, p. 175). Guthrie's analysis is in keeping with that of others in the Scottish Enlightenment, most notably Adam Ferguson, who feared that commerce would destroy the civic virtue of a nation. An important issue here was that standing armies undermined the republican ideal of citizens defending themselves. Guthrie repeated this commonplace of Civic Humanism in the *Geographical Grammar* : 'In a land of liberty it is extremely dangerous to make a distinct order of the profession of arms. In such, no man should take up arms, but with a view to defend his country and its laws: he puts not off the citizen when he enters the camp' (*Grammar 3*, p. 263).[14]

If luxury, as an excess of personal acquisitiveness, could destroy the nation, so could the expansion of empire, the same excess expressed geographically. As such Guthrie warned:

> GREAT BRITAIN is at present that kingdom in Europe which enjoys the greatest prosperity and glory. She ought to be the more attentive therefore to preserve so brilliant an existence. The spirit of conquest neither suits with her physical situation, nor with her political constitution. ... the augmentation of armies must, in some future period, be attended with a consequence of another kind, that is, the establishment of absolute monarchy in Great-Britain. (*Grammar 3*, p. 62)

This argument again echoed Scottish Enlightenment historians such as Robertson and Ferguson on the perils of empire, but also tapped into a wider unease about the virtues of the British empire which was beginning to surface in the 1770s in the context of the discontents expressed in the American Revolution.[15]

Guthrie's overview of the English constitution, then, sees it as a triumph of liberty, as Echard had, and as perpetually threatened, as Salmon had, but sees both of these observations from a Scottish perspective. Historians such as Echard were precisely those against whom the Scottish historiography of which Guthrie was a small part was directed, attacking insular triumphalism. Similarly, the fears over Whig luxury damaging England which Salmon raised were generalised and apparently freed from factional politics by Guthrie's analysis.

A similar narrative and authorial position was established with respect to the Church of England in the *Geographical Grammar*. The trajectory of the Anglican church, like the English constitution, was towards liberty and tolerance. In this case the nadir was represented by the Laudian attempt to impose High Church piety (*Grammar 2*, p. 241). Guthrie viewed modern Anglicanism as both more Erastian and more humane. Since Laud, the Archbishops of Canterbury were said to have been: 'men of very moderate principles and of very inoffensive abilities. This ... has been attended with excellent effects, with regard to the public tranquility of the church, and consequently of the state' (*Grammar 2*, p. 121).

But the Anglican church is not seen as ideal even in its modern form. It approaches 'nearer to the primitive Christianity', but still demonstrates too much of 'the pomp of worship' (*Grammar 2*, pp. 121–2), a phrase which clearly links Anglicanism with Catholicism. The reason Anglicanism is perceived as too ostentatious is that for Guthrie the true genealogy of the Protestant church is through the radical tradition of medieval heresies of Wycliffe *et al.*, not through the Anglo-Catholic myths of the primitive foundation of the English church by Joseph of Arimathea or St Paul pedalled by Heylyn and Wells. For Guthrie, a truly Christian form of church government came from consulting Scripture alone, as in the 'radical' Protestant tradition of Luther and Calvin which the Scottish church drew upon. For this reason the moderate Presbyterian Church of Scotland receives greater praise, being both tolerant and more securely within the Protestant tradition:

> The established religion in Scotland is Presbyterianism. It was formerly of a rigid nature, and partook of all the austerities of calvinism and intolerance of popery, by its persecuting spirit. At present it is mild and gentle; and the most rational christian may accommodate himself to the doctrine and worship of the national church. (*Grammar 2*, p. 73)

Where geography books from Camden and Heylyn on had mostly adopted some form of Anglo-Catholic viewpoint, the *Geographical Grammar* completely rewrites the ecclesiastical history of Britain from the perspective of the Scottish Protestant tradition. The result is a reversal of Heylyn's argument in *Cosmographie*: where for Heylyn a Laudian establishment is the aim and all other systems are a deviation, in Guthrie no ideal establishment is specified in doctrinal terms, but all Protestant churches are to be judged by their success in distancing themselves from a crypto-Catholic, politically destabilising Laudianism.

The *Geographical Grammar* makes one final revision of the understanding of British politics which is the most important of all to the structure of the book, its most visible debt to the Scottish Enlightenment and its most significant innovation in the British geographical tradition. The whole trajectory of English history towards liberty and commerce, as well as the emergent threats of luxury and imperial overreach are not pictured as unique, but merely as one exemplification of the stages through which all societies develop over time. For Guthrie's greatest debt to the Scottish Enlightenment lies in his use of stadial theories of social development. The prefatory material in the *Geographical Grammar* gives an exposition of this theory in two parts. First, the section 'Of the Origins of Nations, Laws, Government, and Commerce' is designed 'to place before his [the reader's] eye a general view of the history of mankind, from the first ages of the world, to the reformation in religion during the 16th century' (*Grammar 3*, p. 36). In Scottish stadial theories, all societies were argued to proceed from a pastoral stage, through an agricultural stage to a commercial stage, with all its attendant threats of luxury and degeneration.[16] Guthrie's 'Origins of Nations' recapitulated this sequence (*Grammar 3*, p. 39), a key role being accorded to towns, which created islands of freedom in a feudal system. Secondly, the third edition added a section 'Of the Origin and Progress of Religion'. As this was not in the first or second editions, it seems most likely that this piece was not written by Guthrie, although its sensitive nature may have led him to have scruples about publishing such a discussion which his posthumous editors did not share. This briefly suggested that the same methods of analysis deployed with respect to society in general by stadial theory could be applied to give a natural history of religion (to use Hume's phrase):

> In deducing the history of religion ... we must make the same distinction which we have hitherto observed in tracing the progress of arts, sciences, and of civilization among mankind. We must separate what is human from what is divine, what had its origin from partic-

ular revelations from what is the effect of general laws, and of the unassisted operations of the human mind. (*Grammar 3*, p. 62)

The argument then given was that polytheism was a natural response of 'rude' peoples to the awesome powers of nature, which they personified and attempted to placate by prayer (*Grammar 3*, p. 63). The true Christian monotheism, however, is the Revelation of God to his chosen people (*Grammar 3*, p. 63). This does not, however, exempt Christianity from naturalistic explanation, in that its doctrine is seen as more amenable to a more civilised stage of development, which explains its gradual spread and purification from superstition over time (*Grammar 3*, p. 64). In this section Guthrie (or his anonymous editor) was treading dangerous ground, as all such naturalistic explanations faced the ire of orthodox clergymen who saw them as atheistic in tendency, and, given the interconnections between church and state, as potentially destabilising to the governmental structure of a Christian country.

Three possible sources for the *Geographical Grammar* 's twin-pronged exposition of stadial theory can be suggested: Adam Ferguson's *Essay on the History of Civil Society* (1767); the prefatory 'View of the Progress of Society in Europe' in William Robertson's *History of the Reign of the Emperor Charles V* (1769); and David Hume's *Essays, Moral, Political and Literary* (1741 and subsequent updates) and *Natural History of Religion* (1757). It would appear from textual comparison that their degree of importance to Guthrie was in the order I have listed them. Indeed, much of the material which had Humean roots could have been cribbed from Ferguson's *Essay*. The basic trajectory and interrelation of liberty, commerce and luxury appears to derive from Ferguson, especially the pronounced fears of luxury and imperial expansion which Ferguson emphasised more than most. Guthrie's chronological span from the first ages to the beginning of the Reformation seems to derive from Robertson's immensely popular *History*, [17] as may the discussion of towns and feudalism. Finally, as I shall show below, certain phrases in Guthrie make it at least likely that he moved beyond Ferguson's citations of Hume's opinions concerning the development of states to their actual sources in Hume's *Essays*. Guthrie's discussion of the natural history of religion, with its adherence to the orthodox belief in revelation and the truth of Christianity, must derive from Ferguson and Robertson and not from Hume's *Natural History of Religion*, something which would accord with the moderate Presbyterian stance we have already seen the *Geographical Grammar* to adopt.

Having prefaced his geography with a discussion indebted to stadial theory, Guthrie went on to deploy it in his discussions of the world in the body of the text. We have seen in detail that England is an exemplification of all the stages the Scots had theorised, but other places were also fitted into the schema. Restricting my analysis to the general summaries of the characters of the continents taken as a whole, Guthrie's debt to stadial theory is clear. The discussion of Europe is worth quoting at length:

> Geography discovers to us two circumstances with regard to Europe, which perhaps have had a considerable tendency in giving it the superiority over the rest of the world. First, the happy temperature of its climate, no part of it lying within the torrid zone; and secondly, the great variety of its surface. ... The immense number of mountains, rivers, seas, &c which divide the different countries of Europe from one another, is likewise extremely commodious for its inhabitants. These natural boundaries check the progress of conquest or despotism ... [and] the seas and rivers facilitate the intercourse and commerce between different nations. (*Grammar 3*, p. 65)

This derives ultimately from Hume's essay, 'Of the rise and progress of the arts and sciences', which Ferguson's *Essay* had used to explain why Europe exhibited all the stages of stadial theory where no other continent had yet to develop beyond feudal despotism.[18] That Hume's essay had actually been used is suggested by the parallel between his continuation of this thought and Guthrie's, which is not in Ferguson:

> *Guthrie*: 'There is no part of Europe so diversified in its surface, so interrupted by natural boundaries or divisions, as Greece ... What Greece therefore is with regard to Europe, Europe itself is with regard to the rest of the globe' (*Grammar 3*, p. 65).
>
> *Hume*: 'mankind, having at length thrown off this yoke [of Papal tyranny], affairs are now returned nearly to the same situation as before, and EUROPE is at present a copy at large, of what GREECE was formerly a pattern in miniature.'[19]

Both see Europe as a continental-scale re-creation of a stage of development ancient Greece went through by virtue of a geography encouraging both separation and commerce. The self-confident assertion of European superiority was hardly new to Guthrie, being a commonplace in geography books throughout the early modern period. What had changed was the grounds for this assessment: where for Heylyn European superiority was a divine judgment, for Guthrie it is naturalised in terms of geography encouraging societal development. Obviously, this may simply be seen as the displacement of the providential sanction of European superiority onto geography, but such a move is entirely in accord with the outlook of the moderate Presbyterian Enlightenment in Scotland.[20]

In the view of the stadial theorists, there was little of real interest outside Europe, as the world was only tracing the same steps Europe had already followed.[21] Guthrie seems to endorse this, his treatment of the other continents being much briefer and in accord with the continental stereotypes propagated by the Scottish Enlightenment.[22] Thus, where Europe occupied 440 pages in the third edition of the *Geographical Grammar*, Asia occupied only 32 pages, Africa 26 and the Americas 116 pages. Asia, as throughout the period under discussion, is viewed as the continent of quasi-feudal luxury and effeminacy (*Grammar 3*, p. 502). More interesting are the discussions of America and Africa, which are both seen as at a less developed stage than Asia. The native Americans are portrayed in a pastoral stage, prior to the development of settled agriculture:

> As they have no particular object to attach them to one place rather than another, they fly wherever they expect to find the necessaries of life in greatest abundance. Cities, which are the effects of agriculture and arts, they have none. The different tribes or nations are for the same reason extremely small, when compared with civilised societies, in which industry, arts, agriculture, and commerce, have united a vast number of individuals, whom a complicated luxury renders useful to one another. (*Grammar 2*, pp. 552–3)

At this stage, and in accord with Ferguson again, the people's ideas are 'too confined ... to entertain the sentiments of general benevolence' (*Grammar 2*, p. 554) associated with more complex societies which allow for the development of more abstract ideas. Related to the simplicity of their ideas, at this stage the Americans can create the purest poetry, and 'their orators, like those of Homer, express themselves in a bold figurative stile' (*Grammar 2*, p. 553). The Scots literati in general argued that sublime poetry was a product of peoples with simple, bold ideas, who expressed them in naturalistic metaphors rather than affected word play. Thus many of this group believed in the authenticity of Macpherson's Ossianic poems, which we have already seen impressed Pennant.[23] Guthrie was simply dismissive of Africa, which was in the same stage:

> The history of this continent is little known, and probably affords no materials which deserve to render it more so. We know from the antients, who sailed a considerable way round the coasts, that the inhabitants were in the same rude situation near 2000 years ago in which they are at present, that is, they had nothing of humanity about them but the form. (*Grammar 3*, p. 596)

In both cases, Guthrie argued that the physical geography of the continent would facilitate the development of commerce and therefore of society. Thus for Africa he wrote:

> The situation of Africa for commerce is extremely favourable, standing as it were in the centre of the globe, and having thereby a much nearer communication with Europe, Asia, and America, than any of the other quarters has with the next. ... In short, Africa, seems to be almost entirely neglected, not only by the natives, who are quite unsollicitous of reaping the benefits which nature has provided for them, but also by the more civilized Europeans. (*Grammar 3*, p. 578; for a similar comment on America, see *Grammar 3*, p. 623)

It should be emphasised, however, that this was not an appeal for colonisation on the basis of the logic of geography of the sort which was to become popular in nineteenth-century geography. As has been shown in the discussion of England, Guthrie not only followed stadial theory in seeing civilised countries as superior, but also in viewing the fabric of their achievement as imperilled by excessive imperial expansion. Guthrie was certainly advocating the development of commercial connections in areas of potentially handsome profits for Europeans, but his vision looked back towards the world of the commercial trading companies, rather than forward to the control of foreign lands by European governments in the nineteenth century, an enterprise which used stadial theories of European superiority, but jettisoned the theme of virtue and its decay which was so important to the first generation of Enlightenment social theorists.

William Guthrie has been seen within the geographical tradition 'not as an original thinker but at least as a populariser of geography'.[24] As will be apparent, a reappraisal of his work from a political perspective must alter that judgement by denying the antithesis: Guthrie was an original contributor to geography by virtue of popularising concepts developed in other contexts. He brought Scottish Enlightenment thought in general, and the moderate Presbyterian strain of it in particular, to bear upon the assessment of the standard information contained in geographical gazetteers and compendia. As with Heylyn, the specificity of his viewpoint is often overwhelmed by the bulk of standard information conveyed, but the *Geographical Grammar* amounts to a wholesale reworking of the intellectual and political presuppositions which had informed Heylyn's *Cosmographie* and the works of many of his English successors. Above all, Guthrie shifts the centre of interpretative gravity away from the Oxbridge–London axis in which it had operated previously. As such, he gave a distanced perspective on the political issues which had been central to the Anglocentric geographical tradition: Guthrie is indifferent to arguments about the intellectual foundations of dynastic allegiance, except insofar as the moral and economic well being of the people is concerned; he is also indifferent to the antiquity and structure of the Church of England, with most of the liturgical and doctrinal matters which Heylyn and Wells discussed being adiaphorous to him. The overall effect of this distancing is a geography book which undercuts the English exceptionalism and patriotism on which geography books had been grounded since Camden. England is unusual in the distance it has progressed along the path sketched by stadial theories, but this has no ontological or providential grounds, and its geographical grounds, following Hume, are continental and European not merely insular and British.[25]

This distancing effect, of course, is not to be confused with the disinterested elimination of politics from geography. On the contrary, Guthrie's text stands for a specific political stance. As with the moderate literati of the Scottish Enlightenment more generally, the greater abstraction at which the arguments operate should not mask a political programme defending a moderate and Protestant form of Christianity, supporting the liberties of 'the people' (albeit not the 'mob'), and working for toleration within the established structures of church and state. Phrases such as 'conservative Whiggism' and 'tolerant conservatism'[26] have been used to express these elements of Scottish thought, which modern academia perhaps finds harder to reconcile than did lowland Scots constructing their identity as North Britons.[27]

Guthrie's own synthesis of this blend of toleration and conservatism, of detached analysis and political persuasion, is summarised in the 'Preface' to the *Geographical Grammar*: 'By comparing together our accounts of the European nations, an important system of practical knowledge is inculcated, and a thousand arguments will appear in favour of a mild religion, a free government, and an extended, unrestrained commerce' (*Grammar 3*, p. vii). Taking this as the political argument framing Guthrie's geography, we can see the distance between him and Heylyn. The discourse of polite sociability is central to the ideal vision of church, state and commerce for Guthrie, the manner in which these bodies operate being at least as important as their content. While religion and indeed providence remain vital to Guthrie, they do not form the framework of his thinking as had been the case for geography books from Camden to Echard.[28] Yet one link remains between Heylyn's Laudianism and Guthrie's Enlightenment: they both continue to support the establishment in church and state, and to fear civil turmoil as the greatest of political evils.[29]

John Walker's *Elements of Geography*: a dissenting excursus

Seen from one perspective, Guthrie's *Geographical Grammar* can be taken as a criticism of the values of a metrocentric English geography forged in the milieu of Oxbridge and London. Such criticisms emerged with particular force in the later eighteenth century, and were by no means confined to Scottish Enlightenment thinkers. As moderate Presbyterians questioned the framework established by Anglican geographers, so too did Dissenters. The best known dissenting educationalist was Joseph Priestley, but his comments on geography are in fact conventional. Priestley argued that geography was an essential prerequisite for the historian, and emphasised the utility of the subject as a collection of commercial information to the training of the middling ranks.[30] The historical framework into which a geographical education was inserted was one Heylyn would have recognised:

> What is truly valuable in the history of past ages is every day cleared from more and more of the obscurity in which it has been involved. In consequence of which, the series and connexion of events may be more strictly traced, so that we may say, the plan of this divine drama is opening more and more, and the grand catastrophe growing nearer and nearer perpetually.[31]

Of course for Priestley it was believers in an established church who had helped to create the obscurity which was gradually being removed under the guidance of providence, but the eschatological framework in which history and geography were emplaced was similar to Heylyn's.[32]

Clearly Priestley's Dissenting vision did not lead him to dissent from the orthodox approach to geographical instruction. Yet we can find just such a 'dissenting geography' – a geography book by a religious dissenter which also dissents from many of the elements normally present in the geographical tradition – in John Walker's *Elements of Geography*, published in Dublin in 1788. Walker was born in Cumberland and trained as a blacksmith. His interests lay in natural science and he was one of the many working men who educated themselves by engaging in fieldwork.[33] Walker moved to Dublin and worked as a teacher in a Quaker School. It was here Walker published *Elements of Geography*, a work which shows the deep impact that the provincial dissenting tradition could make on a geography book. The *Elements* has a remarkable 'Advertisement' which undercuts many of the English geographical tradition's assumptions in the space of one page. Listing the authorities he has drawn on, Walker gives an array of Enlightenment intellectuals: Clarkson, Ferguson, Gibbon, Guthrie, Millar, Priestley, Robertson and Voltaire. He goes on to complain that the *Elements* has been printed incorrectly:

> Men in high stations of life have flattering titles bestowed on them, insomuch that these at length are accounted a part of their real names ... Convinced as I am, in my heart of the evil and indecency of bowing to or flattering a fellow-worm ... it would be wrong in me to address men by their titles rather than their proper names ... Thus where the word Sir occurs in this work I had written 'the Sir', meaning the Sir of the Herald's Office or the Sir of the people at large who have agreed to call the man by that title.[34]

As Walker makes clear, this is not a radical levelling of status, but a theological levelling derived from Quaker spirituality. This is a very different mental world from that in which Edward Wells dedicated his *Historical Geography of the New Testament* to Queen Anne. But it reminds us that the most seemingly innocuous elements of geography books contained a language of inclusions and exclusions, which could be criticised from other theological and political perspectives. The same point is made in another objection to the language of geography books, which also derived from Walker's Quaker heritage, and had much in common with the mentality satirised in the *Spiritual Quixote*:

> Days, Months, Planets, &c. have received names which originated in idolatry; and in describing the solar system I have expressed them thus, the Saturn, the Jupiter, the Mars &c not at all thereby meaning the men Saturn, Jupiter, Mars, &c who lived many ages ago and were afterwards idolized.[35]

Here was a very different natural history of polytheism from that offered by the Scottish Enlightenment and propagated by Guthrie. Walker had to naturalise the explanation of idolatry and thus distance it from Christianity, which he did not put in the same framework (unlike Hume, Robertson and Ferguson). Idolatry was categorically divided from belief, and the natural-historical approach could only be applied to the former. Of course, Walker was really using the 'Advertisement' to prove his Quaker credentials, but it does give an insight into how standard geography books could be read against the grain in a different intellectual tradition.

The text of the *Elements* itself is a miscellany, but some parts of it do show the changes wrought by Walker's different authorial location, intellectually and geographically. The balance of the geographical coverage is unlike any other geography book surveying the whole world. Where most geography books gave more information on Europe than the rest of the world combined, Walker expended only four pages on Europe compared with thirteen on Africa.[36] Most of the African section was an attack upon the slave trade, a theme which surfaced repeatedly in the book. This in turn was part of a broader Quaker pacifist manifesto in the book, which was first sign-posted when Walker discussed the natural and civil divisions of the earth: 'there are the political boundaries that separate kingdoms ... they have generally been marked out by the sword of the conqueror, at the expence of the blood and carnage of his fellows'.[37] Geographers had never been squeamish about the role of war in the construction of their subject matter, but Walker brought this to the foreground to an unprecedented extent and with a new sense of moral outrage.

Building on this Quaker pacifism, Walker's view of Britain removed it from the position of unique virtue accorded to it by so many English geographers as surely as did Scottish stadial theory, but for very different reasons. On slavery, the bulwark of British liberty was anything but that: 'The British Parliament ... is acting with respect to the Slave Trade in a way which seems utterly indefensible on the principles of Humanity.'[38] Slavery was not only a reality in the overseas empire, however, as Walker's discussion of the present state of Britain made clear:

> It has now become a maxim in the constitution, that the air of these islands is too pure for a slave to breathe. Two considerations, however, obviously offer to repress the Englishman's boast. First: The liberty he enjoys is not extended to the settlements abroad; there slavery is maintained and encouraged ... Second: Unless he be a man of property, whatever may be his capacity or integrity, he is unqualified for the legislature ... His religious profession also affects his privileges as a citizen.[39]

Salmon had written in 1749 of the limits of English representative government, but only to point to the impossibility and undesirability of democracy. Walker goes beyond this to point out the limits of the Revolution Settlement which we have seen defended by such a variety of arguments in eighteenth-century geography books. Walker wrote at a time of rising demands for the repeal of the Test and Corporation Acts which limited the participation of Dissenters in public life, particularly disbarring Quakers who refused on point of principle to take such oaths. Walker was obviously writing from a religious perspective, but it is plain to see how such sentiments fed into the rising tide of secular radicalism and rational Dissent.[40]

On the eve of the French Revolution, Walker's *Elements of Geography* represents a strong attack on the politics of England seen as an oligarchical confessional state. Strongly influenced by the Scottish Enlightenment, the possibilities contained in his text are altogether more radical than those in Guthrie or the moderate Presbyterians. The stadial theorists had portrayed Britain as an advanced society, leading the way on a path all would follow. This reduced the singularity of the British achievement to a matter of timing; the pattern would be repeated. For Walker, however, Britain was politically backward, as it hedged its democracy with confessional and socio-economic qualifications. It was also unchristian in its imperial rapacity. Where luxury, tyranny and imperial overreach were threats in Guthrie's analysis of British political life, they were realities in Walker's.

John Pinkerton's *Modern Geography*, the French Revolution and the waning of the Scottish Enlightenment

John Pinkerton's cultural position had much in common with William Guthrie's: a Scot in London, increasingly driven to hackwork for a living, he was a marginal figure in the Scottish Enlightenment who applied its findings in historical works. Pinkerton's historical writings are more thoroughly imbued with Scottish Enlightenment historiography than Guthrie's, yet his geography displays less of this framework than Guthrie's *Geographical Grammar*. This 'retreat' from the Scottish Enlightenment can be seen as part of a general crisis of confidence in the face of the French Revolution.

Pinkerton's early writings showed his affinities and disagreements with Scottish historiography. Writing under the pseudonym Robert Heron, Pinkerton's *Letters of Literature* (1785) praised the 'philosophic spirit' which 'hath induced latter historians to blend the interesting history of society and manners with the dry detail of the true facts'. Yet he felt that many of the Scottish writers were too speculative, comparing Millar and Ferguson unfavourably with Montesquieu.[41] This last point led Pinkerton into a life-long campaign for the publication of primary documents to facilitate a more empirical type of historical enquiry.[42] Pinkerton's own series of histories of Scotland demonstrate just such a mixed relationship of indebtedness to and criticism of Scottish historiography. His histories contained long appendices of historical documents, but also deployed characteristic features of conjectural stadial theories.[43] In particular, Pinkerton used the notion that unintended consequences are vital motors of historical change which had been a commonplace to Hume, Ferguson and Robertson, but which Guthrie did not use. This comes across most forcefully in his description of the emergence of the English parliamentary tradition, which he, like most defenders of the Revolution Settlement, saw as the crowning glory of the English political system:

> [the] slight circumstance, that in England the members were too numerous to meet in one room, decided the destinies of the two kingdoms. England had a free parliament for discussion, and a house of commons: Scotland an influential parliament for assent, and her commons were overpowered by the preference of the superior orders.[44]

Pinkerton, then, was more fully versed in the Scottish historiographical tradition than Guthrie had been, and also possessed a more independent and critical perspective on that tradition. He appeared to be ideally placed to further the deployment of Scots Enlightenment techniques and concepts in the British geographical tradition. That Pinkerton's *Modern Geography*, published in 1802, failed to take up the challenge of further questioning the politics of the English geographical tradition through Scottish Enlightenment concepts can only be understood as part of the trajectory away from reform that Pinkerton described in the aftermath of the French Revolution. Pinkerton was one of many in both Scottish Enlightened and English Whig circles to gradually drift from enthusiasm to disgust with the events in France.[45]

Before the French Revolution, Pinkerton was firmly associated with the reformist Whig wing of British politics. In *Letters of Literature* he made his Whiggery clear:

> Can I pass the subject of Government and Laws without expressing an ardent wish for a reform of our yet happy country ... But God forbid it should be attempted during this tory reign, in which we behold judges carrying politics into the bench of impartial justice, as violently as in the reign of James II ... [I]t is a maxim in the law, as tories tell us, that the king can do no wrong. By the constitution of this country, according to two precedents, the king can not only do wrong, but be put to death as Charles I or banished, as James II.[46]

The Tories attacked could either be the newly-appointed Pitt administration or Lord North and the government which had led the war with America, or perhaps both. The reference to the people's right to depose their kings is akin to the contractarianism espoused by Guthrie, yet more radical, in that Pinkerton extended this to Charles I, and not just the events of 1688/89. The brand of Whig reformism espoused by Pinkerton was also specified more precisely in the same work:

> true and old Whig principles are essential to ... [national] welfare ... As for the new Whig principles, they have done as much harm to the constitution as toryism. An old whig wishes a reduction of prerogative: a new whig wishes to extend it, if it will serve his faction.

In this analysis, the new Whigs were the group led by Charles James Fox, who had joined in coalition with Lord North to form a short-lived ministry in 1783. This ministry was seen by many as the epitome of corrupt politics, being portrayed as an alliance between formerly implacable enemies purely for the sake of power. The old Whigs were the remnants of the Rockingham Whigs, whose most talented representative and publicist was Edmund Burke.[47] Pinkerton was thus allying himself with a reforming Whig tradition which sought power, but only to return the English constitution to the purity of the Revolution Principles of 1688/9. In this connection Pinkerton was involved in a Scottish Revolution Club in 1788 which used the centenary of the Glorious Revolution to remind the public of its aims and tenets.[48] For Pinkerton the causes of the departure from the line prescribed by the events of 1688/9 were to be found, as for Pennant and Guthrie, in 'great wealth and luxury ... enervating both body and mind'.[49]

Given this reformist cast of mind, it is unsurprising that Pinkerton's initial reaction to the French Revolution was positive: wishing to revive the true principles of 1688/9 in England, he was one of many who initially saw the French Revolution as creating the same sort of libertarian constitution in France.[50] While privately he had doubts about the French Revolution in his correspondence by 1794 in the aftermath of the Terror,[51] Pinkerton was unusual in continuing to defend a more positive interpretation in print as late as 1797. This view, couched in the language of Scottish stadial theory, came in Pinkerton's *History of Scotland*. The 'Preface' to the History alleged that the historian was not 'to pollute his pages with the faction of the day', thereby distancing himself rhetorically from 'the present opinions and commotions in Europe'.[52] The text, however, was more engaged for the Jacobin cause than the 'Preface' would lead one to expect. Writing of the reign of Robert II of Scotland (1371–90), Pinkerton clearly had a modern context in mind, as the shifting tense of his argument suggests:

> But the period is at length arrived, when the violence both of religious, and of political, party was to suffer a considerable abatement. When universal right of conscience shall order governments to refrain from the private business of the bosom; and shall render even the name of toleration ridiculous by tearing down the veil, which obscured the most sacred prerogatives of mankind. When the interested aims of antiquated parties were become objects of equal neglect to enlightened nations, who began to see that one party alone could serve them, an union of the people against their corrupters and oppressors.[53]

Pinkerton here updates his old Whig language of 1785 in a way which is sympathetic with the Jacobins. He goes beyond the Revolution Settlement of 1688, to which the term 'toleration' was a mantra, to envisage a complete liberty of conscience, rather than toleration from the vantage point of a hegemonic church and state. Pinkerton's optimism was due to a conjecture on his part about a further stage in stadial theory, in which Montesquieu's maxim that republicanism was only appropriate for small states would be transcended by the greater political sophistication of the populace:

> Such have been the uniform steps of civilization, in all ancient and modern nations. The democracies of Greece had passed through the ordeal of monarchy: and if the republican form of government be ever found convenient and durable, in a large country, even after the invention of representation, it must be when the national character had been enlightened, and improved to a superlative degree.[54]

Pinkerton was obviously prepared to countenance such a political experiment in France and America, unlike Burke whose reading of stadial theory suggested such a form of government should emerge over time, rather than being created *de novo*. This experiment would test whether the vaunted British constitutional balance of crown, lords and commons did not have a redundant element in the form of the monarchy. Pinkerton had travelled a long way from the 'balanced absolutism' theorised by Heylyn and Bohun, the *sine qua non* of their balance being an ordeal to be jettisoned as soon as possible in his version.

The Terror certainly shook Pinkerton's faith in the French experiment even before his *History* was published, but the final decay of optimism into hostility was only publicly aired as late as 1802 in his *Modern Geography*. In the 'Preface', Pinkerton conveyed the impression that while the uncharted areas of the world were at last becoming known to Europe through the voyages of La Perouse and Vancouver, the old world itself was becoming an uncharted chaos through which no geography book could steer: 'Whole kingdoms have been annihilated; grand provinces transferred: and such a general alteration has taken place in states and boundaries, that a geographical work published five years ago may be pronounced to be already antiquated.'[55] As becomes clear in the text, it was not blissful to be alive at this moment, but a very hell. In his discussion of France, Pinkerton no longer equated universal freedom of conscience with the purest form of religion as he had in discussing Robert II:

> The religion of France was the Roman Catholic, till the recent revolution established freedom of conscience, or rather gave an undue

> ascendancy to concealed atheism, which any superstition remarkably absurd has a tendency to produce. But the strongest minds remained deistical [i.e. Christian, believing in a God], instead of flying from one extreme to another. (*MG*, I 253)

This might be seen as still weakly supportive of France, in that the blame is laid at the feet of *ancien régime* Catholicism rather than revolutionary excess. Yet his account draws on the traditional English Francophobe imagery of France as the seat of luxury and inhumanity,[56] transferring this from the absolutism of the monarchy and church to the Directory:

> Even the violent changes which have taken place seem to have little affected their characteristic gaiety, and Paris continues to be one of the happiest cities in the world: while the screams of massacre resounded in some parts of the city, in others the theatres were crowded, and nothing was heard but the sounds of pleasure. (*MG*, I 257)

Plus ça change, plus c'est la même chose? An immutable French national character is invoked to reveal a continuity beneath the revolutions in church and state, which confirms French moral depravity and inferiority to Britain.

The results of the experiment with a large republic show the French national character has not been enlightened to a superlative degree, as Pinkerton had still hoped in 1797. This led Pinkerton to rethink his entire conception of republicanism as the *ne plus ultra* of constitutional systems, the taint of the French experience holding good for America:

> The most impartial travellers have been impressed with regret and astonishment at the spirit of selfishness and avarice [in America] ... If the spirit of a monarchy, according to Montesqueiu, be glory, the spirit of a commonwealth, by the same authority, is virtue. ...[I]t might perhaps be argued, from the example of Holland, Switzerland, and the French Directory, not to mention many republics of antiquity, that the spirit of a commonwealth is money ... It is however to be hoped and expected that this character will not, like that of the Jews, become indelible. (*MG*, II 563–4)

Even if not indelible, Pinkerton was clear that the spirit of the French Revolution was contagious. His characterisation of Toussaint

L'Ouverture's slave uprising in French Haiti mobilised Swift's imagery of the Houhynhms for conservative purposes:

> [the French colonies] have been convulsed and ruined for a season by intestine commotions, arising from the wild theory of the rights of man being extended to the negroes, who feel that they have a right to ruin and destroy, but none to build and improve. Perhaps the right of horses may next be discussed. (*MG*, I 254)

From pushing the philosophical generalisations about societal progress in a more optimistic direction than even the stadial theories of the Scottish Enlightenment, Pinkerton had reversed his course and was using the 'lessons' of geography to enforce a scepticism about constitutional change and imperial liberty.

In treating Britain in *Modern Geography*, Pinkerton also reverted to a tradition of religious nationalistic imagery he had eschewed a decade previously. While the French colonies succumbed to madness, 'by the protection of all-ruling Providence the British empire rose superior to the struggles' (*MG*, I 255). Internally, Britain was also portrayed in a manner which suggested but slim grounds for the sort of reformist alarms Pinkerton himself had raised in 1785. Anglican tolerance is no longer ridiculous, but a *via media*:

> The church of England is established upon a most peculiar basis, and truly characteristic of a moderate and judicious nation. As in the political system, extremes ... are carefully avoided, and despotism or anarchy, from whatever source ... prevented ... so in the church, while papal power, and the other catholic chains are proscribed, the other extremes, tending to loose democracy, are equally avoided. (*MG*, I 33)

This *via media* has a decidedly post-1789 character, avoiding Papal excess as ever, but with the other extreme being democracy, rather than fanatical Presbyterianism. Indeed, the middle way is now between a religious and a political extreme, rather than between two religious extremes. When Pinkerton rounds this discussion off by saying that Anglicanism has retained episcopalianism 'in its ancient splendour', his overall image of the English church–state system is more unstintingly positive than Guthrie's or even Salmon's. Clearly, the language of corruption and reform was now impracticable, given the real danger Britain faced from a militarily-successful France: a reversion to praise of

English providential moderation in religion and government was the only option. The moment for more radical questioning of English political values in geography books from the geographical, intellectual, religious and political margins had been ended at least temporarily by changing political languages and realities in the aftermath of the French Revolution.

It is worth tracing the trajectory of Pinkerton's political opinions to their close. In his *Recollections of Paris* (1806) Pinkerton moved a stage beyond the simple defence of the English and attack on the French contained in *Modern Geography*. *Recollections of Paris* attempts, *inter alia*, to show that geographical realities had doomed the political project of the Revolution to failure from the outset. In the 'Preface' Pinkerton pictured Rousseau as the idol of Jacobinism, and argued he was in fact 'an angel of darkness' due to his ignorance. Rousseau's ignorance manifested itself in his arguments about the state of nature. The true state of nature was displayed in 'innumerable books of voyages' which Rousseau ignored. The result was catastrophic:

> the savage rout and uproar of Sans Culotterie drowned the sober voice of wisdom and moderation, proscribed what even the wildest savages value as they do their own eyes, being in truth the very eyes of the body politic, the sacred and eternal aristocracy of talents.[57]

Pinkerton reverted to the imagery and conceptual translations Heylyn had deployed a century and a half earlier: geography was indispensable as a source of information for political theorising, essential to clear scholarly vision, and any viable theory would enshrine hierarchy as the eye of the body politic, where those theorists who were ignorant of geography were lost in darkness and would plunge political life into a similar state.

The language of enlightenment and darkness, then, could be used against the *lumières*. For Pinkerton, the turmoil of the Revolution had been engendered by a geographical and social illusion, a mirage before the eye of the body politic, concerning the state of nature and the South Sea islands, to the depiction of which we have already seen Wesley object on Christian grounds. Invoking his authority as a geographer, Pinkerton gave an alternative picture of savage societies, allegedly built from the facts of travellers:

> Among the rudest savages we find several classes of men superior to the rest. The chief or leader is a man of superior strength and

> courage; and as in that rude existence the qualities of the father, as among horses and other animals, commonly descend to the son, we find a kind of hereditary government among some of the most savage tribes. Among savages too, we always find what may be termed nobility .... The next class to the nobles are the priests ... Such are the established degrees of inequality experimentally found among all savage tribes, under whatever climate, in whatever region of the habitable globe.[58]

Pinkerton gives his savages a settled monarchy, aristocracy and church, such that the French Revolution, by emulating Rousseau, had moved away from the true state of nature. This polity is also naturalised by the example of animals; clearly if the French did liberate horses after the Haitian uprising, their natural instinct would have been to found a moderate church and state polity according to Pinkerton, with the English as the Houyhnhms of political theory and the French Revolutionaries as the Yahoos. What is most obvious is that Pinkerton, despite his scientific rhetoric of experiment, cited no evidence for his assertions about the savage state in the *Recollections*. In this context, Pinkerton's publication of a seventeen volume *Collection of Travels* (1808–14),[59] while it was certainly hackwork to eke out a living, was not without political significance.

Pinkerton's aim in writing *Modern Geography* had been to supersede Guthrie's *Geographical Grammar*.[60] Coming from a comparable authorial position, Pinkerton's geography book shows how the reformist element in Scottish thought was eroded in the face of the French Revolution.[61] As was said of Guthrie, the stability provided by established political structures was less dispensable than reform of those structures. In Guthrie this was a latent prioritisation, as it was in Pinkerton's early writings, but it was activated as the French Revolution failed to deliver the values of sociability and moderation on which the Scottish Enlightenment had been premised. In the process, the role of the Scottish Enlightenment in British intellectual and political life was also altered:

> Burke established a rhetoric of prescriptive conservatism with such intellectual power and religious conviction, and found so large a public willing to endorse it ... that he bade fair to displace Scottish social theory from its role as the chief ideological support of the Whig order. This does not mean, however, that he was either indifferent or hostile to it.[62]

Pinkerton had started his writing career by identifying himself with Burke and the (post) Rockingham Whig position in 1785. Whilst he had diverged from Burke's position in the immediate aftermath of 1789, by 1802 and the start of his engagement with geography, he had come to a synthesis using Scottish theory to defend the established order which brought him back towards Burke. In describing this course, Pinkerton also signalled a retreat from Guthrie's reformist perspective on English political life, let alone from Walker's impassioned condemnation. Whilst the modern forms of conservative and radical political language started to emerge at this time, in geography books only the conservative position was represented, and the window of opportunity for reformist political questioning of the 1770s and 1780s was closed.

10
On the Cusp of Modern Geography: Fieldwork and Textuality in the Career of James Rennell, 1764–1830

The personal politics of Major James Rennell were similar to those of John Pinkerton. Taking a critical line against Lord North's handling of the discontents of the Americans, he came to a more conservative conclusion about the French Revolution:

> The illustrious geographer was always an advanced Liberal in politics, but he loved his country's greatness above all things, and imbibed a strong feeling of horror for the excesses of the French Revolution. He consequently attached himself to those Liberals, such as Lord Spencer and Sir W. Windham, who joined Burke in supporting the government of Pitt in a war policy.[1]

Rennell's pamphlet on the French situation, *War with France the Only Security of Britain* (1794), was written in the aftermath of the Terror, and pointed to this as the moment of transition between regarding French 'struggles for liberty, with a sympathy worthy of freemen', and the realisation that their 'national character' revealed them to be 'incapable of receiving rational liberty'.[2] Clearly, Rennell's break with the Revolution came earlier than Pinkerton's, at least in published form, but effected the same shift from viewing the events of 1789 in the light of 1688/89 to seeing the French national character as antithetical to freedom. Despite describing a similar trajectory politically in the period 1770–90 as Pinkerton, and writing geography at the same time as Pinkerton, Rennell points to a very different politicisation of geography.

The way in which politics informed the texture of his geographical writings was one which bespoke a profound shift in the politicisation of geography which would grow in strength in the nineteenth century as a new conception of the geographer began to develop.

Rennell as a modern

Many elements of Rennell's biography seem to mark him out as working within a qualitatively distinct tradition of geography from that to which all the individuals discussed so far conformed. To the modern reader, it is hard not to see Rennell as working in the nexus of a formalised and institutional geography and deploying that geography for imperialist purposes. Rennell seems to verify the contention that the 'new' geography was solely a product of and for the grandiose imperial ambitions of the nineteenth century.[3]

First, there is a sea change in the character of geography in Rennell's writings, away from a textual conception and towards fieldwork. None of the geographers I have discussed up to this point saw any need to conduct fieldwork, and although some of them (like Salmon) had travelled this did not inform their geographical writing. Rennell, by contrast, was among the first to betoken the 'formal empire of professional knowledge', including geographical knowledge, which characterised the 'second', post-American, British empire.[4] From the outset Rennell's career as a sailor and then, from 1764, as Surveyor General to the East India Company, had led him to recognise the practical difficulties of mapping and gathering geographical information.[5] Rennell retained this awareness when writing geography books, his basic interest being the construction of accurate maps and descriptions based on the collation of all available surveys and travel journals. This also led him to revise his constructions in the light of new data, demanding carefully updated editions.[6] This painstaking method led Rennell to change the scale of geographical enquiry in two senses. First, his works were about regions, notably Africa, Asia Minor and India, rather than summaries of the world as a whole.[7] Secondly, Rennell concentrated on specific thematic enquiries in physical and human geography – the oceanic currents of the Atlantic, the hydrography of the Nile, Ganges and Brahmaputra, and so on[8] – rather than claiming to summarise all the salient physical and social characteristics of a country, as had the early modern textual approach.

This altered conception of geography is mirrored in a different career pattern for the geographer. All the writers discussed so far wrote one or more geographical texts as part of a far wider *œuvre*, normally centring on historical and theological works, and saw geography as merely ancillary. By contrast, Rennell wrote a large number of geographical works which, while designed to illuminate other areas of enquiry, also intimated an independent role for geography. Rennell saw the improvement of geographical knowledge as a task in its own right. His works also form a series of interconnected investigations about the modern and ancient geography of one area, Asia Minor and India. Here we see the emergence of a different conception of geographical authorship, in which the geographer is an individual writing a series of related works about one area, based on their own fieldwork. This contrasts with the early-modern conception, where a geography book is a one-off summary of geographical knowledge, functioning as a groundwork for other and more significant forms of enquiry, and built by the collation of a miscellany of texts by travellers.

In writing a series of geography books, Rennell also forged a different concept of the scholarly and institutional relationships between geographers. Rennell clearly saw himself as part of a community of enquiry composed of past and present geographers, and referred to their work to correct and discuss it. As an example of this, Rennell's discussion of the voyage of Hanno, a disputed circumnavigation of Africa in ancient Greek times, concluded with 'some strictures on the opinions of some celebrated modern geographers, on the subject of this voyage'. These strictures were in fact a careful discussion of Rennell's agreement with, and divergence from, the opinions of D'Anville, Bougainville, Gosselin and other eighteenth-century writers on ancient geography.[9] Rennell was unfailingly polite in his discussion of these and other geographers, acknowledging his debts and cautious in his conjectures. His politeness, moreover, was not restricted to the English, as is shown by his admiring references to D'Anville in particular. The scholarly ethos here is very different from that under which most geographical discourse operated, 1650–1800. Most geography books were compiled by plagiarism and the attribution of indebtedness was sporadic at best. Moreover, geographical discussion was often less than courteous, and designed to discredit a writer rather than correct their scholarship, let alone improve the state of geographical knowledge, as we saw in Bohun's case. Such discussions remained largely within the ambit of factional English/British church–state politics. The eighteenth century has been viewed as an era in which politeness led to less acidic exchanges in the European 'republic of letters', but such a process could only begin to operate where the writers saw themselves as a community of like-minded individuals contributing to a larger intellectual project.[10] In geography, such a perception was delayed by its lowly and propædeutic status in the humanist educative structure. The scholarly engagement with other geographers evidenced by Rennell marks the formation of such a community among geographers.

This community was a personal and not simply a textual phenomenon for Rennell. Rennell was a member of the African Association, founded in 1788 'as no species of Information is more ardently desired, or more generally useful than that which Improves the Science of Geography'.[11] The club collected subscriptions and funded Park, Horneman and others to explore the interior of Africa. Rennell was a member of the Association from 26 May 1792,[12] and wrote his illustrations of Africa based on the reports sent to the Association for the perusal of its members. Eventually the African Association would merge with the Royal Geographical Society, but it marked an important albeit less formal stage in the creation of a geographical community.

Rennell and the politics of modern geography

James Rennell's writing career as a geographer, then, takes on a very different shape from that of any of the individuals discussed to date. Given this, it is unsurprising that the ways in which his geographical texts are politicised are also very different. Above all, geography is now an activity which serves the state practically rather than textually.[13] Heylyn *et al.* defended the English church and state by narrative methods which valorised specific constitutional, ecclesiological and dynastic structures. Rennell, by contrast, rarely makes any *explicit pronouncements* about the structures of governance, but his entire practise as a geographer *assumes* the legitimacy of those structures. In his mapping and surveying, and thence in his geographical 'constructions', Rennell supported the British state.[14] This can be seen most clearly from his approach to India.

Rennell was quite candid that war and imperial conflict were essential to the progress of geographical knowledge: 'In a national point of view, the improvement of geography is always taken up very late at home; since a coarse approximation to the truth answers most purposes; and abroad it cannot be well done, until a subjugation of the country has taken place.'[15] The legitimacy of such a subjugation is not a point at issue (or, as we shall see, only rarely), and it is in the effacement of such political questions that Rennell signals a change in the intersection of geographical and political discourse. Indeed, the whole question of resistance by indigenous people is constructed at several points by Rennell as a resistance to geographical progress. Thus he confesses that the construction of his Indian map is sketchy in several places, adding:

> It is very much to be regretted that there are no observations of longitude, to correct the bearing of a coast of so large an extent as that between Cochin and Bombay; or that a measured line was not drawn across some part of the peninsula; as for instance, between Madras and Mangalore ... This, however, can hardly be expected, whilst Hyder Ally continues master of Mysore and its dependencies.[16]

Where most early modern geography books showed a marked concern for legitimacy, in Rennell's hands geography is markedly silent as to legitimacy. This is not simply Rennell's public discourse, but also his mode of expression in his Indian surveying journals, which were written to the East India Company in Calcutta. Rennell's surveys were carried out in 1764–67, which was also a time of heightened tensions between local rulers and the Company.[17] In his *Journals*, local resistance was always described in terms of its deleterious impact on the survey, until Rennell at last joined forces with a detachment of English Sepoys (i.e. Indian soldiers in English pay) and was seriously wounded in 1766.

Rennell's configuration of the Indians as obstacles to the progress of geography is part of a broader system of national and ethnic stereotypes, which builds from an assumption of European superiority. Rennell's life's work was directed towards illuminating the geography of Asia, but his summary characterisation of Asians was a stereotypical invocation of Asian effeminacy and changelessness akin to Guthrie's:

> Both the Hindoos and Chinese appear to have carried the arts just to the point requisite for useful purposes; but never to have approached the summit of perfection ... Arrian's Indian history ... shews us how very little change the Hindoos have undergone in about 21 centuries, allowances being made for the effect of foreign conquests.[18]

The ethnic imagery Rennell deployed became more complex in his later writings, spurred by the researches of the explorers sponsored by the African Association. Just as Rennell's geographical method was a comparative one, comparing and collating examples and evidence, so such a method was applied to the peoples of Africa and India.

> The contrast between the Moorish and Negro characters, is as great, as that between the nature of their respective countries ... The Moors appear to possess the vices of the Arabs, without their virtues; and to avail themselves of an intolerant religion, to oppress strangers: whilst the Negroes, and especially the Mandingas, unable to comprehend a doctrine, that substantiates opinion or belief, or the social duties, are content to remain in their humble state of ignorance ... I know of no fitter title to confer on them, than that of the HINDOOS of AFRICA: at the same time, by no means intending to degrade the MAHOMEDANS of INDIA, by a comparison with the AFRICAN MOORS.[19]

As will be apparent, Rennell used some version of stadial theory, the Indians having progressed to a certain point and been changeless ever since, and the most advanced of the Africans having reached the level of civilisation of the Indians. Both groups, however, were seen as inferior to the Europeans, a message of considerable (albeit unspoken) political importance woven into the texture of Rennell's analysis. Far more than for a hack writer such as Guthrie, the direct political importance of such a theorisation to political practice must have been apparent to Rennell.

What we see with Rennell, then, is the elimination of direct political argument by textual interventions in defence of specific church and state structures. In its place, a series of assumptions about the legitimacy of subjugating other nations is developed, these being expressed in a language of geographical data collection. The legitimacy of subjugation is in turn bolstered by imaging ethnic groups as inferior to the Europeans, a finding itself given authority by its sanction in geography books. While one way of explaining European superiority might be providential, Rennell was in fact notably reluctant to introduce theological argument into his writings. As far as I am aware, he only once deployed the theological argument from design, and that in the context of ocean currents: 'It appears scarcely any portion of the ocean remains still: and, doubtless this is amongst the wise dispensations of THE CREATOR, to preserve its purity'.[20]

This, of course, again marks Rennell off from most of his predecessors, who had developed theologically and denominationally specific arguments within the course of their discussions. The state of Rennell's own faith cannot be determined, but he clearly had important doubts about the Mosaic chronology and saw its defenders as 'superstitious', as his letter to the geologist John Hunter showed. Rennell had read a manuscript by Hunter and advised greater tact in theologically sensitive matters:

> At page 3 you have used the term 'many thousands of centuries' ... Now, although I have no quarrel with any opinions relating to the antiquity of the globe, yet there exists a description of persons very numerous and very respectable in every point, except their pardonable superstitions, who will dislike any mention of a specific period that ascends beyond 6,000 years. I would therefore, with submission, qualify the expression by saying 'many thousand years', instead of 'centuries'.[21]

That Rennell's own theological position is undecipherable from his voluminous geographical writings is in itself significant. He clearly divorced theology from geography to an extent which none of his predecessors had. As politics was sublimated to the level of presuppositions, so theology was effaced, the result being what appeared a more 'pure' form of geographical writing. In fact, of course, such a shift was itself a politicised change.

Rennell as early modern

Rennell's geographical writings cannot be treated, however, simply as a break from the early-modern British tradition of political argumentation in geography books. His work also showed strong continuities with that textual tradition.

While Rennell did remove most of the specific political argument from his work as a geographer, there were eruptions of anti-Jacobinism in his texts. That these interventions were incongruous to the sort of geographical persona he had established for himself was clear to Rennell, but the French Revolution was too momentous to be effaced as questions of imperial legitimacy had been. Thus Rennell cited Herodotus, prefacing that:

> we cannot resist the temptation of inserting the following remarks, at this time, as they shew the strong contrast between the virtuous republican of Greece, and a modern republican, formed on a Gallic model.[22]

Similarly, a discussion of Aboukir Bay and the geography of the Nile led to this apostrophe to Nelson after the defeat of the Jacobin threat:

> this place, under the modern name of ABUKEIR, has received a new, and perhaps more lasting, impression, of 'the stamp of fate', by its overlooking, like SALAMIS, the scene of a naval battle, which, like that of Salamis, may lead to a decision of the fate of Europe. This most brilliant victory, achieved solely by the Britons, Europe felt as her own; and Frenchmen *alone*, mourned the defeat. To this spot, the genius of Britain conducted his favourite NELSON, who at one blow destroyed the fleet of the enemy.[23]

Given the context, it is noticeable that Rennell still avoids any providential ascription of divine aid to Britain, preferring the neutral (or secularised?) reference to 'the genius of Britain'. This contrasts with Pinkerton's invocation of providence to explain the calm of the British Empire in the face of the French example of anarchy, and with the more general importance which Protestantism continued to have in British patriotism.[24]

Rennell's conception of empire is also not one which links him without ambiguity to nineteenth-century imperial theory and practice. In fact, while his geographical writings show many signs of a colonialist mentality, the model of colonialism to which they adhere was on the wane in the later eighteenth century. For while the legitimacy of empire was assumed by Rennell, like Guthrie he advocated a limited empire in India, not total domination:

> I believe there are many who think that the British might have extended their possessions in Hindoostan, *ad libitum*: however, one of the greatest of our Indian statesmen, Lord Clive, thought that the Bengal provinces and the circars, together with a moderate tract of land round Madras, and the island of Salsette, near Bombay, were fully equal to the measure of good policy, and to our powers of keeping possession.[25]

Rennell had been encouraged throughout his time in India by Clive, and in turn supported his vision of empire, and he generalised the point in his survey of the history of the Mogul empire, saying 'its history is one continued lesson to kings, not to grasp at too much dominion; and to mankind, to circumscribe the undertakings of their rulers'.[26] Moreover, Rennell's advocacy of limited empire also spilled over into an acceptance of its limited worth to the colonised. Arguing that Bengal 'enjoyed a greater share of tranquillity' than the rest of Asia, Rennell admitted that this was because it was a 'tributary state' to Britain such that

> the grand object for which the Bengal provinces are held, militates against the ease and happiness of their inhabitants: for there can be no inducement to increase a national income for the purpose of finally enriching another nation.[27]

Rennell's imperial assumptions were modelled in the era of the control of India by the East India Company in the 1760s. The power of the Company was undermined by the British Government from 1773 onwards, with the final transfer of power in 1813. Rennell's practical and intellectual work, then, related to a conception of empire which was to be overtaken before his death. Whilst much of his argument and many of his assumptions about the politics of empire would be continued in nineteenth-century geography, this was in the framework of a different conception of empire.[28]

From a cartographic perspective, Rennell's work for the East India Company was reliant on methods of collation and selection among the evidence of route surveys, as all eighteenth-century cartography had been. As such, while Rennell indeed surveyed in the field, his construction of maps relied on the adjudication of evidence back in the office, unlike the procedures of triangulation which were used (albeit inconsistently) in the Great Trigonometrical Survey of India in the nineteenth century. Despite Rennell's 'rhetoric of science, accuracy, exactness, and precision' his collations in fact 'converted conjecture into knowledge', being reliant on precisely the same procedures of selection and appropriation as the textual tradition of geography books in the 'long' eighteenth century.[29]

The final and most important way in which Rennell was squarely within the early modern tradition of geography can be gleaned from the titles of some of his works: *The Geographical System of Herodotus Examined*; *Observations on the Topography of the Plain of Troy* (1814); and *Illustrations (Chiefly Geographical) of the History of the Expedition of Cyrus* (1816). A high percentage of Rennell's geographical output was designed to illustrate the geography of classical histories. In this, Rennell's practise of geography was entirely traditional, seeing it (as Heylyn had) as one of the eyes of history. That classical histories were prioritised was due to the humanist educative structures that continued to dominate public schools and the universities. While Rennell had very little formal schooling and did not read Greek, he still operated within that scale of values at a time when the monopoly of classical learning was already being attacked by those who sought a form of schooling more relevant to middle-class children destined for commercial careers.

Reading Rennell: on the cusp of modernity

Reading Rennell's geographical works and experiencing their whole texture discloses the ambiguous interconnection of various elements which, retrospectively, appear incongruous. It becomes apparent that to read James Rennell in the context of the practise of geography which he inherited, rather than to take him as a symbol for 'the rise of modern geography', is to uncover a transitional sort of geography and geographer.

Looking at Rennell's methods of illuminating classical histories, we find an exemplification of the melding of methods and objectives which both before and after this time were seen as belonging to separate modes of enquiry. The basic aim of explaining the geography of a classical text was a traditional one, which had also engaged several of the individuals discussed earlier, as for example Edward Wells in his edition of Dionysius's *De Situ Orbis*. Much of Rennell's work was also unexceptional, notably his explanation of myths and fables about places, and his attempts to relate ancient place names to modern geography. Rennell is unusual, however, in his use of a knowledge of geomorphological processes in this work, where the traditional tools of humanism had been etymological and philological. As an example, in his *Geographical System of Herodotus*, Rennell discussed the plausibility of circumnavigations of Africa by the ancient civilisations. Rennell's discussion emphasised that such voyages could have taken place by describing the favourable oceanic and near-shore currents which modern sailors had charted around Africa. These, together with research on the nature of Greek vessels, convinced Rennell that such voyages could have been made by hugging the coastline and using the favourable currents.[30] Rennell, then, used the latest information on oceanography, a subject in which he made pioneering observations, to adjudicate on disputed claims by ancient historians. The methodology involved here, with what now might be called process geomorphology being used in a humanist geographical enquiry, was one Rennell repeated in all his ancient geographical works. Such an interaction of humanism with emergent scientific knowledge was not unusual in the period up to 1800, but the increasing specialisation of scientific knowledge put it under strain with the passage of time.[31] Rennell was one of the few geographers to participate fully in both modes of scholarship.

A similar interweaving (to modern eyes) was created where Rennell used modern military speculations to elucidate the topography of Troy as described in the *Iliad*. Thus, trying to locate the 'Throsmos', a mound where the Trojans encamped after forcing the Greeks back to the shore in Book VIII, Rennell argued for one hill on the plain rather than another as only it would allow for the rapid drawing up of a battle front if the Greeks should try to flee.[32] Rennell here deployed exactly the reasoning he so frequently used in his modern geographical writing, most notably about India, where topographical description was blended with discussion of the strategic qualities of that topography.[33] Here, then, humanism and militarism interacted, unsurprisingly given that Rennell believed much of the elucidation of the theatres of ancient history had only been possible thanks to military operations such as the French invasion of Egypt (resulting in *Description de l'Egypte*) and the naval voyages commissioned by the British Admiralty.[34] The interaction was not only in one direction, either, as Rennell argued Cyrus's expedition should be 'particularly studied by military men; as containing many important lessons, as well of tactics, as of policy'.[35]

These interactions permeate the entire structure of Rennell's geographical enquiries in general, and affect the politicisation of his geography in particular. As a typical example, having exhorted military men to read Cyrus's expeditions, he added that Cyrus's troops had shown remarkable fortitude, and then drew a parallel: 'we regard the courage and perseverance of our countryman, CAPTAIN, now ADMIRAL, WILLIAM BLIGH, and his boat's crew, as one of those remarkable instances'.[36] This was a gesture of loyalty towards the British naval establishment in a controversial case. Similar militaristic parallels were drawn in Rennell's geographies, which reinforced his anti-Jacobinism. His parallel, quoted earlier, between Nelson's victory at Aboukir and the Greek victory at Salamis (described in Herodotus, Book VIII) by extension made a connection between Napoleon and the Persian tyrant Xerxes, perhaps implying that both were oriental despots. Such a reading was confirmed in the posthumously published *Comparative Geography of Asia*, were Rennell likened Napoleon to Alexander the Great, both of whom 'shewed no feeling either for man or beast'.[37] Adding this use of ancient-modern parallels to Rennell's other modes of geographical politicisation discussed earlier, it is clear that his texts operated in several different modes of political argumentation. The total experience of reading Rennell for his political stance is an ambiguous one, with anti-Jacobin outbursts mixed with historical parallels, with ancient and modern geography described in terms of strategic value, with modern and classical imperial geographies in interplay, and with late humanist fears over imperial overreach operating in tandem with the emergent colonialist mentality of expansion as a civilising mission.

Rewriting Rennell: eliminating early-modern geography

The liminal status of James Rennell's geography, and the modes of politicisation that geography facilitated, have been effaced over time. The loss of a sense of Rennell's ambiguous position in the history of geography came in the late nineteenth century, as geography began to press its claims for recognition in schools and universities. At the time of Rennell's death in 1830, Beaufort recognised his humanist achievement:

> Germany can boast of her Cluverius and Cellarius and France of her d'Anville but no eminent geographer has yet adorned this country. Rennell amply redeemed us from that reproach. To the industry of the former, and to the acuteness of the latter he added a sagacity which reconciled the most discordant passages of history, a perseverance which ransacked every source of information.[38]

As the nineteenth century progressed, Rennell was rewritten to conform to a modern image of geography, while continuing to perform the role Beaufort had given him as the first eminent English geographer, even being aggrandised into the first 'real' geographer. This process is well exemplified in Clements Markham's 1895 biography of Rennell. Markham constructed Rennell in the light of the cult of genius developed by the Romantics. The comparison with the poet was explicit as Markham wrote his own *The Prelude: Or the Growth of a Geographer's Mind*:

> He [the geographer] should have been trained by years of land or sea surveying, and by experience in the field in delineating the surface of a country. He must possess the topographical instinct; for, like a poet, a geographer is born – he is not made. The birthplace and

> youthful surroundings of our first English geographer will, therefore, have a special interest for the student of his life-story; for we thus become acquainted with the images which filled his brain at the time when his geographical instincts first began to mould themselves.[39]

The geographical instinct which drove Rennell in this construction was an unambiguously modern one, based on fieldwork which made geography a 'science'. Older notions of science as meaning simply 'knowledge' were jettisoned for science as method, and the pantheon of humanist geographers to which Beaufort had inducted Rennell as an English representative was now replaced by a trajectory in which Rennell eclipsed that tradition.

> [T]he critical judgment of Rennell was superior to that of d'Anville. Both were the most eminent geographers of their respective times, and both used similar methods in considering questions relating to geographical subjects. The difference between these two eminent men was that while one was a student from his boyhood, the other had been a surveyor on active service by sea and land ... The conclusion is that to form a perfect geographer long experience in the field as an explorer or a surveyor must be combined with the close and deep study of a man of letters.[40]

Another way of looking at Markham's conclusion is that he distorted the texture of Rennell's writings and career to form him into a perfect geographer in the late nineteenth century mould which was only beginning to emerge when Rennell lived and wrote. Rennell wrote in the twilight of the early-modern tradition of geography, and his methods remained more textual than Markham was willing to admit. Debating the topography of Babylon, for example, it was Rennell's opponent, Claudius Rich who had done fieldwork, not Rennell. Rennell argued that Rich could not have found the correct site of the city, because he had relied on the relation of the river channel to the city described in antiquity, where it was clear from Rich's own investigation that the channel had meandered over time, and therefore should be in a different positon *vis-à-vis* the site at the present day.[41]

Markham's assessment was part of a more general evaluative pattern extending to the present day which ignores the transitional and liminal position Rennell occupies in the history of geographical thought. Rennell has been co-opted into the historiographical framework analysed in Chapter 1 in which geography is born *de novo*, or at least *de minimis*, at the end of the eighteenth and beginning of the nineteenth century. Far from allowing 'a more telling and accurate assessment' of Rennell this process has *eliminated* his *liminality*.[42] This treatment of Rennell was necessitated by the desire to locate him at the wellsprings of modern geography, where he had initially been placed by Beaufort for very different reasons. The elimination of one aspect of Rennell's *œuvre* to achieve this mirrors the more general effacement of the early-modern textual practise of geography as a consistent and theorised enquiry within the structure of a humanist education.

11
The Scottish Enlightenment and British Geography (II): James Bell and J.R. McCulloch, 1830–50

James Bell's *System* and the rise of statistical geography

James Bell was heir to the traditions discussed in the two preceding chapters, namely, the Scottish Enlightenment geographies of Guthrie and Pinkerton, and the mass of geographical data produced by the fieldwork tradition of which Rennell was so distinguished an early representative. Bell's relationship with the Scottish Enlightenment was, like Guthrie's and Pinkerton's, somewhat skewed by his location as a writer. Where Pinkerton and Guthrie developed their ideas in London, Bell's career centred on Glasgow, which was distinct from Edinburgh in its relationship with Enlightenment ideas, being both more commercial and more rigidly Presbyterian,[1] both facts which (as we will see) are evident in Bell's geographical writings. Writing in the 1830s, Bell could also draw on a wider range of geographical data than any of the writers discussed previously. This was in part due to the tradition of colonial fieldwork, Bell himself paying homage to Rennell as the first British 'geographer of any eminence' for his work in the field.[2] The accumulation of data also had a domestic angle thanks to the nineteenth-century penchant for parliamentary commissions and the institution of national censuses from 1801. Bell, then, lived in 'the new age of blue books, royal commissions and select committees, the age of statistics'.[3]

Bell's task was to compress the vast array of information available into the old genre of the geographical compendium, which purported to give an accurate summary of all the nations of the world. The result, as we shall see, was one of the greatest gazetteers in the English language. But Bell's *System of Geography* (1832) also betokens the breakdown of the humanist conception of geography and the ways in which that tradition developed a political position through textual manoeuvering. Bell's *System*, then, is janus-faced, being both a summation and an undercutting of the textual tradition of geography and its forms of political expression. If Rennell's work moved in two directions thanks to his fieldwork, Bell's represents the same dilemmas transferred to the geographer's study. It is noticeable that Bell's work occupies this liminal position concurrent with parallel ambiguities in the British political system on the eve of the decade of reform in the 1830s which transformed political languages and realities.[4] The breakdown of humanist geography, which in Bell is more at the level of theory than practice, was, as we shall see, to be pushed towards its conclusion by McCulloch, a self-conscious liberal of the Reform era.

Continuities: the textual politics of Bell's *System*

Looking first to Bell's *System* as a continuation and summation of the humanist, early-modern conception of geography, there can be no doubt that the text conveyed a political and theological argument in long-established ways. For while it was frequently claimed by contemporaries that the use of vast quantities of statistics and an array of tabulation made texts of the variety which Bell constructed impartial sallies into the 'science of society', and while Bell himself argued that the *System* would not meddle in contemporary politics (see *System*, III 12), there can be no doubt on the basis of a close reading that the claim to objectivity was merely rhetorical.

Politically, Bell expressed hostility to Lord North, than whose administration none 'could be more impolitic and unfortunate' (*System*, III 10). This was not simply an historical point, given that much landed opinion continued to hold to the species of conservatism which North had propounded. Furthermore, Bell adopted the traditional scepticism of the Scottish Enlightenment towards eulogies of the British political system, akin to that which we have already seen in Guthrie and Pinkerton. For Bell, the British constitution was no clearer on the monarch's powers in 1832 than it had been when James II was ousted in 1688 (*System*, III 12–13). Likewise, Bell's Scottish intellectual heritage left him free to attack the English legal system, suggesting the common-law tradition developed excessive complexities which obstructed justice:

> The laws of Britain, though they be precise, are justly censured for being voluminous and complex, putting it in the power of dextrous

> lawyers to extend their causes to a most vexatious length. (*System*, III 17)

As the threat from the French Revolution subsided, there was a resurgence in the Scottish questioning of English political pieties which had waned with Pinkerton.

Where the material discussed so far is familiar from the first generation of the Scottish Enlightenment, Bell also turned his attention to the political issues of the 1830s, his position being consistent with liberal demands for rational reform of the constitution. In this connection, Bell praised the 1824 repeal of the Combination Laws, which had prohibited trade unions and thereby allowed excessive wealth to accumulate in the hands of the few (*System*, III 29). On similar grounds of economic equity, Bell also attacked the East India Company as 'a society of monopolists whose power and possessions have ... become so exorbitant' (*System*, III 27). Finally, Bell welcomed the Catholic Emancipation Act of 1829 as a constitutional reform which would promote the economic development of Ireland:

> That great and important measure being now carried, there is reason to hope that the discontent, disunion, and strife which have so long distracted Ireland and perverted her best energies, will soon cease to operate. (*System*, III 236)

In all these instances two things are noticeable. First, Bell as an heir to the Scottish tradition is on the reforming side in his comments, far from those to whom toleration of Catholics and unions remained anathema. Secondly, and somewhat distinct from the earlier Scottish tradition, Bell's comments about reform all elide political and economic argumentation, justifying political change to a great extent on the basis of economic expediency.

The political comments from Bell's *System* cited thus far would be accordant with either the Whigs (moderate or Foxite) or the reform-minded liberal Tories of Lord Liverpool's administration in the 1820s.[5] Both groups contemplated and enacted political reforms grounded in economic argumentation derived from Ricardo. Yet the politics of Bell's *System* can be located somewhat more precisely as belonging to the Whig, rather than the liberal Tory tradition by further textual analysis.

As we have already seen for his politics more generally, so Bell's Whiggery expressed itself in both traditional and new ways in the *System*. On the one hand, Bell deployed traditional textual methods of espousing a political position in a geography book in his potted Whig history of England. Both James I and Charles I were criticised for their absolutist notions of monarchical power, and the Glorious Revolution was praised as it 'fixed the English government in that state of freedom and moderation which has since so happily characterised it' (*System*, III 72). Bell's history of England, then, was a Whiggish one of the triumph of liberty 'in consequence of the progress of political and religious knowledge, and the increasing diffusion of commercial wealth' (*System*, III 69). Lest the Whiggish message should be missed, Bell's *System* also included a list of England's great statesmen which was in fact a radical-Whig pantheon: Hampden, Pym, Hollis, Sydney and Russell (*System*, iii 114). On the other hand, Bell's Whig position was also expressed in a modern political context in his support for electoral reform, which is worth quoting at length, as it reveals much about the *System*'s mode of argumentation and the ways in which it started to diverge from the humanist textual politics of geography. Discussing the Scottish electoral system, Bell was led into the following reflection:

> Parliamentary representation ... notwithstanding the great influx of wealth, the increase in inhabitants, and the general diffusion of knowledge, themes upon which political speculators dwell with so much delight, remains as it was fixed 123 years ago! ... Edinburgh, the metropolis of the kingdom, has only one representative in the national senate, while an English hamlet of six old houses sends two representatives to the same assembly! The whole kingdom sends 45 members to parliament; a single county in England, namely Cornwall, sends 44!! Nor is this all: all of the few representatives Scotland possesses are not to be regarded as the representatives of national feeling and the guardians of national interests. Of the two millions and a half of population, not above 3000 individuals possess the legal right of voting for their representatives; and even of this handful nearly one half are false, nominal, and fictitious voters! We have no inclination to enter upon politics, nor though we had, is this the fittest place for discussing them; but we do think that the fact just alluded to such a one as must strike every reflecting person with astonishment. (*System*, III 180)

Here we see a changing mode of politicisation, in that Bell hopes that facts presented will of themselves have the persuasive force to engender political reform. Bell displays a contempt for 'speculatists' in politics, aligning himself as a pragmatist. Allegedly against rhetorical modes of political persuasion, Bell rests his case on numerical statement, also adding a disclaimer that he is not making any political point, as this would be inappropriate in the genre of the geography book. As will be apparent, Bell's position is itself a form of rhetoric using statistics and facts to persuade in a genre which, given the divisions of knowledge of the period, had always been politicised. There can be no doubt that Bell was making a partisan point in this passage: his *System* was published on the eve of the 1832 Reform Act, and supported its basic principles of electoral reform to eliminate the old 'rotten' boroughs and give the franchise to the commercial classes. Bell's support for reform is unsurprising: he himself had been a trader before he turned his attention to geography, and his native Glasgow before the Reform Act shared a parliamentary seat with Dunbarton, Renfrew, and Rutherglen, such that 'political influence and participation, even amongst the merchant elite of Glasgow, was therefore restricted, second-hand and shared with three disparate burghs'.[6] Bell, then, was making a partisan and broadly Whig/liberal point in his comments on electoral reform which supported his own class's competence to act in the political process, but this was clothed in the rhetoric of letting the facts 'speak for themselves' which was to become so important to nineteenth-century political argumentation.

In the realm of church politics, Bell's *System* also acted in a traditional manner to support a specific and partisan position by its textual construction, this being even more apparent and unchanged than was the case for his argumentation about constitutional politics. Unsurprisingly, Bell had no doubt that the Christian religion was the true faith which revealed the word of God (*System*, I cxxix). He was also strongly anti-Catholic, characterising the Reformation as a 'struggle betwixt light and darkness' (*System*, III 132). Furthermore, this struggle was not simply an historical phenomenon, but an ongoing battle, as became clear in Bell's comments on Ireland. As we have seen, Bell approved of Catholic emancipation, but this was on commercial and pacific grounds, his attitude towards Catholicism as an institution and doctrine being uncompromising. The Irish had been 'perverted by the false lights of superstition' such that, for Bell,

> we ought not to be surprised that they have distinguished themselves by acts of ferocity and barbarism, almost totally unknown where knowledge is promoted and government regulated by the principles of justice and the dissemination of religious truth. (*System*, III 251)[7]

Within the Protestant tradition, Bell is complimentary about the Anglican Church, but in a manner which is uninterested in its doctrinal disputes. Thus he says it is not clear whether Anglicanism is Calvinist or Arminian without adjudicating (*System*, I cxxvii), and that the thirty-nine articles give Anglicanism a creed 'the same in all essential respects with those of the other reformed churches' (*System*, III 132), itself a controversial position. This detached attitude to Anglican doctrine, which earlier English geographical writers such as Wells and Gibson could not have contemplated, is coupled with an appeal for a parliamentary commission to eradicate pluralism which suggests Bell saw a need to reform the Church of England (*System*, III 134).

Lest Bell's attitude to the Anglican church be taken as proof of a growing secularisation in geographical writing, his attitude to his own Scottish Presbyterian church shows that the *System* was articulating a preference for a specific theological and ecclesiological position through its textual construction. Even in his discussion of the English Reformation, Bell had traced its genesis through the 'radical' tradition of Wycliffe and the Lollards, rather than adhering to the High Church argument about the primitive and Catholic status of the Anglican church which we have seen in Camden, Heylyn and Wells (*System*, III 132). Unsurprisingly, then, he was far more effusive in his praise of the Scottish kirk as a Calvinist inheritor of the 'radical' tradition, seeing Presbytery as the primitive form of church government (*System*, III 194). This purity was not simply a doctrinal issue, but (as with Pennant) led the Scottish church to a practical evangelism unmatched by Anglicanism, this being reflected in the superior intellectual state of the Scottish peasantry (*System*, III 181). Bell was personally a zealous Calvinist, and the Glaswegian Enlightenment context from which he came was likewise known for its 'commitment to a radical, pious and evangelical brand of Scottish Presbyterianism'.[8] It is unsurprising, then, that while his *System* shared with Guthrie's *Grammar* a redirecting of the denominational politics of geography towards Scottish Presbyterianism, it was distinguished by a fervour unseen in Guthrie, which suggests anything but a linear trend towards secularisation in the religious politics of geography. This fervour was most apparent in Bell's impassioned defence of the early Scottish Reformers who had been attacked in 'this degenerate age': 'though many of them perished in the bold and patriotic enterprise, their blood became the seed of the church of Christ; and their memory will be held in everlasting veneration' (*System*, III 196).

Bell's intellectual context as a writer had something in common with the world of 'evangelical economics' which Boyd Hilton has investigated.[9] Bell's was an extreme form of Calvinism concerned with 'atonement', yet he, unlike those analysed by Hilton, never developed a coherent system linking his statistical and political material with his faith in the *System*. Perhaps the nearest the *System* came was in its support of evangelism in the imperial context (see *System*, IV 26 and IV 494). Indeed, if evangelicalism became central to early nineteenth-century social thought, Bell tends to look in the other direction towards eighteenth-century modes of textual politicisation, with only the occasional foray into a religious social policy.

Reconfiguring the 'political' and the 'geographical' in the *System*

The prefatory 'Address' to Bell's *System*, contributed by his publishers, Fullarton and Co. of Glasgow, reverted in outlining the market for the book to the old commonplace that geography was vital to the politician, giving 'adequate intelligence of the statistics of his own country, the relations of foreign States, and the balance of political powers' (*System*, I vi). While this commonplace is as old as geographical writing itself, this version was anything but a Strabonic vision of geography as the 'science of princes', and implicitly recast the sphere of the political with which geography interacted and to which geography contributed. The key point to note is that the politician is assumed to need to know the 'statistics' of his own country. Here we see the onset of the nineteenth century's characteristic recourse to statistics in the construction of political arguments which has already been alluded to in discussing Bell's view of parliamentary reform. This different conception of the sphere of the political was reflected in Bell's *System* by the proliferation of tables of data and statistics more generally, a subject to which I return below.

The reworking of the sphere of the political was matched by a reconfiguration of the notion of geography which Bell enacted at least at the theoretical level. Again the prefatory 'Address' intimated this, placing Bell's *System* in an emergent tradition of 'scientific geography' whose 'father and founder' was Büsching (*System*, I viii). For 'scientific' we can here read 'statistical' as 'one of Büsching's outstanding contributions was his incorporation of statistics into geography'.[10] Bell elaborated on this positioning of his work in his 'History of Geography' in volume six of the *System*. Bell's 'History' pictured geography as a modern subject and a 'science' in the modern sense of the term. Thus, after Varenius's *Geographia Generalis*, Ptolemy was no longer of use, 'a mere skeleton' (*System*, VI 492). This, of course, was antithetical to a humanist conception of the subject, wherein the ancients were reverenced, their texts updated by means of notes, and their worth not judged simply in terms of being in accord with the latest discoveries. As if to reinforce the point, Bell's admiration of Samuel Bochart's *Geographia Sacra* was counterbalanced by his frustration with Bochart's humanist textual scholarship:

> the reader is amazed at the extent of learned research shown by the author, and the immense store of oriental and classical quotations brought to bear upon the subjects discussed. The work, however, has imperfections. Strained and fanciful etymologies are not seldom adopted, inconclusive reasonings as frequently introduced, and illogical conclusions drawn. (*System*, VI 493)

Interestingly, Bell was particularly dismissive of the British geographical tradition before Rennell, identifying its commercial Grub Street milieu of production as disabling it from serious scholarship. Where the French government had patronised the science of geography and the collection of data, in Britain 'any who shall undertake a system of universal geography, must, in order to ensure its reception, and remunerate his labour, accommodate his work with the taste of the public' (*System*, VI 501). The result was that most British geography books were not 'pure', the public being insufficiently 'enlightened' to demand such works (*System*, VI 502). Bell, then, was advocating either a more enlightened commercial or a more interventionist governmental geographical tradition which would concern itself with the collection and presentation of data about the surface of the globe. He wrote this some time between 1828 (when the publishers' 'Address' is dated) and 1832 (when the *System* appeared). Concurrent with this, of course, was the foundation of the Royal Geographical Society in 1830, which was to push geography in the direction Bell desired, as geography shifted from being the Strabonic and humanistic 'science of princes' to being the statistical and explorational 'science of empire' collecting information of service to the government.[11]

How these shifts in the spheres of politics and geography which Bell's *System* set out at a programmatic level would influence the construction of geography books is partly realised in Bell's work. This can be seen in the sheer textual preponderance of tables and numerical data in the *System*, which is unfamiliar to anyone accustomed to eighteenth-century geography books. But it is in Bell's posthumous

Gazetteer of England and Wales (1834–35) that he gave the clearest indication of the changes which new conceptions of politics and geography wrought in the established genres of geographical writing. The *Gazetteer* was explicitly pitched as a product of the era of legislative reform and parliamentary data collection:

> independent of all other considerations, the fact of a new census having been completed within the last few months [i.e. that of 1831], and of a new principle of franchise having been established in the kingdom, would of themselves warrant and call for the appearance of a new Gazetteer, in which the results of the new population-returns, and the other arrangements consequent upon the new system of election should be accurately noted and described. And it is this consideration which induces the Editor to think that the time is also propitious to his undertaking.[12]

True to this positioning, the *Gazetteer*'s construction is a marked break with previous generic conventions as far as politicisation is concerned. Above all, there is an absence of textual politics in Bell's *Gazetteer*, when compared with such works as Bohun's *Geographical Dictionary* or with chorographical works such as Gibson's editions of *Britannia*. As an example, we can compare the treatment of Banbury in Bell's *Gazetteer* with the parallel treatment in Gibson's 1695 edition of *Britannia*. In his entry for Banbury, Bell gives its population from the censuses of 1801 and 1831, the rateable value of property, and details about local trades. He further lists vicarages, their patrons and value to the incumbent, and then enumerates chapels and meeting houses. This is the method Bell adopts for all the towns and villages in the *Gazetteer*. Beyond this, he gives brief historical comments and occasional details of church history from Tanner's *Notitia Monastica*, in the case of Banbury noting that its inhabitants 'were almost universally puritans'.[13] The contrast with *Britannia*'s entry on Banbury is striking:

> There is a credible story, that while *Philemon Holland* was carrying on his English edition of this *Britannia*, Mr Camden came accidentally to the Press, when this sheet was worked off; and looking he found that to his own observation of *Banbury* being famous for *Cheese*, the Translator had added *Cakes* and *Ale*. But Mr Camden thinking it too light an expression, chang'd the word *Ale* into *Zeal*; and so it pass'd to the great indignation of the *Puritans* of this town.[14]

While the counterpoint may be somewhat extreme in this instance, it does highlight the change in geography as statistics took over from narrative as the textual mainstay. The 1695 *Britannia* engages in gentle mockery of Puritan earnestness, and at the same time undercuts a previous editor in order to boost its own credentials. By contrast, Bell's *Gazetteer* is an unembellished recital of facts and figures. This is not, of course, a depoliticisation, since statistics are seen as vital to the politician, but it does relegate the *Gazetteer* to the role of political raw material, rather than being a political participant as *Britannia* had been.

Conclusion

It is hoped that my analysis of Bell's geographical works makes it clear why I opened by characterising him as a liminal figure in the textual tradition of geography. On the one hand, Bell aspires to a new definition of the spheres of the political and the geographical which links the two by statistics and data collection, rather than by textual annotation, editing and discursive polemic. This aspiration comes closest to realisation in the *Gazetteer*, which in one respect amounts to a denial of political intent in a geographical genre which had previously operated within a political ambit, and in another presages a politics of data presentation and the rhetoric of objectivity promulgated by nineteenth-century sciences of society. Yet Bell's main contribution to geography, his *System*, was in fact rather traditional in its textual politicisation, despite its programmatic statements of disciplinary redefinition. Despite some flattening of intradenominational Protestant disputes, the *System* was strongly supportive of Scottish Presbyterianism coupled with moderate political reform, and as such perpetuated the nexus of geography and Scottish Enlightenment thought forged in the work of Guthrie and Pinkerton. The role of statistics in public and political life was not fully settled in the 1830s, and Bell's *System* reflects just such an uncertainty in its disjunction between innovative ambition and traditional textual construction in its politicisation.[15]

J.R. McCulloch's geography and political economy

John Ramsay McCulloch's contribution to geography came in a similar context to James Bell's, namely as a representative of a later generation of Scottish Enlightenment thinkers coping with the proliferation of statistical data.[16] McCulloch, however, was more successful in negotiating this situation than Bell had been, bringing together programmatic shifts in the realms of geography and politics with a changed textuality for geography books. McCulloch's achievement was based on referring his geographical work back to a system of values derived from classical political economy.[17] As we shall see, McCulloch's conception of political economy gave him a vision of a new role for geography in the division of knowledge which tied it to a new conception of politics as a political-economic science of governance. In short, political economy provided an Ariadne's thread for McCulloch of the sort which Bell conspicuously lacked. With McCulloch, then, we can reverse the order we adopted with Bell, moving from his reconfiguration of the realms of politics and geography to his execution of that programme in his geographical compendia.

Geography as the eye of political economy

McCulloch was and is known for his work as a political economist. Yet his vision of political economy drew geography within its ambit, a fact reflected by his own *œuvre*, with McCulloch writing three broadly geographical compendia,[18] and by his library, which boasted a host of geographical works and the catalogue to which showed a considerable interest in discussing preferable conventions for the presentation of geographical knowledge.[19]

For McCulloch, both geographical and statistical science were recent developments,[20] and at several points he argued that it was only with Conrad Malte-Brun's *Universal Geography* (English translation, 1822–33) that geography came of age.[21] The late rise of a scientific geography was a point on which McCulloch agreed with James Bell (although he was implicitly relegating Bell's own work to a pre-scientific status), but he did not blame this on the Grub Street commercial publishing environment, as Bell had, but on the continued hold of 'scholasticism' – by which he meant late humanism – on English education as defined by Oxbridge. McCulloch's most important exposition of this point came in an 1829 article for the immensely influential *Edinburgh Review*, 'the house organ of moderate liberalism'[22]:

> What with Nonsense verses at school, and Novel-reading, Apocrypha controversies, and Phrenology afterwards, we have no time to attend such matters [as political arithmetic and statistics]. Notwithstanding all that is said about the march of intellect, and the efforts to multiply sixpenny systems, it is a fact that even the science of geography, interesting and important as it is, is at this moment, and has long been, at a lower ebb in Great Britain than in

> any country in Europe, Spain not excepted. In despite of the superior means of information at our command, we have not published a single complete treatise on geography during the last hundred years, that is not utterly contemptible.[23]

Particularly important in this tirade is the contemptuous reference to 'apocrypha controversies', a clear attack on humanist scholarship of the sort which Wells or Gibson introduced into geography books, and a clear distancing by McCulloch of modern and scientific geography from such a tradition.

McCulloch's attack on humanistic education was essentially a political-economic one, as his discussion of Oxford University in his *Geographical Dictionary* shows:

> essentially a scholastic establishment ... well fitted to make good Greek and Latin scholars, and perhaps good divines. But it is obviously quite unsuitable, as a place of instruction, for the *élite* of the youth of such a country as this. It must be entirely changed before it either send forth legislators capable of appreciating the various interests of this vast empire, or individuals capable of promoting and extending those manufacturing and commercial pursuits, to the success of which we are mainly indebted for our unparalleled increase in wealth and population. (*Dictionary*, II 451)[24]

A modern industrial and commercial economy demanded an altered form of education, which prioritised 'useful' knowledge in the natural and (emergent) social sciences. That this attack was resolved by McCulloch from the general level here to the level of geography specifically is apparent from comments in his *Catalogue* concerning Camden's *Britannia*. McCulloch owned a copy of Richard Gough's 1789 edition of *Britannia*, a work which he believed had reached the end of its life as a work of contemporary geographical scholarship after two hundred years:

> it may be doubted, considering the extraordinary growth of manufactures and commerce ... whether another edition would be desirable. We incline to think that the historical and antiquarian portions of the work can no longer be advantageously combined with an exact description of the physical and industrial condition of the country.[25]

The very forces which McCulloch studied as a political economist made humanist scholarship in general and its geographical manifestations in particular quite obsolete.

What would a modern geography suitable to a commercial and manufacturing age look like, and how would it fit into a revised division of knowledge? McCulloch was clear and consistent in pivoting his answers to these questions around political economy as the science of society. This is apparent in McCulloch's history of the emergence of political economy, where statistics, geography and history are configured as the handmaidens of an inductive political economy: 'he [the political economist] takes the facts furnished by the researches of the statistician, and after comparing them with those furnished by historians and travellers, he applies himself to discover their relation'.[26] While McCulloch mentions travellers' accounts here, he normally expressed a preference for geographical abridgements rather than travel books *per se*, as they organised material by topic, thereby allowing the political economist to find the relevant information with more facility than a travel narrative. This point was clearly expressed in discussing Richarderie's *Bibliothèque Universelle des Voyages* (1808):

> A work of this sort, were it reasonably well executed, and made applicable to the present time, would be invaluable ... it would give the general reader sufficient information with regard to the state of different countries at different periods, stripped of the verbiage and useless detail [of travel accounts].[27]

Accounts from across time and space, then, were subordinate in the division of knowledge to the science of society, such that geography and history became the eyes of political economy, where for most eighteenth-century geographical writers geography and chronology had been the eyes of history. McCulloch was clearly relegating the milieux in which political controversies had been played out in the eighteenth century in favour of new discursive sites. (That McCulloch was only interested in geography insofar as it contributed to political economy is apparent from his lack of interest in the attempts of Rennell *et al.* to increase contacts with Africa, as he could see no economic benefits springing from this.[28]) Geography being the eye of political economy in McCulloch's system did not make it political as such, since he firmly divided politics from political economy, the former being concerned with the constitution, the latter only with the acts of a government viewed through the lens of their contribution to the prosperity of the nation as a whole rather than specific classes.[29] But even politics, for McCulloch, was a sphere to be judged by the political–economic criterion of utility not the traditional one of liberty: 'freedom is not, as some appear too think, the end of government: the advancement of the public prosperity and happiness is its end; and freedom is valuable only as it contributes to bring it about'.[30]

McCulloch's recasting of the role of geography in the division of knowledge, then, was away from a humanistic – or as he preferred 'scholastic' – conception towards envisaging it as a study propaedeutic to the scientific, political–economic analysis of a modern commercial and industrial society. As previously, geography was to be a preparatory subject, but now the master enquiry was political economy, where previously it had been divinity and history. Political economy occupying the central role for McCulloch also led to an altered conception of the sphere of the political. Utility to society as a whole took over from polemic battles between rival (and elite) denominational and constitutional positions as the defining feature of political questions and analyses in the era after the Reform Act.

Political economy in McCulloch's geographical works

At one level, which will be analysed in this section, McCulloch's geographical works faithfully reflect the reworking of geography's position in the division of knowledge which he advocated at a theoretical level. His three geographical dictionaries ordered their material and understood it through political economic theory in a way and to an extent which no geographical writings had previously. What is more, these dictionaries were influential, their sales figures being very impressive in the context of the era, the commercial dictionary selling some 28 000 copies and the geographical dictionary some 4000.[31]

Adopting a well-established genre, the geographical dictionary/gazetteer, McCulloch kept within its norms in terms of construction but altered their function. This is noticeable in his historical sections, McCulloch's sketch of English history in the *Geographical Dictionary* centring on the emergence of the modern economy and on the role of economic forces in that history to an unprecedented extent (*Dictionary*, I 792–802). Thus McCulloch argued that 'by far the most important source of change' in Elizabeth's reign was the depreciation of currency and consequent rise of prices (*Dictionary*, I 798). Similarly, Henry VII's reign was judged 'one of the most important' in English history because it destroyed feudal power, ushering in a more commercial society (*Dictionary*, I 796). None of this, of course, was unprecedented, Scottish Enlightenment historians in particular linking economic and political change in their narratives, and this being transferred to geography books by Guthrie and Pinkerton, but McCulloch was more unrelentingly economistic in his narrative than his predecessors.

True to his argument that political economy was concerned with the welfare of the whole population, McCulloch's geography books also displayed a concern for what would come to be called 'condition of England' questions, asserting at two points that 'the condition of the great bulk of the people of Britain, and especially of the labouring classes, has been signally, and indeed astonishingly, improved since the middle of last century' (*Dictionary*, I 457; see also I 791). This concern was by no means restricted to the home context: in discussing the state of India under British rule, McCulloch balanced a recognition that agricultural improvement had been encouraged with acceptance that 'there has been but little, if any, improvement in the condition of the people of Bengal under our government ... It is the landlords who have been the great gainers' (*Dictionary*, II 23).

More generally, McCulloch's geographical dictionaries reinforced at numerous points doctrines from his version of classical economics. For example, the Corn Laws were attacked as being less useful in securing a supply of food than free trade (*Dictionary*, I 770). McCulloch also argued that imperial possessions were a drain on the British economy, and should only be founded as free and autonomous units which could engage in trade free from colonial restrictions imposed by the mother country (*Dictionary*, I 469). As will be apparent from the examples chosen, behind these specific views lay a general defence of free trade against the rival economic doctrines of mercantilism and physiocracy:

> Had he [Colbert] acted on the maxim, *Laissez-nous faire*, of the merchants he consulted, though his measures might not have been equally successful at the outset, they would have been incomparably more so in the end. The *prestige* that attached to the name of Colbert has tended materially to prolong the mania for regulating in France; and consequently to enfeeble and paralyze the industry of the country. (*Dictionary*, I 856)

Standing behind this was a belief in the Mandevillian doctrine of private vices, public benefits, modulated through the economic lens of

Adam Smith's 'hidden hand' of the market (albeit without Smith's civic humanist fears over the effects of luxury):

> as all commerce is bottomed on a principle of reciprocity, and is sure to stimulate the industry, and to add to the wealth of all who engage in it, we may safely affirm, that while the whole people of Britain are pursuing only their own interests, they are contributing in the most effectual manner to diffuse the blessings of civilisation and a taste for luxury and refinement. (*Dictionary*, I 455)

Clearly, while retaining the generic conventions of a geographical dictionary, McCulloch did act to alter the overall meaning of that genre, such that it both provided data from various nations as the raw material for economic analysis and reflected the results of such analysis in the arguments it espoused. In his 'Preface', McCulloch had said that his *Geographical Dictionary* would 'not scruple to commend and censure, as well as to describe' (*Dictionary*, I viii). As will be apparent, the grounds of commendation and censure resided in the doctrines of political economy.

The politics of McCulloch's political economy

So far, we have followed McCulloch's application of a programmatic division of knowledge in the geographical realm, taking it at face value. But it must be pointed out that McCulloch's separation of politics from political economy was disingenuous to say the least. The classical economics of which he was a representative was distinguished by its 'concern with formulating analytical propositions which had more or less immediate policy implication'.[32] McCulloch was identifiably a liberal in politics thanks to his political economic doctrines, and a liberal of a radical bent. McCulloch's geography books did not simply develop political positions through their use of political economy, but also deployed more traditional textual methods, his *Statistical Account*, indeed, eschewing an exclusively statistical presentation in favour of including historical and descriptive elements.[33] The political positioning which McCulloch's geographical writings evidence was of a piece with that he expressed elsewhere in his *œuvre*.

Although McCulloch's political economy agreed in its Ricardian policy implications (if not its intellectual bases) with the actions of the liberal-Tories led by Lord Liverpool and then Peel,[34] he was hostile in general to the Tory position. Thus McCulloch's *Geographical Dictionary* was unusual in attacking the popular memory of George III: 'this prince … [was] ill-educated, obstinate, and strongly imbued with anti-popular prejudices' (*Dictionary*, I 800). He renewed this attack on George in his *Catalogue*, and coupled it with criticism of Samuel Johnson's Tory politics as 'bigotted'.[35] McCulloch also attacked this politics in its Scottish guise of the so-called 'Dundas despotism', the Tory connection based around the Dundas family which controlled political power in McCulloch's youth (from about 1784 to 1827). His *Commercial Dictionary* attacked Dundas for defending the East India Company, intimating that Dundas had deceived parliament as to the financial health of the company so as to secure the renewal of its charter, which would be profitable to him if not the nation as a whole.[36] McCulloch's own career had been blocked by Dundas, as had those of many of his fellow whig/liberal contributors to the *Edinburgh Review*.[37] In McCulloch's case, a proposal to create a chair of political economy at Edinburgh University for him was quashed by Dundas,[38] which reduced McCulloch to a life of publishing in order to make ends meet. The brief attack on Dundas's character in the *Statistical Account* was McCulloch evening the score after the Dundas hegemony had collapsed in 1827.

McCulloch's hatred of Tories was almost matched by his dislike of the aristocratic Whigs in the tradition of Charles James Fox, centred on Holland House.[39] Fox (and his successors) was disliked by McCulloch on the predictable grounds that he was 'wholly ignorant of everything that had any reference to matters connected with commerce and finance'.[40] All this locates McCulloch's own political position fairly precisely as what Mandler calls a 'liberal' as opposed to a 'Whig', or as what Brent terms a 'new Whig'.[41] This group centred their understanding around political economy, prioritising this over legislative reform which was the core of the older Foxite Whig tradition. Indeed, legislative reform was only favoured by the liberals if it improved the economic functioning of the nation. This positioning as a liberal explains why, despite his hatred of the Tories, McCulloch could praise Pitt the Younger as he understood political economy and sought to implement its findings.[42] It also explains McCulloch's scepticism about the 1832 Reform Act as expressed in his *Geographical Dictionary*, which stands in marked contrast with Bell's position in the *System*. Where Bell saw a Reform Act for Scotland as a statistical necessity, McCulloch believed it had had a disastrous effect, the property qualification leading to rapid sales of land and the subdivision of property to create voters, the result of which was that improvement and therefore national prosperity were threatened (*Dictionary*, II 657). McCulloch's views in the English context were slightly different but equally hostile to the effects of electoral reform. Here, he believed that the franchise had been extended to those 'little better than a beggar' who could not be so 'interested in good government, and in the maintenance of that security so essential to all great undertakings, as the owner of an estate or of a cotton-mill' (*Dictionary*, I 462). 'Beggars' who were uneducated in the science of society and could be corrupted by bribery were not the basis of a rational electorate which could institute good government. In both cases, McCulloch judged legislative reform exclusively in terms of its economic consequences. His was certainly not the position of a Tory opponent of electoral reform; as the passage on the effects of the Reform Act in England cited above makes clear, he opposed stasis, wanting the commercial and industrial elites ('the owner of a cotton mill') to have electoral influence.[43] More generally, he thought the English fixation with landed property was misleading and unfavourable to economic efficiency (see *Dictionary*, I 853). McCulloch, then, while doubting that the poor had the independence to vote judiciously, was not a civic humanist, or was at least one of the most attenuated kind, as he did not ground virtue in the land, tending instead to relocate it in commerce.

Likewise in the sphere of church politics, McCulloch's geographical texts did adopt a position, and one which was highly controversial, this again being driven by his subordination of matters of intellectual principle to the dictates of national prosperity and productivity. In its traditional form, McCulloch was simply uninterested in church politics. As such, he could praise toleration in general and Catholic emancipation in particular in the *Geographical Dictionary* as aiding the economy (*Dictionary*, II 45–6). Further, and more controversially, McCulloch advocated the redistribution of church income in Ireland away from the Protestant established church towards Catholics on the economic grounds that 'it is in truth little better than a contradiction and an absurdity to make the church of a small minority the national church, and to assign to it funds that might amply provide for the religious instruction of the whole people' (*Dictionary*, I 463). McCulloch showed none of the proselytising zeal of Bell's *System* over the conversion of the Irish Roman Catholics. That McCulloch was out on a limb here can be seen by the fact that these views as expressed in the *Geographical Dictionary* were published in 1841, the same year which saw a conservative government elected in 'defence of the Church against the designs of the Dissenters and the Irish Roman Catholics'.[44] Finally, and to show that McCulloch's was basically a secular and strongly anti-clerical geography book, he went beyond advocating the redistribution of Irish church funds to apply the same economic logic in the English context, viewing Anglicanism in Erastian terms as a mere government department:

> there is no principle or right of any kind to hinder parliament, should it be so disposed, from dealing with church property as it would deal with any thing else. An established church ... is a mere human institution, with functionaries appointed and paid by the state; and should parliament be honestly impressed with the conviction that the interests of religion and morality will be better promoted by diverting a portion of the church property to other purposes, it is not entitled merely but it is its bounden duty so to divert it. (*Dictionary*, I 788)

Where Walker's *Elements* (see Chapter 9) was a rare example of an anti-clerical geography book, this was for denominational reasons connected with his Quakerism. By contrast, McCulloch's was an anti-clericalism which was essentially secular, as it was driven by non-religious concerns. McCulloch, then, produced what may well be the first secularised geographical dictionary in the British geographical tradition.

McCulloch's geographical output had a politics and a stance on church politics, those of an extreme liberal wedded to the principles of political economy. The overall effect is the diminution of historical and narrative modes of politicisation (although as we have seen they are by no means eliminated), and a total flattening of denominational and ecclesiological textual politics. The 'apocrypha controversies' which McCulloch so despised were effaced from his geographical writings. In McCulloch, then, the humanist tradition of textual politics reaches the most attenuated form it can without disappearing altogether, something which Bell had signalled as a desideratum but failed to achieve. McCulloch recast the existent genre of the geographical dictionary, making it the eye of political economy and placing it under the aegis of this science of society.

Geography at mid-century

There is an important sense in which McCulloch (and indeed Bell) were representative of the nature of geographical writing at the mid point of the nineteenth century. Both exemplify a more general reorientation of geography towards statistics and political economic analysis, jettisoning the historical and humanist modes of politicisation which the bulk of this book has analysed. As Fontana has observed, political economy was 'a style of analysis rather than a theory ... not a single doctrine, but an aggregate of more or less converging views'.[45] Geography books in the period display a movement towards just such a 'style', McCulloch only being unusual in the rigour and thoroughness with which he tied geography to a doctrinally defined version of classical political economy. When the *Gazetteer of the World* (1851–56), commenced by Bell as the successor to his *System* and finally published under the auspices of the Royal Geographical Society some 20 years later, said of the pre-Union mode of governing Ireland that it was 'now the proper subject, not of statistic statement, but of history', the assumptions behind this statement summed up the transition which had occurred in geography.[46] The *Gazetteer* assumed that the realm of geography was statistic statement, and that to classify a fact as historical was to consign it to oblivion insofar as a work was geographical. This, of course, was a view diametrically opposed to that of Heylyn, Wells and many other geographical authors this book has studied, for whom the interrelationship of geography, history and politics was a seamless one articulated by a humanist framework.

Yet there is another sense in which it would be misleading to end my narrative with McCulloch, as the geography books of the period 1650–1850 might then appear to trace a smooth and thoroughly Whiggish trajectory from the royalist, denominational and humanist geography of Peter Heylyn to the electoral (albeit not fully democratic), secular and social scientific geography of McCulloch. In fact, McCulloch's geographical work, like John Walker's *Elements of Geography*, marks a radical high water mark, unrepresentative in important ways of the political tide of geographical writings in the mid-nineteenth century. For there can be no doubt that McCulloch's extreme advocacy of political economic reasoning made him, by the standards of the period, a 'philosophical radical'.[47]

In two important ways Bell's and McCulloch's geographical work did not reflect the nexus of political ideologies which geography books of the mid-nineteenth century displayed. First, at the dawn of the 'golden age' of British empire and of Victorian pride in nation and empire, both Bell and McCulloch, coming from the Scottish Enlightenment tradition, did not reflect the patriotism which most geography books exhibited. Secondly, and applying to McCulloch only, geography books continued to adopt Christian viewpoints throughout the Victorian period, there being no progressive secularisation in geography. A flavour of the ideology of geography books in the mid-nineteenth century can be gleaned from Walter Blackie's *Imperial Gazetteer* (1855). Blackie was a member of the RGS, and his work included a patriotic paean concerning the British empire:

> in many respects the greatest now existing, or that has ever existed, in the world; ... encircling the globe, and ruling its destinies, pre-eminent alike in wealth and population, manners and religion, law, literature, and arts – an empire, in short, in which the great problem of human government is practically solved, and the power of the sovereign so happily reconciled with true freedom in the subject, that other States, in reforming themselves, look to the British Constitution as their perfect model.[48]

The loyal and adulatory tone here is a long way from McCulloch's reflections on the weakness of a constitution which gave the vote to 'beggars', or from Bell's asides about the failure of the constitution to delimit the power of the monarch. It is noticeable that Blackie's and the *Gazetteer of the World*'s pride in Britain was expressed in a religious idiom: for Blackie, the perfection of the British constitution was 'the result of providential arrangement'.[49] Similarly, the Scottish critique of the Anglican church in Bell and McCulloch was a long way from Blackie's praise of England: 'nowhere are the great truths of Christianity more deeply pondered, more thoroughly understood, more cordially believed, more sedulously practised'.[50] Taken together, the constitution and the Protestant church (unlike McCulloch, Blackie expressed a hostility towards Catholicism which remained widespread in Victorian Britain)[51] gave the British, in a characteristically Victorian idiom, a 'superiority of character'.[52] Such a 'superiority', of course, could easily translate into a belief in the 'white man's burden' of imperial control, and the geography books of the mid-nineteenth century start to display such a belief, which is an attenuated and imperial version of Scottish stadial theories. The RGS was to propagate such views into the twentieth century, and the *Gazetteer* it sponsored in the 1850s could already speak of 'the vitality, energy, and reproductive power of the coloured race' as being 'decidedly inferior to that of the Anglo-Saxon, Anglo-American, and other fair branches of the Caucassian race'. The consequences of such racial and imperial reasoning were already presaged in the same work, the Australian aborigines being said to be 'destined, like all other civilized races, to disappear before the progress of the white race'.[53]

In Blackie and the *Gazetteer*, then, we can see the ideologies of patriotism, imperialism and race added to the supposedly analytical tools of statistics and political economy which Bell and McCulloch had deployed. It was in the nexus of these discourses that the Victorian conception of geography, so different from its humanistic predecessor, was defined.[54]

12
Coda: Halford Mackinder and the Empire of 'New' Political Geography, c.1887–1919

Placing Mackinder

Halford Mackinder is normally credited with taking a leading role in the disciplinary codification of geography in Britain, and is further seen, together with Alfred Mahan and Isaiah Bowman, as the instigator of 'political geography' or 'geopolitics' in the English-speaking world. Mackinder, then, is traditionally placed as one of the founding fathers whom histories of geographical thought and of political geography discuss in their opening chapters. By contrast, he forms a coda to this book, which is in itself a critique of simplistic notions of geography's emergence being coupled to an independent disciplinary structure and of resultant misconceptions concerning when and how politics and geography have been linked. My purpose in this chapter is both contextual and comparative: contextual, in that I wish to place Mackinder's geopolitical project in the historical context of the crisis of confidence in the status of Britain as a world power which affected late-Victorian and Edwardian British political discourse; and comparative in that I wish to play Mackinder's conception of the link between politics and geography – so crucial to twentieth-century political geography – off against the humanist textual politicisation of geography books which the bulk of this study has analysed. Such a method allows for a closer scrutiny of Mackinder's claims to have forged a 'new' geography, on which subject conventional histories of geography tend to have taken Mackinder at his own word, and should highlight some of the discursive disjunctions and continuities between the intertwining of geography and politics which humanist methods established and that which has become dominant under the rubric of political geography.

Before turning to Mackinder, it is important to sketch the political context in which his work emerged. Mackinder's career as a thinker coincided with the late-Victorian and Edwardian crisis of confidence over the future of Britain, a crisis which affected both Liberals and Conservatives.[1] This crisis came in the aftermath of the golden era of Victoria, Palmerston, Gladstone and free-trade Liberalism, for which McCulloch's reconfiguration of geography as the handmaiden of political economy was ideally suited and with which the previous chapter closed. Mackinder himself was a political actor, being a member of political clubs such as the Liberal League, the Co-Efficients and the Compatriots,[2] and becoming a Conservative member of parliament from 1910 until 1922. Mackinder started his career as a 'Liberal Imperialist', switching to the Conservative party in the wake of Joseph Chamberlain's 1903 speech which heralded the radical Conservative/Unionist campaign for tariff reform to preserve British supremacy, and which the bulk of Mackinder's Liberal Imperialist colleagues opposed.[3] The significance of this switch in political allegiance should not be overemphasised since, except on the subject of tariffs, the continuities between Liberal Imperialists and Conservatives were substantial, the two groups comprising a 'unified political formation' whose political programmes were in substantial agreement as to the problems Britain faced and the sources of viable solutions.[4] Mackinder, unsurprisingly, saw the overlap between the two positions, arguing that in his conception of manpower lay 'the reconciliation of Colonial Liberalism with protection'.[5] It is worth looking at the elements in the analysis of the political state of Britain shared by Liberal Imperialists and tariff-reforming Conservatives, as it is here we will find much of the inspiration behind Mackinder's conception of geography as a discipline, his understanding of the role of geography in the division of knowledge, and his political doctrines as espoused in his geopolitical writings.

For both Liberal Imperialists and tariff-reforming Conservatives, the key problem for *fin de siècle* Britain was to maintain her status as a great power. This was in large part a geographical issue: both groups took pride in Britain's insular position,[6] but feared that the large emergent nations – the United States, Russia and above all Germany – would usurp British supremacy, this being enabled by the abundant resources available to such massive continental powers.[7] For Britain to counteract her small size required a two-pronged policy, with domestic and foreign aspects. On the one hand, the nation state itself had to maximise its 'efficiency', this being the 'political catchcry' of the Edwardian era, which 'cut completely across the conventional distinctions between "left" and "right", "liberals" and "conservatives"'. The political programme of the campaigners for national efficiency was: 'an attempt to discredit the habits, beliefs and institutions that put the British at a handicap in their competition with foreigners and to commend instead a social organisation that more closely followed the *German* model'.[8]

A key part of this programme to foster domestic efficiency, and the most relevant aspect to Mackinder, was a desire to update the English educational system and curriculum in line with the German structure to meet the needs of the modern commercial and political world.[9] But efficiency alone could not maintain Britain's status as a world power. Instead, and turning to the foreign side of their programmes, Conservatives and Liberal Imperialists agreed on the need for Britain to protect her empire, which made her *de facto* one of the large states. For both groups, this meant a rejection of free-trade economics of the sort which McCulloch had espoused and which had become the backbone of Gladstonian Liberalism, in favour of some form of neo-mercantilist 'national economics' which defended the interests of the British empire as a trading bloc against rival states which used protectionist economic tariffs.[10] One consequence of this was the Liberal Imperialist rejection of Irish Home Rule, the policy with which Gladstonian Liberalism had become inextricably identified in its later years, in favour of retaining a larger British geographical (and therefore economic) entity.[11] Farther afield, this economic bloc had to be bolstered by a strong merchant and naval fleet if Britain's insular position was not to generate the potential for a blockade by a large rival nation and if her empire was to be united and protected. As such, Liberal Imperialists and tariff-reforming Conservatives agreed on the need for the maintenance of British naval supremacy in the face of the expanding German fleet.[12]

Both the Liberal Imperialists and the tariff-reforming Conservatives, then, developed programmes, rhetorical and practical, designed to combat a perceived deterioration in Britain's position as a world power, which involved improved efficiency at home coupled with a drawing together of the British empire, the aim being to form a trading bloc capable of competing with the Germans in a struggle between national economies which was frequently described in pseudo-Darwinian language.[13] For both groups, therefore, a domestic politics was unavoidably an imperial politics, and such a national politics of integration provided a collectivist *via media* between the old excesses of High-Victorian free trade and the new threat of a socialist collectivism based on the politics of class, which the rise of the Labour party seemed to prefigure.[14] That Mackinder's own political career was a series of variations on this *via media* is well known,[15] but I wish to argue further that it informed all the elements in his construction of a 'new' geography.

Geography's educative empire: an imperial humanism?

The world system as analysed by Liberal Imperialists and tariff reformers was one where large blocs struggled with one another in a zero-sum game for survival and supremacy, far removed from the expansionist free-market doctrine of international trade based on comparative advantage. Mackinder's sense of the need for a 'new' geography sprang from the same basic observation: with the era of discovery over,[16] the world was now a 'closed system' to be imagined as a whole,[17] wherein all parts were known and owned. Geography, then, could not merely chronicle new discoveries, which (according to Mackinder) had always been its role to date. Mackinder's well-known work as an educationalist, trying to carve out a space for geographical education at all levels from primary school to university, derived, then, from a sense that the world system had shifted in a way which demanded a changed function for geography, just as this shift was perceived to necessitate a new conception of the foundations of British politics.

Mackinder was clear about what the new geography must turn its back on, consistently deriding the old descriptive geographical grammars with which this study has been concerned. For him, the 'old' geographies were 'irrational', 'that is ... [their] main function is not to trace causal relations ... [but to provide] a body of isolated data to be committed to memory'.[18] At one level, this attack on 'old' geography had a specific factional resonance in late-Victorian and Edwardian political discourse. Both Liberal Imperialists and tariff-reforming Conservatives, in their rejection of liberal free-market economics, were sceptical about its fixation with trade statistics, which they saw as giving a false impression of the health of the nation. Mackinder shared that scepticism, developing in his *Money-Power and Man-Power* (1906) the notion of 'manpower' as a better and more holistic measure of economic performance than trade statistics.[19] Mackinder's dismissive comments concerning descriptive geographies, with their 'mere mapping of frontiers and detailing of statistics',[20] then, transferred this political scepticism into the sphere of geographical education by attacking the statistical tradition of geography which originated with Büsching in the mid-eighteenth century and was particularly intertwined in the British geographical tradition with free-market political economy thanks to the efforts of Bell and McCulloch.

At a second level, however, Mackinder's target in criticising 'old' geography was far broader than the statistical/political–economic variety of geography books. For Mackinder there was a consistent thread in geography books and geographical instruction from the ancient Greeks to the previous generation: 'there piled up a chaos of miscellaneous information, which was until recently referred to as "descriptive geography"'.[21] Note, then, that Mackinder lumped together all previous descriptive geographies, irrespective of their methodological variations, up to and including the statistical/political–economic reworking of the genre. For him, their generic continuity as gazetteers outweighed any changes in approach. This amounts, of course, to an attack on the entire early-modern humanist tradition of textual geography. But interestingly, Mackinder attacked the geography of the humanist tradition in precisely its own terms, arguing that the new geography must develop a genre with 'no apparatus of notes and memoria technica to de-humanise it. ... No more devilish means of sterilising young minds was ever invented than the old-fashioned text book of geography'.[22] The generic conventions of geography as codified in the humanist educative structure, then, dehumanised, where a post-humanist geography book, by contrast, must make human interest its prime concern.

What textual format was the 'new' geography to adopt in order to give geography a function in the era of a closed world system? Contrary to the structured paragraphing of gazetteers, geography books should be written in 'ordinary literary form', that is, as continuous prose, preferably with a strong narrative line, engaging the reader.[23] Further, geography had to be restructured to be 'progressive in method'. Where old gazetteers and the classroom instruction they fostered led students to 'learn of successive regions by precisely the same methods at all stages of mental development', the new geography would expand outwards from the home to broader regions and gradually increase the amount of theoretical material the pupil was expected to absorb from a narrative.[24] Mackinder certainly deployed this progressive method in constructing his six-volume series, *Elementary Studies in Geography* (1906–14).[25] This method was designed to illuminate in ever-increasing richness the main causal preoccupation of the scientific geographer in Mackinder's 'new' geography: 'the interaction of man in society and so much of his environment as varies locally', a nexus which integrated physical and political (or human) geography.[26]

Given that this was the format which the geography book should adopt to promulgate a progressive form of geographical instruction centred on the causal relation between man and the environment, what was the outcome of such instruction and how was it of use in the closed world system of the late-nineteenth and early-twentieth centuries? For Mackinder, the 'very essence of geographical power' as instilled by the 'new' geography was the ability 'to visualize ... until it becomes possible to think of the whole World's surface at once in all its complexities'.[27] Since this power to visualise the whole world was 'essential to an Imperial people', for Mackinder such a 'sense of geographical and historical perspective goes far to make the statesman', a point which updated the Strabonic commonplace in an Edwardian context.[28] Mackinder's new geography, then, was designed to produce people who, by visualising the whole world, could 'think imperially', a much-touted concept in the Liberal Imperialist and tariff-reforming circles in which Mackinder travelled.[29] 'Thinking imperially' was contrasted with the 'Little Englandism' which the free-trade wing of the Liberal party was said to foster by not recognising the need to coalesce into a larger trading bloc in order to survive.[30] Mackinder explicitly set his new progressive geographical method against such intellectual myopia:

> 'Little they know of England who only England know'. Still more truly may it be said, little they know of the British Empire who only the Empire know. ... Let us first teach children to read the map and to think geographically, then let us regulate and broaden their imagination through wider and wider fields until at last they can grasp the globe in the background of a thought, and place a given detail in its world setting.

Mackinder's geography, like the perspective of the Liberal imperialists and the tariff reformers, was to be nationalist by the very fact of taking a global and imperial perspective: 'let our teaching be from the British standpoint, so that finally we see the world as a theatre for British activity'. This was the training appropriate to a clash of national powers, wherein Britons were 'citizens of an empire which has to hold its place according to the universal law of survival through efficiency and effort'.[31] Such a geographical training, furthermore and in accord with the mantra of 'efficiency', would close the 'lead of Germany in Atlas cartography and in scientific and philosophical Geography'.[32]

Mackinder's new geography, then, necessitated a new genre of geography book of unprecedented narrative and literary ambition, and a new, progressive structure of geographical instruction, in order to create an ability to visualise the world as a whole, this being the key to creating imperial citizens aware of the struggle in which they were engaged for national survival. In short, in its textual and didactic specifications the new geography was a response to the crisis of confidence into which Britain was plunged by the rise of rival nation states, being firmly allied with those groups of both a liberal and a conservative disposition who looked to the empire as Britain's salvation. In such a context, the old humanist geography had to be jettisoned.

Complementary to Mackinder's new form of geographical text and instruction was a new vision of the role and position of geography in the broader structure of education. If the old geography of humanism was dehumanising, Mackinder's vision was of a post-humanist geography being humane: 'My high ambition ... has been to supply for the masses of our children that kind of writing which in another sphere is known by the name of humane letters.'[33] This ambition was to be achieved by rejecting the propaedeutic role consistently accorded to geography in the humanist structuring of knowledge from Heylyn to McCulloch. Instead, Mackinder saw geography as an independent discipline, and indeed as a key one to a student's intellectual development. Geography, together with history, theology and philosophy, was one of the 'outlook subjects' which trained the imagination, its role being to provide a spatial perspective. Furthermore, it was perhaps the key outlook subject, as it contained elements of art, of philosophy and of science.[34] Symptomatic of the change Mackinder sought to effect in the status of geography, it was no longer an 'eye of history', the humanist commonplace, but at least equal to history (as in Mackinder's *Scope and Methods* paper)[35] if not superior 'because for the vast majority of people the present must ever be more important than the past', which meant there was a danger in 'the sterilizing tendency of too strong an historical sense'. Geography, then, was not simply one of the outlook subjects, but 'the chief outlook subject'.[36]

This elevation of geography's status from being a preparatory subject to being a mainstay in the curriculum led Mackinder to a series of grand claims for geography's role. Geography, by bridging the arts and the sciences, was said to be the new classics, able to train statesmen and citizens in a common culture:

> In the days of our fathers the ancient classics were the common element in the culture of all men, a ground on which the specialists could meet. The world is changing, and it would seem that the classics are also becoming a speciality. ... [I]t is ... our duty to find a substitute. To me it seems that geography combines some of the requisite qualities.[37]

A post-humanist geography, then, was envisaged by Mackinder as taking over from classics as the linch pin in a new form of education, appropriate to educating imperial citizens in the 'changing world' of a closed political system. Mackinder envisaged a vast educative empire for his new geography, which would become the centre of what might be called an 'imperial humanism', replacing the humanism of the classics, which had flourished in the era of geographical discovery from the Renaissance on. If Mackinder's initial problematic was the reconfiguration of geography for an era after the age of discovery, his solution was a grand assertion of the indispensability of a 'new' geography in the closed system context: in an age of competitive national blocs in which British survival depended on empire, geography would educate imperial citizens into the sort of awareness advocated by Liberal Imperialists and tariff-reforming Conservatives. Accordant with this position, it would have *imperium* within the structure of knowledge. Mackinder's response to the crisis of British confidence, then, involved the destruction of the humanist conception of geography by its reversal, in order that the discipline might be born, phoenix-like, from the ashes of the body of 'old' geography: geography was to be written in narratives, not the formulaic paragraphs of descriptive gazetteers; geography was to be a progressive causal enquiry, not an accretion of information; geography was to be central, not propaedeutic. Such a reborn geography would produce humane imperial citizens.

Political geography and empire

If geography was to be the key subject in an imperial humanism, for Mackinder geography's own 'crowning chapter' was in the realm of political geography.[38] How, then, did Mackinder put his claims about the nature of geography into operation in his own political geography?

Mackinder's political–geographic argumentation

In accord with his own dictates as we have seen them outlined above, and in a clean break with the generic conventions of the humanist geographical gazetteer, Mackinder's political–geographic works are complex narratives of some literary ambition. Further, these texts are accompanied by a wealth of figures which visualise the political–geographic argument: thus *Britain and the British Seas* has 132 illustrations in its 358 pages; Mackinder's *Geographical Pivot* paper contains perhaps the most lasting image which the geopolitical tradition has produced in its depiction of 'the natural seats of power'; and his lectures on India are a text to accompany and elucidate photographs.[39] This congruity between programme and practice also applied to Mackinder's geographical works for children, the *Elementary Studies in Geography* series. As Mackinder's biographer summarises, 'these principles [of geographical presentation] he put into practice. ... [His works] abound in graphic descriptions enlivened by the use of strong adjectives, picturesque phrases, and striking analogies'.[40]

Initially, then, it would appear that Mackinder is allowing geographical facts presented in narrative and visual forms to drive his political analyses, promoting geography to a sovereign role as the 'new' geography advocated. This would accord with his recorded belief that the facts of geography had some causal influence (though not a fully deterministic one) over political actions.[41] Such an approach would completely alter the balance between the spheres of politics and geography, giving the latter an active role in constituting the former. Of course, this would also amount to a wholly new discursive connection between the spheres of geography and politics from that developed by the traditional geographical gazetteer, wherein the genre of the geography book had been the container for a textual politics generated by 'high' political debate, independent of geographical considerations. At a first glance at Mackinder's work the role of geographical data in the construction of a political analysis would appear to be far greater than previously, or indeed to be emergent for the first time.

Yet Mackinder's rhetoric should not be taken at face value. From a deconstructive perspective, Ó Tuathail has pointed out that Mackinder's alleged recourse to visual facts is undermined by the need for textual support to explicate those facts.[42] From an historical perspective, a parallel scepticism can be developed about Mackinder's visual facts, since those 'facts' seem to adhere rather closely to the doctrine of a Liberal Imperialist or a tariff-reforming Conservative, that is to Mackinder's personal politics. As an example, take Mackinder's visual facts concerning Ireland's relationship with the British mainland in *Britain and the British Seas*. He presents two maps (see Figure 12.1), and adds the following textual gloss to elucidate their visual meaning:

> The seas which divide Ireland from Great Britain are truly inland waters. They penetrate through the mountainous oceanic border of Britain to the plains of the interior, and in certain parts present long stretches of flat shore, as in Lancashire and to the north of Dublin. The Irish Sea is a British Mediterranean, a land-girt quadrilateral, wholly British, whose fours sides are England, Scotland, Ireland, and Wales. The mountains of all four are visible from Snae Fell.[43]

The point Mackinder wishes to make is that the seas between Ireland and Britain are not a divisive feature at all when their visual geography is viewed correctly. Rather, the natural geographical unit is that formed by Britain and Ireland, the sea between them being a connective inland one in its true nature, just as North Africa and Southern Europe form a unit given focus by the unifying influence of the Mediterranean. Of course, the 'facts' Mackinder constructs textually from his visual material could all be embedded in different arguments: St George's Channel could be taken as a funnel for trade, rather than an inland sea; more simply still, the Channel could be seen as a geographical divider, a view reinforced by Figure 12.2 in *Britain and the British Seas*, which shows a deep sea trench separating Britain from Ireland;[44] and that all the mountains of Britain and Ireland can be seen from Snae Fell is no more a visual argument for geographical unity than the fact that Dover can be seen from Calais is an argument for a French annexation of Kent. Further, Mackinder's language presupposes what it alleges the visual facts demonstrate: that the Irish sea is 'truly' an inland water and is 'wholly British'. If that were self-evident, no supporting text would be required.

This analysis of Mackinder's use of textual methods to support his visual geopolitical facts could be repeated for numerous instances of his political–geographical argumentation. Rather than multiply examples, however, two general points can be made, the first contextual, and the second, following on from that, comparative. First, Mackinder's discussion of Ireland's geographical relationship with Britain clearly supports the view that the two are linked, which in contemporary terms would support the view of the Liberal Imperialists and the Conservatives that Gladstonian Liberalism's obsession with Home Rule for Ireland was a mistake. Throughout his political life, and

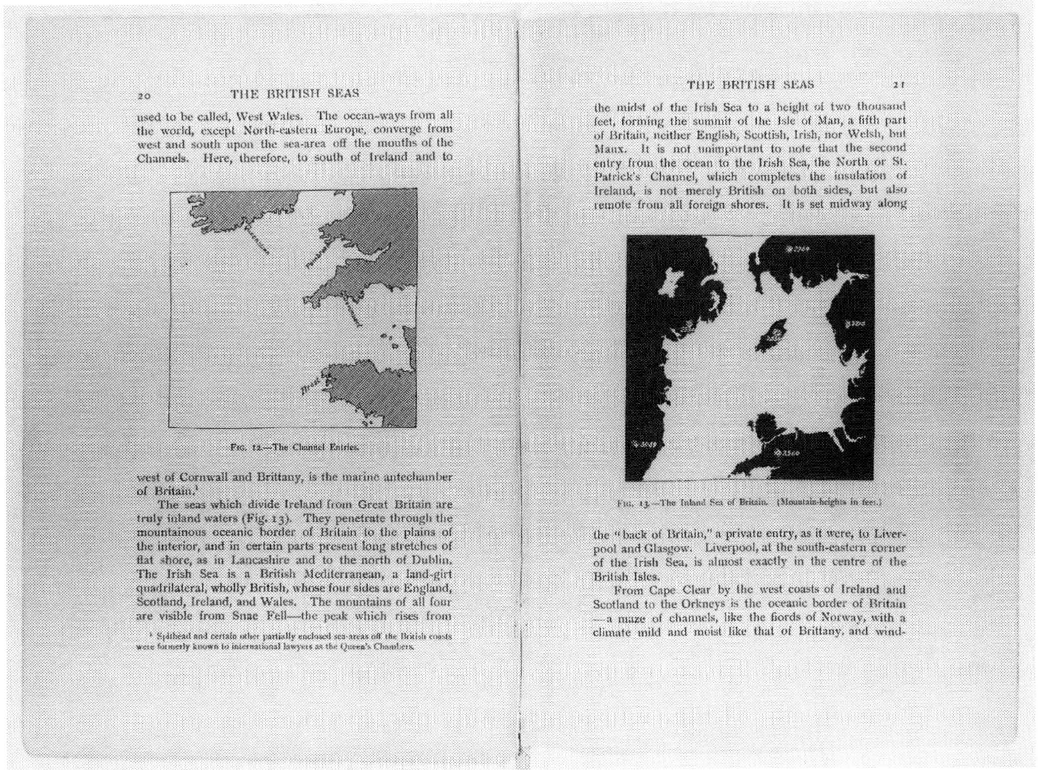

20 THE BRITISH SEAS

used to be called, West Wales. The ocean-ways from all the world, except North-eastern Europe, converge from west and south upon the sea-area off the mouths of the Channels. Here, therefore, to south of Ireland and to

FIG. 12.—The Channel Entries.

west of Cornwall and Brittany, is the marine antechamber of Britain.[1]

The seas which divide Ireland from Great Britain are truly inland waters (Fig. 13). They penetrate through the mountainous oceanic border of Britain to the plains of the interior, and in certain parts present long stretches of flat shore, as in Lancashire and to the north of Dublin. The Irish Sea is a British Mediterranean, a land-girt quadrilateral, wholly British, whose four sides are England, Scotland, Ireland, and Wales. The mountains of all four are visible from Snae Fell—the peak which rises from

[1] Spithead and certain other partially enclosed sea-areas off the British coasts were formerly known to international lawyers as the Queen's Chambers.

THE BRITISH SEAS 21

the midst of the Irish Sea to a height of two thousand feet, forming the summit of the Isle of Man, a fifth part of Britain, neither English, Scottish, Irish, nor Welsh, but Manx. It is not unimportant to note that the second entry from the ocean to the Irish Sea, the North or St. Patrick's Channel, which completes the insulation of Ireland, is not merely British on both sides, but also remote from all foreign shores. It is set midway along

FIG. 13.—The Inland Sea of Britain. (Mountain-heights in feet.)

the "back of Britain," a private entry, as it were, to Liverpool and Glasgow. Liverpool, at the south-eastern corner of the Irish Sea, is almost exactly in the centre of the British Isles.

From Cape Clear by the west coasts of Ireland and Scotland to the Orkneys is the oceanic border of Britain—a maze of channels, like the fiords of Norway, with a climate mild and moist like that of Brittany, and wind-

Figure 12.1 Maps of Britain and Ireland from Halford Mackinder, *Britain and the British Seas* (1902)

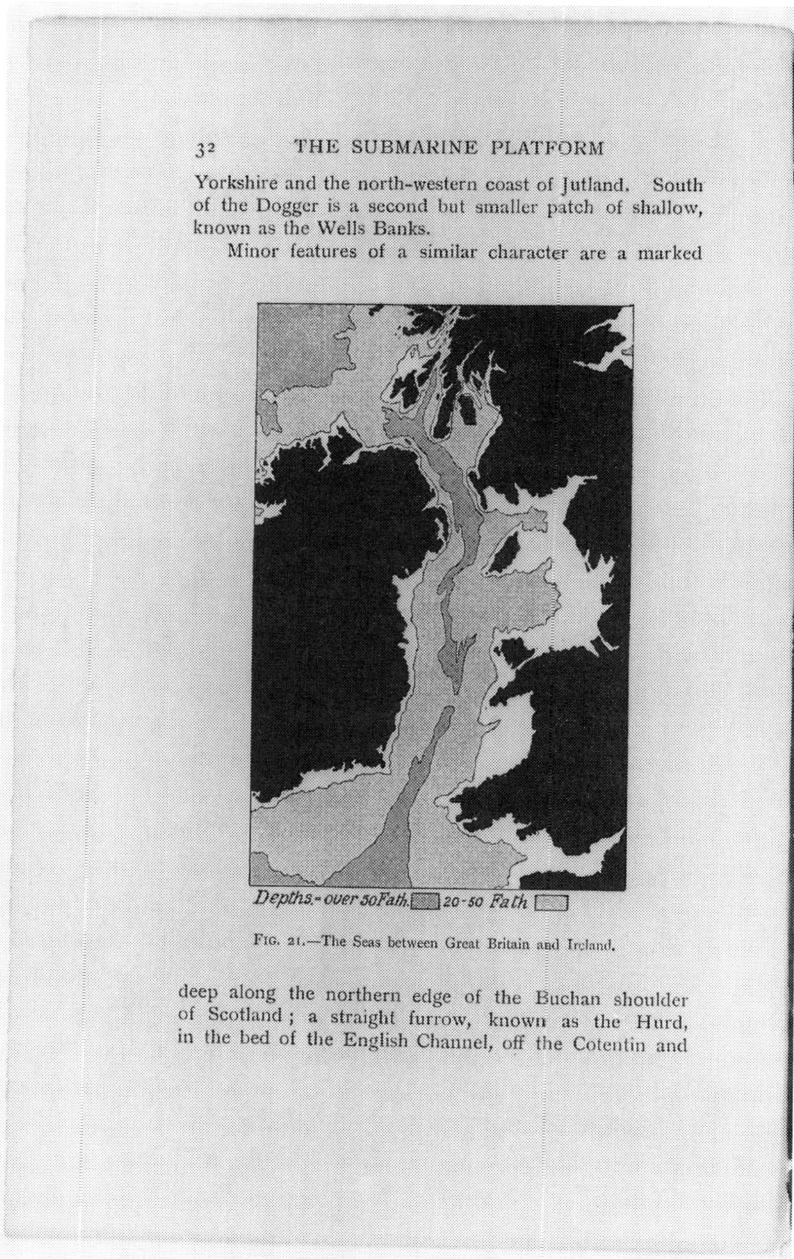

32 THE SUBMARINE PLATFORM

Yorkshire and the north-western coast of Jutland. South of the Dogger is a second but smaller patch of shallow, known as the Wells Banks.

Minor features of a similar character are a marked

FIG. 21.—The Seas between Great Britain and Ireland.

deep along the northern edge of the Buchan shoulder of Scotland; a straight furrow, known as the Hurd, in the bed of the English Channel, off the Cotentin and

Figure 12.2 Map of Britain and Ireland from Halford Mackinder, *Britain and the British Seas* (1902)

despite his shift in political allegiances, Mackinder was a committed Unionist, and it is Unionism which the analysed discussion in *Britain and the British Seas* supports. In other words, just as Mackinder's conception of the 'new' geography was forged in the crucible of his imperial and nationalist politics, so was his practice of political geography. The extent to which Mackinder's political geographic work supported his political project will be detailed below. Following on from this, the second and comparative point is that the congruity between Mackinder's political beliefs and his geographical argumentation makes his 'new' geography a different and far less radical break with the humanist politicisation of geography than his rhetoric and programmatic statements claimed. Akin to the predecessors he derided, Mackinder relied on careful textual construction to support a pre-established political position through an ostensibly geographical text. (Indeed, Mackinder was criticised for his deductive method, selecting facts in accordance with a pre-determined argument, in H.R. Mill's contemporary review of *Britain and the British Seas*.[45]) It was only the format of Mackinder's geographical works which interrupted this fundamental continuity in the lines of force between geography and politics. Despite Mackinder's determinist/possibilist rhetoric about geography driving political actions, his geographical argument is still subordinate and subsequent to a political stance. The shift which Mackinder does effect is more modest: he uses *geographical facts* about depths, distances, geology, topography and the like to support his political position, in contrast to the humanist politicisation of geography, which used the *generic conventions of geography books* containing paragraphs concerning the constitution, history and church affairs to embed a political position in a geography book. Humanist geography and Mackinder's 'new' geography, then, have the same lines of force, with politics driving geography, but different connections, Mackinder shifting from the generic level to the level of discrete facts, thereby giving the appearance of a far greater role for geography in political analysis. Herein lies the distinction between the humanist politicisation of geography books with which this study has been concerned, and the twentieth-century discourse of political geography.

Mackinder's political–geographic argument

We have already seen how Mackinder's political–geographic analysis of the Irish Sea was accordant with his Unionist politics. I now wish to extend the contextual analysis, to show in greater detail that Mackinder's political geography argued that geography, considered strategically, supported the imperialist and nationalist sentiments which united Liberal Imperialists and Conservatives.

For Mackinder, it was the 'Columbian epoch' following the discovery of the Americas which had made Europe geographically central to world trade and relations, by giving sea-power an advantage over land-power in a geopolitical trading system centred on the North Atlantic basin.[46] Within Europe, it was Britain which had made herself pivotal: 'The existence of the world centre is, of course, a corollary of that oneness of the [Atlantic] ocean which is the basis of British power, whether that power be naval or commercial.'[47] Yet Mackinder thought that he was writing as the Columbian era was coming to a close: the age of discovery was over, and the advent of the railway meant that the balance of power would swing back towards land powers, making the large nation states like Germany, Russia and America prime candidates to become the new central powers.[48] Mackinder's problematic for a geopolitical analysis, then, was precisely the crisis of confidence over Britain's ability to compete with larger states which dominated late-Victorian and Edwardian political discussion.

The 'geographical logic' which Mackinder discerned for Britain in this geopolitical situation is politically locatable as precisely that of a Liberal Imperialist and/or tariff reforming Conservative. In other words, for Mackinder the facts of geography supported his own political position and undermined that of free-market Liberals and Conservatives as well as the class politics of socialism and the emergent Labour party. For Mackinder free market economics, 'the dream of the sixties of the last century,' had been appropriate to the expanding world of the Columbian era, 'but when competing countries seek to monopolise markets by means of customs tariffs, even democracies are compelled to annex empires'.[49] Britain was losing the competitive race with Germany and the United States, especially with the former, which 'occupies a central position on the Continent of Europe'.[50] Britain was also threatened by the continental power of Russia in the era of the railways, that nation developing a protectionist national economics of the sort Britain was exhorted to adopt by Liberal Imperialists and tariff-reformers: 'Russia, by her tariff system and in other ways, is steadily hastening the accomplishment of what I may call the non-oceanic economic system.'[51]

Mackinder's political-geographical writings made it clear that in this era of competition between national blocs, Britain was in a struggle for her very survival, having recourse to Darwinian imagery:

> the most important facts of contemporary political geography are the extent of the red patches of British dominion upon the map of

> the world, and the position of hostile customs frontiers. They are the cartographical expression of the eternal struggle for existence as it stands at the opening of the twentieth century.[52]

In this situation, the British had to update their conception of the empire and then protect it. This was most apparent at the close of Mackinder's school text, *The Nations of the Modern World*, which supported the tariff-reformers' conception of an imperial federation of self-governing dominions: 'we must grow used to the thought that the Empire is no longer based on a Mother Country and her Colonies. ... The new view, acceptable to all the Britains, is that the Empire rests on a group of allied nations'.[53] These nations should form a bloc which would trade within itself on the principles of national economics rather than the bankrupt ideals of free-market economics, the latter destroying communities and creating a divisive politics of class rather than a unifying one of nation.[54] To protect this imperial federation required that Britain retain her naval supremacy, a point on which tariff reformers and Liberal Imperialists agreed, and which Mackinder ably encapsulated in his flagrantly Liberal-Imperialist conclusion to *Britain and the British Seas*:

> The whole course of future history depends on whether the Old Britain besides the Narrow Seas have enough of virility and imagination to withstand the challenge of her naval supremacy, until such time as the daughter nations shall have grown to maturity, and the British Navy shall have expanded into the Navy of the Britains.[55]

To summarise, Mackinder claimed that the facts of geography supported a national economics built around the trading bloc of a reconfigured British empire. Supported by a strong navy and the selective use of tariffs, such a bloc could compete with the emergent great land powers which had been facilitated by the development of the railways. It is all too apparent that Mackinder's geopolitical analysis supported the types of position which Liberal Imperialists and tariff-reforming Conservatives held in common. In other words, Mackinder's visual facts of geography supported the political position he himself held concerning the ways out of the perceived crisis of Edwardian Britain. Just as surely as for the old humanist geographers, but at the level of facts not generic conventions, Mackinder allowed his geographical writings to act as a container for a pre-established political position.

Conclusion: how new was the 'new' political geography?

Halford Mackinder advertised his work as a 'new' geography[56] which included a new approach to the interconnection of geography and politics. By and large, he has been taken at his own word: Mackinder's biographer thinks his work is 'justifiably entitled "The New Geography"';[57] conventional histories of geography have repeated that Mackinder was the founder of modern British geography;[58] and his place is similar in Livingstone's more sophisticated Darwinian reading in *The Geographical Tradition*, Mackinder at least introducing an 'alternative discourse'.[59] Similarly, Mackinder's work linking politics and geography has been taken as the originating moment of political geography and geopolitics by conventional summaries of political geography and by more radical approaches.[60] Even Ó Tuathail, from a perspective influenced by a Nietzschean/Foucauldian genealogical scepticism about the analysis of historical origins, sees Mackinder as one of a group retrospectively constructed as the founders of a geopolitical tradition 'for certain ostensibly good reasons'.[61]

From a contextual-historical perspective informed by an analysis of the previous centuries of engagement between geography and politics, this assessment of Mackinder's geography in general and his political geography in particular as 'new' must be qualified. In one sense, Mackinder certainly does mark a final break with the early-modern humanist geographical tradition which this book has investigated. Mackinder's repositioning of geography in the division of knowledge in order to 'humanise' it amounted to a jettisoning of the humanist framework which had been transmitted from the Renaissance onwards (albeit in an extremely attenuated form by the time it had been reinterpreted in line with political-economic dictates by McCulloch). The common thread in the humanist educative structure was that geography was a propaedeutic enquiry whose knowledge could be conveyed in single-volume structured summaries of information about the globe, arranged nationally. Mackinder sought to reposition geography as the key subject in a new 'imperial humanism' which would train British citizens to be patriotic by virtue of thinking imperially. In this connection, he dismissed old geography books as 'arid', a label with which they have been stigmatised by histories of geography ever since, preferring instead to construct illustrated narratives.

Yet in another sense, an historical analysis must see Mackinder's language of the 'new' as a rhetorical ploy. For just as in the humanist geographical tradition, so in Mackinder, geography sanctions a political position which has been determined before recourse has been made to any geographical enquiry. In this specific case, Mackinder's belief in a strong imperial solution to the late-Victorian and Edwardian crisis of political confidence drove the geographical facts which he chose to deploy. Mackinder's rhetoric of the 'new' suggested that geographical facts would now drive political analysis, in implicit contrast to the humanistic tradition where geography books were simply sites in which political positions were articulated. In fact, Mackinder's political geography effected no such reversal in the lines of force between politics and geography. The shift Mackinder did enact was, rather, that geographical facts were organised according to a political argument, whereas previously the organisation of geography books had facilitated the wholesale emplacement of political arguments little inflected by geography. In this shift, Mackinder set the tone for twentieth century political–geographic and geopolitical analyses, and further he enacted the final effacement of the already-emasculated humanist modes of politicising geography, but he in no way established a relation of equals between politics and geography.

13
Enlightenments and Geography: Continuity and Change in the Politics of Early-Modern British Geography, c.1550–c.1850

Attempting to provide an overview of the 'historical event' which linked political languages and geographical texts between 1550 and 1850 requires a less minute focus than has been deployed in the preceding biographical analyses. At this more general level, a fairly consistent characterisation of the political positioning of early-modern British geography between William Camden and J.R. McCulloch can be developed.

Geography's role was the textual collation of information provided by travellers and scholars, normally at a world scale, subdivided by nation. Given this textual conception of the subject, politicisation was achieved by the inflections geographical compilers gave to their material during the process of abridgement. By the presentation of the constitution, the ecclesiastical polity, and the religious and political history of England and Britain, specific visions of the church–state relationship could be promulgated. Most of the contentious issues revolved around the dynastic legitimacy of monarchs, the role of the king-in-parliament, the nature of the Revolution Settlement of 1688/89, and the true origins, ecclesiology and doctrines of the established church. Issues of political representation, mass political participation and of extending religious toleration did not surface frequently in the geographical tradition studied here until its later years.

Given this general manner of introducing politics into geography books, what varieties of political positions did the books espouse? The politics of most geography books veered towards orthodoxy, supporting a strong interpretation of the monarch's role and a High Church interpretation of Anglicanism. The nature of the orthodox position changed over time, of course, from the Laudian rigour of Heylyn, through the orthodox or clerical Whig version of 1689 in Echard to the social Whiggery of Guthrie. Furthermore, what at any one time counted as adherence to an orthodox political position could be debated, as we saw in the case of Tory Anglican travel writers. For all these shifts, however, the fear of faction and the spectre of the Civil War remained the most active historical exemplar, leading all these varieties of orthodoxy to an aversion to any radical breaks in the British governmental structure. The term 'orthodox' is used advisedly, because the politics of British geography books was a theological politics. The nature of the church settlement advocated differed between cases, but historical arguments about Joseph of Arimathea, John Knox or William Laud were as important as those about the monarchs. The vision of the ideal church structure – promoted via discussions of such individuals – was coupled with arguments about political structure, orthodox churchmanship combining with the defence of monarchical prerogative in most cases.

Dissonant voices were few in geography, but where such a possibility was perceived or actualised, heterodoxy worked along the axes of dynastic politics and dissenting churchmanship, rather than manifesting itself in appeals for democracy – that is, until the early nineteenth century. Bohun, in the second edition of his *Geographical Dictionary* was accused of being subversive for supporting the exiled House of Stuart. Whether the allegation was true or false, it is noteworthy that the sensitive issue definitive of radicalism was dynastic not democratic. Similarly, John Walker's *Elements of Geography*, while not alone, was at the extreme end of the political spectrum established in geography books because of its questioning of the reality of English liberty. Once again, however, the radicalism is not of a proto-democratic variety, but is in this case a matter of denominational affiliation. This fits well with the suggestion that 'the roots of reform lay in theology', and that this remained true until the end of the period discussed:

> What was evident in England in the half century after the American Revolution was not any massive conversion of the population to toleration (Repeal, Emancipation) or democracy (Reform), but the largely negative phenomenon of the stready erosion of Anglicanism, and the related growth of Protestant Dissent and unbelief.[1]

Geography books espoused theological politics, from Heylyn's Laudian version, through Wells's High Churchmanship to Guthrie's and Bell's questioning of the Church of England from a Presbyterian perspective. Certain political structures were unavoidably preferred given the theological position from which the author wrote.[2] It is only as the tradition of early-modern geography comes to a close with J.R. McCulloch in the mid-nineteenth century that we start to see a geography book uninterested in theological and ecclesiological affairs in its politicisation.

Geography books took up very specific political languages, normally relating to the immediate actualities of practical politics. Salmon's *History of Queen Anne*, for example, developed a Tory narrative of that reign, designed to encourage criticism of the actions of Robert Walpole as prime minister under George I and II by a twofold process of contrast with Anne's Tory ministries and conflation with her Whig ministries. This in turn was transferred to the historical section of his *Geographical and Historical Grammar*, the attack by this time being extended from Walpole to all Whigs in the light of the record of Walpole's successors. Salmon, then, was adopting a specific 'Patriot' version of the language of virtue and corruption known as Civic Humanism and conveying this message through a geography book. All geography books until the later eighteenth century show such deployments of historically precise political languages. As such, generalised characterisations of geography as 'bourgeois', 'liberal' and the like are not of great use, as they assume transhistorical or at least macrohistorical generalisations about the nature of politicisation in the early-modern period which are unrealistic, fixing as they do subsequent political meanings on a previous era in the passage of differences which constitutes the language of politics. To be a Tory geographer in the reign of Queen Anne, as Edward Wells was, required a very different set of textual moves from being a Tory geographer under the early-Hanoverian Whig hegemony writing about Queen Anne, as Thomas Salmon was.

If we do wish to provide a larger characterisation of the politics of British geography in the period, the tentative status of which must be clearly understood before it is attempted, the best starting point is the disjunctions between the English, Scottish and French Enlightenments. The English Enlightenment was built around the defence of church and state, England being seen as the home of the politeness codified by Addison's *Spectator*. Given this, the English Enlightenment was:

> conservative and in several ways clerical, a response and indeed a reaction to the traumatic experiences of the seventeenth century, part of the establishment of that post-Puritan ruling order which made England at once the most modern and the most counter-revolutionary of European societies. ... [It] was the instrument of ruling groups, aristocratic and clerical.

Pocock argues that this Enlightenment set itself against three strands of thought: 'the republican patriot joined the sectarian enthusiast and the Oxford High-Churchman as the three opponents of the Whig regime whom the ideology of politeness sought to exorcise.'[3] English geography books certainly avoided the charge of Republicanism by virtue of their monarchism, and their Anglican bias tended to leave them free from any charge of enthusiasm. The third excess, Oxford High-Churchmanship, was far more common in geography books, Camden, Heylyn and Wells all coming from just such a background, which also manifested itself in Johnson's travel writing. It is clear, then, that even in the intellectual context established by 'the most counter-revolutionary of European societies', geography remained a conservative discourse, tending to defend the powers of the monarch and the Catholicity of the Church of England with a degree of warmth beyond that recommended by the 'politeness' forged by Addison and Shaftesbury.[4]

Several reasons for the positioning of geography at the conservative end of the spectrum of a politically conservative English Enlightenment can be adduced from the genre of the geographical compendium. First, most geography books in the period were fairly miscellaneous compilations of useful knowledge, aimed at a wide audience. As a consequence, they offered a bowdlerised version of 'politeness', coupled with a traditionalist politics, being aware that the more refined versions of Civic Humanism and politeness had a limited elite constituency.[5] Secondly, most geography books which had a more scholarly market in mind were aimed at a milieu in which Renaissance humanism continued to guide the curriculum. The stable role of the geography book, as illustrative of historical texts, perhaps encouraged stasis in its form and politics. As the conservative elements of English intellectual life were frequently more concerned for precedent than radicals who invoked the law of nature, it is possible that the twinning of geography and history which led many historians to write geography books, also led those books to trend towards the conservative end of the political spectrum.[6] Finally, it must be remembered that most geography books were produced by direct plagiarism from past books, such that many of the ideological elements from previous political situations continued down the decades. Given, for example, the continued reliance on Camden and Heylyn (either directly or through further intermediary appropriations) in the construction of sections on Britain in later geography books, elements of their Elizabethan and Laudian politics could not but remain visible long into the eighteenth century, although alert editing could alter their political resonance in accord with changed situations, as we saw in the case of Gibson's editions of Camden's *Britannia*. The result was geography texts of a fragmentary and conservative politicisation, in which the transmission and appropriation of political ideas was far more important than innovation and criticism. This marks the politicisation of geography books off from the political tradition *per se*, as studied by Pocock and others, wherein 'counter-moves' and critical appropriations bulk far larger than in the early-modern geographical tradition.

The elements of continuity, however, can be overplayed. There were certain trends in the politicisation of geography over the period as a whole leading to the breakdown of the 'historical event' of early-modern geography and its politicisation and necessitating a recasting of the connections between the two discursive realms. First, there was the gradual and partial decoupling of theological from political argument. To start with Heylyn is admittedly to start with an extreme in the theological understanding of politics, but it is noticeable that geographers from Salmon onwards seem less directly concerned with matters of church government. Salmon defends episcopacy just as Guthrie takes a Presbyterian angle, but the interaction between church and state is less thoroughgoing: Salmon sounds like a pale imitation of Edward Wells, and Guthrie's moderate Presbyterian approach leads to very respectful criticism of the Church of England. If one of the driving forces in the English Enlightenment was, as Pocock has argued, the Latitudinarian Anglican tolerance developed in the later seventeenth century, its impact in geography books was delayed until the mid-eighteenth century. At this point, it tended to be introduced into British geography by figures connected with the Scottish Enlightenment, as they themselves forged a more Latitudinarian or 'moderate' form of Presbyterianism.

Secondly, there was some weakening of the preoccupation with dynastic issues over time. Issues of legitimacy were vital to Heylyn, Ogilby and Bohun, each envisioning a balanced system with the monarch as an absolute head. They conceive of liberties as protected not jeopardised by such a constitutional structure. By the later eighteenth century, the language of balanced power was being used in a sense with which we are more familiar today. The balance, however, was not seen as in need of any change, and the representation of the people was rarely discussed. If there was widespread discontent leading up to the Reform Act of 1832, it was not expressed in geography books until the very time of the Act (in the case of Bell's *System*) and franchise reform only became important in retrospective discussion of its merits and flaws (in McCulloch's work). In this respect, the politics of geography books is at least accordant with Clark's interpretation of the reform movement as a late stage in the secularisation of radicalism. Demands for increased religious toleration were given a voice in geography books from the mid-eighteenth century. By the time Reform became an important issue in geography books, there had been a notable diminution in the intensity with which religious issues were discussed, as exemplified in McCulloch, although the argument should not be pressed too far, given the discernibly Presbyterian approach which James Bell could still adopt.

Finally, there was a major transformation in the conception of empire in geography books, mirroring that in political argument as a whole. For Camden and his post-Reformation contemporaries,

> this Kingdom was an 'empire' in the sense that it contained within itself absolute sovereignty over itself, so that its king (in parliament or out of it) was absolute and sovereign head of all jurisdiction both spiritual and temporal, and therefore head of the English Church as inseparable from the English body politic.[7]

The generations of geographers we have seen defending the church and state remained within the same conceptualisation of empire, wherein the absolute and indivisible nature of power had to be demonstrated. The fears expressed over the treatment of the American colonies and their subsequent secession wrought fundamental changes in British theory about empire. Guthrie's *Geographical Grammar* had expressed doubts about the wisdom of British policy in America, thus countenancing the separate political status of the colonies rather than assuming they formed part of the absolute empire of Britain. After American secession, 'new imperial discourses' had to be developed as the 'first', North American, empire was replaced by a predominantly Asian 'second' empire. In these discourses, colonies were considered as distinct from Britain and therefore needing different forms of government. Multiple constitutional systems in local contexts had to be considered. One of these was 'the Anglo-Indian, which debated whether empire in South Asia was necessarily despotic because exercised over subjects naturally or historically incapable of liberty'.[8] Guthrie reflected this belief in his deployment of stadial theory and James Rennell was clearly implicated in the emergence of these concepts in his Asian geographies. New conceptions of empire would continue to tax geographers throughout the nineteenth century and give them a new political role as specialists in analysing the appropriate form of government for each colony, their accreditation coming from their personal acquaintance with the colony in question. This emergent conception of the geographer as an imperial field-worker also affected the nature of geography books: we see a proliferation of statistics and information about imperial lands in the early nineteenth century, which altered the nature of the evidence presented in geography books and made the humanist ideal of a single work conveying all available information about the globe increasingly impractical.

As we characterised the politicisation of early-modern British geography in terms of the English Enlightenment, so we can characterise the aforementioned factors which led to its breakdown in terms of the Scottish Enlightenment's effect upon British intellectual life. There can be no doubt that those who forged the Scottish Enlightenment were loyal to the British monarchy and parliament, nor that they mainly adhered to the established church of Scotland. Yet at a discursive level their impact was more subversive of the pieties of English political discourse. First, the Scottish Enlightenment was not, any more than the English, a secular phenomenon (and in this it diverged from the French Enlightenment, with which it has all too often been conflated): 'their endeavours ... were aimed at the effective reconciliation of a presbyterian morality with fashionable enlightened values'.[9]

Yet the church which the moderate literati of the Scottish Enlightenment supported was the Presbyterian Church of Scotland. This loyalty could, by its very existence, act as a solvent to English church–state discourse, questioning it from an external perspective:

> In so far as religion was concerned, North Britons remained vigorously Scottish, committed ... to their established traditions of kirkmanship. A rigid ecclesiology of distinct temporal and spiritual spheres, fundamentally different from the Erastian basis of the Church of England, maintained in Scotland a very strong sense of ecclesiastical distinctiveness.[10]

We have seen in the writings of Guthrie, Pinkerton, Bell and McCulloch that a Scottish Enlightened perspective began to question the Anglican discourse within which geographical books had functioned until the mid-eighteenth century. With the exception of McCulloch, all these authors engaged in such a questioning from a position more sympathetic to Presbyterianism. As is clear from the case of Thomas Pennant's travel writings, an Englishman who agreed with the Scottish evaluation of Presbyterianism could also act to undermine the Anglocentric religious values structuring geographical knowledge. Further, John Walker's attack from an Irish Quaker perspective shows that other components of the British archipelago and other denominational perspectives, each in conflict with the values of the English Enlightenment, could significantly undermine the basic parameters within which the politicisation of early-modern English geography had operated.[11]

Secondly, while the Scottish Enlightenment championed the virtues of English government, it also undermined Anglocentric narratives about those virtues. Thus, while Scottish political discourse from the 1730s acted 'to orientate Scots through a powerful anti-feudalist critique of Scottish society towards a political and institutional identification with England' it was also apparent that 'the desire for anglicization did not stem from anglocentricity'.[12] On the contrary, Scottish Enlightenment history questioned English patriotic accounts of the rise of its political system, seeing it within a more general model of the stages through which all societies developed:

> The Scottish Enlightenment presented English Whiggism with an unusual challenge. From a standpoint at once within and outwith the English political world, the Scots set themselves to rid English Whig politics and history of its parochial prejudices and ancient constitutional myths, and to interpret them afresh in a properly cosmopolitan perspective.[13]

Such a perspective, of course, cut against any interpretation of English history as unique, thereby draining dynastic and constitutional debates of much of the vigour they had maintained from an Anglocentric perspective. It is just this Scottish Enlightenment positioning, questioning from within and without, which Guthrie pioneered in the geographical tradition and which successive Scottish geographers maintained, thereby relativising what had been taken as political absolutes in the English early-modern geographical tradition.

Thirdly, shifting conceptions of empire were related to a Scottish political–economic critique of the 'first' British empire, based as it was upon mercantilist and protectionist principles. For Hume:

> Whig doctrine was used as a cover for the promotion of a dangerously misconceived idea of commercial empire. Not only did the British exaggerate the threat of France to the liberties of Europe; British commercial policy represented a threat still greater, and more immediate.[14]

The 'second' British empire which emerged after the loss of the American colonies was one wherein racism and the need to propagate civility became central justifications for British rule. Just as the critique of the first empire was articulated by Scottish Enlightenment literati, so the justification of the second empire drew upon their theorisations, albeit in ways which Hume and Smith would not have accepted: 'The theories of "moral sentiments" and human development through stages were ... easily uncoupled from their individualist and historical underpinnings, to be taken up and employed as the dominant public discourse of a new Christian imperialism.'[15]

Furthermore, the 'second' empire saw a drive towards the collection of information and its presentation in statistical forms, to improve the efficiency of imperial government. This again related to the discursive frameworks of the Scottish Enlightenment, with its advocacy of an empirical political economy as, in Adam Smith's words, 'the science of the legislator'.[16] In other words, if the altered geography and governance of the British empire after the loss of the North American colonies played a major role in changing the political discourse of geography, then both the analysis of that loss and the new forms of imperial governance developed in response to it were tied up with the Scottish Enlightenment and with the subsequent popularisation of its ideas. We can see in the career of Rennell and in the texts of Bell and McCulloch the outcomes for geography of the empirical turn in political discourse: the geographer is driven towards a fieldwork which produces results of use to the legislator; and geography books are pushed towards the display in tabulated form of the outcomes of that fieldwork.

It is worth noting, then, that the collapse of the early-modern geographical tradition and of its politicisation was itself a geographically-driven process in two senses. First, in the transformation of the English Enlightenment's politicisation of geography which the Scottish

Enlightenment effected, we see one space of knowledge creation altering a discourse which, while based on a different space of knowledge creation, namely the English, assumed itself to be a universal 'view from nowhere'. The geography of political discourse, then, altered the politics of geography. Secondly, early-modern English geography and its politics emerged as the 'first' English empire started to develop in the mid-sixteenth century.[17] It, then, was predicated on a certain spatial arrangement of the extended British archipelago centred on the North Atlantic basin. It is as this 'first' empire breaks down to be replaced by a new one that the politics of English early-modern geography itself breaks down under the weight of information about the new, 'second', empire. The changing political geography of imperial Britain, then, helped to alter the discursive construction of geography's politics.

Taking these trends together, there are good reasons to argue the early nineteenth century witnessed a significant break in the interaction of geographical and political discourse. First, in political argumentation, the detachment of theological from political language has been seen as ending the imperial conception of the English church and state established at the Reformation and reformulated after 1688.[18] This rupture was codified by parliamentary reform, Catholic emancipation and the repeal of the Test and Corporation act: 'the eighteenth-century historian ... sees in the events of 1828–32 a dissolution of the social order with which he is familiar'.[19] These events altered the grounds for political debate, and the presuppositions about political inclusion and exclusion under which the geographers discussed here operated were rendered obsolete.

Secondly, in the intellectual realm this period saw increasing specialisation in the sciences, and the increased inability for arts and sciences to speak at more than a generalised level.[20] Science as its own community of discourse demanded new self-images, and in geography this process of reconfiguration revolved around the transition from a textual to a practical self-image which we have seen being retrospectively fixed on Rennell. A decreasing percentage of geographers had a university education,[21] and the best training for a geographer came to be seen as fieldwork. The result was a proliferation of geographical data, unmanageable within the old textual conventions of early-modern geography. This linked geography, governance and the armed forces more closely than in the early-modern period, while decoupling geography from humanistic enquiry, classical, historical and literary.

Taken together, these changes in the structure of political discourse and the political system which it discussed, in the nature of the intellectual community and in the geographer's education and function within both that community and the broader public realm, bring the engagement between geography and politics which I have discussed to a close. To generalise, we can perhaps see the decline of the early-modern conception of geography (and its link with politics) and the subsequent recasting of geography as a two-part process spanning the nineteenth century. In the first phase, c.1800–70, which this book has covered in detail, the generic conventions of early-modern geography remained in place, as did a textual form of politicisation. Yet these continuities were progressively weakened as geography was allied to statistics and then to political economy. Quantitative material began to swamp and replace prose argument as the mode of geographical expression. It is this which can be witnessed in the work of Pinkerton, Bell and McCulloch, which simultaneously relied on and undermined the textual tradition of early-modern geography. More generally put, the first half of the nineteenth century saw an uneasy compromise between the older textual modes of the English Enlightenment and the criticisms and inflections of those modes wrought by the intellectual and political currents of the Scottish Enlightenment. In the second phase, the reformulation of geography and its politics occurs in the late nineteenth century as an institutional space is carved for the subject in a modern, post-humanist curriculum. At this point the founders of the 'new' geography, led by Mackinder in Britain, attacked as arid and dull the early nineteenth century statistical reworking of early-modern geography which had accommodated the English and Scottish Enlightenment traditions. Mackinder sought to forge a new form of geography and a new link between geography and politics. As we have seen, much of this was a rhetoric of novelty rather than a reality, but it was backed by an evangelising historiography which argued that only with an institutionalised form of geography did geography reach maturity, and that all previous engagements called geography had been 'precritical', worthy only as anticipations of what the late nineteenth century achieved. The history of geography has remained within the conceptual space established by this polemic until the present day. As will be apparent, the present work agrees with the chronological framework which Mackinder formulated, seeing his era as the key one in the formulation of geography as it is practised today. It disagrees, however, with Mackinder's effacement of the coherence of the early-modern geographical tradition, postulating instead that geography's identity is a passage of differences, in which the historian will be as interested in the early modern as in the modern tradition, since not the least part of the intellectual politics of modern, fieldwork geography has been the effacement of its early modern textual predecessor by historiographical sleight of hand.

Notes

1 Effacing and Recovering the History of Geography

1. Michael Oakeshott, *Experience and its Modes* (Cambridge: Cambridge University Press, 1933), p. 34.
2. Ibid., p. 76.
3. Michael Oakeshott, 'The Activity of Being an Historian', in his *Rationalism in Politics and Other Essays*, Timothy Fuller, new edn. (Indianapolis: Liberty Press, 1991), pp. 151–83, at pp. 170–1.
4. Oakeshott, *Experience*, p. 106.
5. Oakeshott, 'Activity', p. 161.
6. Oakeshott, *Experience*, p. 107.
7. Ibid., p. 108.
8. Michael Oakeshott, *On History and Other Essays* (Oxford: Blackwells, 1983), p. 53.
9. Ibid., p. 60.
10. Ibid., p. 62.
11. Ibid., p. 112; see also Oakeshott, *Experience*, pp. 120–1.
12. See also the reflections on historical 'traditions' in Mark Bevir, *The Logic of the History of Ideas* (Cambridge: Cambridge University Press, 1999), pp. 200–13.
13. Oakeshott, *On History*, pp. 54, 65 and 93.
14. Oakeshott, 'Activity', p. 180.
15. Oakeshott, *On History*, p. 117.
16. Ibid., p. 28.
17. Oakeshott, *Experience*, p. 102.
18. Ibid., p. 103.
19. Oakeshott, 'Activity', p. 181.
20. Oakeshott, *Experience*, pp. 157–8.
21. Oakeshott, 'Activity', p. 159.
22. Ibid., p. 169; see also Oakeshott, *Experience*, p. 142.
23. Richard Hartshorne, *The Nature of Geography: a Critical Survey of Current Geographical Thought in the Light of the Past* (Lancaster, Pa: AAG, 1939), p. 35.
24. Carl Sauer, 'The Morphology of Landscape' in John Leighty, ed., *Land and Life: a Selection from the Writings of Carl Ortwin Sauer* (Berkeley: University of California Press, 1963), pp. 315–50 at p. 319.
25. Ibid., pp. 320 and 321.
26. David Stoddart, *On Geography and its History* (Oxford: Blackwells, 1986), pp. 30 and 28.
27. Ibid., pp. 33 and 37–8.
28. Margarita Bowen, *Empiricism and Geographical Thought: From Francis Bacon to Alexander von Humboldt* (Cambridge: Cambridge University Press, 1981), p. 10.

29. Ibid., p. 11.
30. Ibid., pp. 11 and 274.
31. Ibid., p. 90.
32. Ibid., pp. 72 and 75.
33. Q. Skinner, 'Meaning and Understanding in the History of Ideas', in James Tully, ed., *Meaning and Context: Quentin Skinner and his Critics* (Cambridge: Polity, 1988), p. 294, n. 41; and Bowen, *Empiricism*, p. 146.
34. Bowen, *Empiricism*, pp. 98 and 161.
35. Geoffrey Martin and Preston James, *All Possible Worlds: a History of Geographical Ideas*, 3rd edn. (New York: John Wiley, 1993), p. 8.
36. Ibid., p. 2.
37. Ibid., pp. 96–7.
38. Ibid., p. 109.
39. For more detail on Mackinder's position and its subsequent history, see below, Chapter 12.
40. Martin and James, *All Possible Worlds*, p. 95, emphasis added.
41. See for example David Harvey, 'On the History and Present Condition of Geography; an Historical Materialist Manifesto', *Professional Geographer*, 3 (1984), pp. 1–11; and Michael Eliot Hurst, 'Geography has Neither Existence nor Future', in R.J. Johnston, ed., *The Future of Geography* (London: Methuen, 1985), pp. 59–91.
42. See for example Neil Smith and Anne Godlewska, eds, *Geography and Empire* (Oxford: Blackwells, 1994) which has one chapter on Elizabethan England and no chapter on the eighteenth century.
43. See for example David Miller and Peter Reill, eds, *Visions of Empire: Voyages, Botany, and Representations of Nature* (Cambridge: Cambridge University Press, 1996).
44. David Harvey, *The Condition of Postmodernity: an Enquiry into the Origins of Cultural Change* (Oxford: Blackwells, 1989), p. 249.
45. Henri Lefebvre, *The Production of Space* (Oxford: Blackwells, 1991), esp. p. 48 and p. 53; see also Denis Cosgrove, *Social Formation and Symbolic Landscape* (Beckenham: Croom Helm, 1984).
46. Derek Gregory, *Geographical Imaginations* (Oxford: Blackwells, 1994), pp. 16–33.
47. Charles Withers, 'Encyclopaedism, Modernism and the Classification of Geographical Knowledge', *Transactions of the Institute of British Geographers*, N.S. 21 (1996), pp. 275–98 at p. 275.
48. See Charles Withers, 'Geography, Science and National Identity in Early-Modern Britain: the Case of Scotland and the Work of Sir Robert Sibbald', *Annals of Science*, 53 (1996), pp. 29–73; and *idem.*, 'How Scotland Came to Know Itself: Geography, National Identity and the Making of a Nation, 1680–1790', *Journal of Historical Geography*, 21 (1995), pp. 371–97.
49. Withers, 'Encyclopaedism', pp. 279–80.
50. Ibid., p. 289.
51. Miles Ogborn, *Spaces of Modernity: London's Geographies, 1680–1780* (New York: Guilford Press, 1998), p. 1.
52. Ibid., p. 28.
53. David Livingstone, *The Geographical Tradition: Episodes in the History of a Contested Enterprise* (Oxford: Blackwells, 1992), p. 102.

54. Ibid., p. 112.
55. Withers, 'Sibbald', p. 63.
56. See Yvonne Sherratt, *Adorno's Critique of Enlightenment: a Positive Interpretation* (Cambridge: PhD thesis, 1999), Introduction and Chapter 1.
57. Oakeshott, *On History*, p. 33; Oakeshott, 'Activity', p. 166.
58. Oakeshott, *On History*, p. 3.
59. Ibid., p. 31.
60. O.F.G. Sitwell, *Four Centuries of Special Geography: an Annotated Guide to Books that Purport to Describe all the Countries of the World Published in English before 1888, with a Critical Introduction* (Vancouver: University of British Columbia Press, 1993), Figures 1 and 2.
61. Conal Condren, *The Language of Politics in Seventeenth-Century England* (London: Macmillan, 1994), p. 3.
62. See Alan Downes, 'The Bibliographical Dinosaurs of Georgian Geography (1714–1830)', *Geographical Journal*, 137 (1971), pp. 379–87.
63. Roger Chartier, *The Order of Books: Readers, Authors and Libraries in Europe between the Fourteenth and Eighteenth Centuries* (Cambridge: Polity, 1994), pp. 2–3.
64. While I put the two together for my present purposes, I am aware that Skinner and Pocock are not theoretically identical in approach: see Bevir, *Logic*, pp. 32–52.
65. Skinner, in Tully, ed., *Meaning and Context*, pp. 66, 44 and 47.
66. Ibid., p. 43.
67. Ibid., pp. 63–4 and 77.
68. Robert Darnton, 'History of Reading' in Peter Burke, ed., *New Perspectives on Historical Writing* (Cambridge: Polity, 1991), pp. 140–67 at p. 161.
69. Donald McKenzie, *Bibliography and the Sociology of Texts* (Cambridge: Cambridge University Press reprint, 1999), p. 61.
70. Jerome McGann, *A Critique of Modern Textual Criticism* (Chicago: Chicago University Press, 1983); Michel Foucault, 'What is an Author?', in Paul Rabinow, ed., *The Foucault Reader* (Harmondsworth: Penguin, 1984), pp. 101–20; Elizabeth L. Eisenstein, *The Printing Press as an Agent of Change: Communications and Cultural Transformations in Early-Modern Europe* (Cambridge: Cambridge University Press, 1979); and Mark Rose, *Owners and Authors: the Invention of Copyright* (Cambridge, Mass.: Harvard University Press, 1993).
71. McGann, *Critique*, pp. 48 and 122.
72. Chartier, *Order of Books*, pp. viii–ix, summarizing McKenzie, *Bibliography*.

2 The Sphere of Geography and the Realm of Politics in Britain, c.1650–1850

1. Much of the material in this chapter appeared as 'The Character of English Geography, c.1660–1800: a Textual Approach', *Journal of Historical Geography*, 24 (1998), pp. 385–412.
2. For an exemplary study of literature which starts from contemporary understandings of politics rather than vague evocations of the 'politics of literature', see Howard Erskine-Hill, *Poetry and the Realm of Politics: Shakespeare to*

Dryden (Oxford: Oxford University Press, 1996) and *Poetry of Opposition and Revolution: Dryden to Wordsworth* (Oxford: Oxford University Press, 1996).

3. See J.G.A. Pocock, *The Ancient Constitution and the Feudal Law: a Study of English Historical Thought in the Seventeenth Century: a Reissue with a Retrospect* (Cambridge: Cambridge University Press, 1987); *idem*, *The Machiavellian Moment: Florentine Political Thought and the Atlantic Republican Tradition* (Princeton: Princeton University Press, 1975); and Quentin Skinner, *Liberty before Liberalism* (Cambridge: Cambridge University Press, 1998).
4. For the persistence of the historical approach to politics, see J.A.I. Champion, *The Pillars of Priestcraft Shaken: the Church of England and its Enemies 1660–1730* (Cambridge: Cambridge University Press, 1992); and R.J. Smith, *The Gothic Bequest: Medieval Institutions in British Thought, 1688–1863* (Cambridge: Cambridge University Press, 1987).
5. See Howard Erskine-Hill, *The Augustan Idea in English Literature* (London: E.J. Arnold, 1983); and Philip Ayres, *Classical Culture and the Idea of Rome in Eighteenth-Century England* (Cambridge: Cambridge University Press, 1997).
6. See J.C.D. Clark, *English Society, 1688–1832: Ideology, Social Structure and Political Practice during the Ancien Régime* (Cambridge: Cambridge University Press, 1985); *idem*, *The Language of Liberty 1660–1832: Political Discourse and Social Dynamics in the Anglo-American World* (Cambridge: Cambridge University Press, 1994); A.M.C. Waterman, 'The Nexus between Theology and Political Doctrine in Church and Dissent', in Knud Haakonssen, ed., *Enlightenment and Religion: Rational Dissent in Eighteenth-Century Britain* (Cambridge: Cambridge University Press, 1996), pp. 193–218; James E. Bradley, *Religion, Revolution and English Radicalism: Non-Conformity in Eighteenth-Century Politics and Society* (Cambridge: Cambridge University Press, 1990); Condren, *Language of Politics*, pp. 33–4; and Champion, *Pillars of Priestcraft Shaken*.
7. Skinner, in Tully, ed., *Meaning and Context*, p. 77. See also p. 106.
8. See Lesley Cormack, '"Good fences make good neighbours": Geography as Self-Definition in Early-Modern England', *Isis*, 82 (1991), pp. 639–61; Lesley Cormack, *Charting an Empire: Geography at the English Universities, 1580–1620* (Chicago: Chicago University Press, 1997), Chapters 3–5; Sitwell, *Four Centuries of Special Geography*, pp. 2–8; and David Woodward, *Maps as Prints in the Italian Renaissance: Makers, Distributors and Consumers* (London: British Library, 1996), pp. 7–20.
9. Samuel Johnson, *A Dictionary of the English Language* [1755] (London: Longman facsimile, 1990), *sub*. world. This was sense 1 in Johnson's *Dictionary*, where sense 3, 'the earth; the terraqueous globe' is the sense that has come to predominate. See also Thomas Blount, *Glossographia: or a Dictionary, Interpreting all such Hard Words of Whatsoever Language, now used in our refined English Tongue* (London, 1661); and Robert Cawdrey, *A Table Alphabeticall* (London, 1604).
10. Ephraim Chambers, *Cyclopædia*: *Or, an Universal Dictionary of Arts and Sciences*, 2nd edn. (London, 1738), *sub*. Cosmography.
11. Chambers, *Cyclopædia*, *sub*. Mathematics, Mixed.
12. The general geography was not always written by the same person as the descriptive part, the astronomical section of Guthrie's *Geographical Grammar*, for example, being written by James Ferguson, FRS.

13. Chambers, *Cyclopædia, sub.* Chorography. For the history of English chorography, see Stan A.E. Mendyk, *Speculum Britanniae: Regional Study, Antiquarianism, and Science in Britain to 1700* (Toronto: Toronto University Press, 1989); and Graham Parry, *The Trophies of Time: English Antiquarians of the Seventeenth Century* (Oxford: Clarendon Press, 1995).
14. Blount, *Glossographia*; and Cawdrey, *Table Alphabeticall.*
15. Ptolemy, *Geography*, Edward Stevenson, ed. (New York: Dover Press, 1991), Bk. 1.1, p. 25. For the enduring importance of this image, see Svetlana Alpers, *The Art of Describing: Dutch Art in the Seventeenth Century* (Harmondsworth: Penguin, 1989), pp. 133–6.
16. Thomas Dyche and William Pardon, *A New General English Dictionary*, 2nd edn. (London, 1737), *sub.* Chorographer.
17. See Roy Porter, *The Making of Geology: Earth Science in Britain 1660–1815* (Cambridge: Cambridge University Press, 1977), p. 12.
18. J.N.L. Baker, *The History of Geography: Papers by J.N.L. Baker* (Oxford: Blackwells, 1963), p. 103. This complaint aside, Baker has done more to elucidate eighteenth-century geography as the eighteenth century defined it than any other scholar.
19. See, for example, the practise of the 'father of American geography', Jedidiah Morse, who relied on correspondence and travel books, as described in Ralph Brown, 'The American Geographies of Jedidiah Morse', *Annals of the Association of American Geographers*, 31 (1941), pp. 144–217 at pp. 147–8, pp. 160–2 and 210.
20. Thomas Salmon, *A New Geographical and Historical Grammar* (London, 1749), p. 1.
21. This division is to be found in Bernhard Varenius's *Geographia Generalis* (1650), and, as Baker pointed out (*History of Geography*, p. 113), was not original to him. While Baker cites Keckermann, the distinction really goes back to Ptolemy and Strabo: for Ptolemy (*Geography*, 1.1) 'geography looks at the position rather than the quality' (i.e. at the mathematical), where for Strabo (*Geography*, H.L. Jones, trans. and ed., Cambridge, Mass.: Harvard University Press, [1917–32], 1.1.1) geography will also inculcate knowledge 'of things on land and sea, animals, plants, fruits, and everything else to be seen in various regions'.
22. Geography was defined as engaged in description by Nathan Bailey, *An Universal Etymological Dictionary* (London, 1721); and by Dyche and Pardon, *Dictionary*, amongst others.
23. Edward Phillips, rev. John Kersey, *The New World of Words: Or, Universal English Dictionary*, 6th edn. (London, 1706).
24. Patrick Gordon, *Geography Anatomiz'd: or, the Geographical Grammar*, 7th edn. (London, 1716), p. 3.
25. See Rachel Laudan, *From Mineralogy to Geology: the Foundations of a Science, 1650–1830* (Chicago: Chicago University Press, 1987) and Martin Rudwick, 'Minerals, Strata and Fossils', in N. Jardine *et al.*, eds, *Cultures of Natural History* (Cambridge: Cambridge University Press, 1996), pp. 266–86.
26. Noah Webster, *An American Dictionary of the English Language* (New York, 1828).
27. William Guthrie, *A New Geographical, Historical, and Commercial Grammar*, 3rd edn. (London, 1771), p. iv.

28. William Barrow, *An Essay on Education* (London, 1802), II 59–60.
29. Gordon, *Geography Anatomiz'd*, pp. 2–3.
30. Johnson, *Dictionary*. On the meaning of 'science', see also Andrew Cunningham and Perry Williams, 'De-centring the "big picture": *The Origins of Modern Science* and the Modern Origins of Science', *British Journal for the History of Science*, 26 (1993), pp. 407–32 at pp. 420–1. Bowen, *Empiricism and Geographical Thought*, p. 146, notes that geography was 'outside the orbit of the physical sciences', but does not recognise that geography in the understanding of the period was nevertheless scientific because 'science' as a term had a different domain of application.
31. William Cunningham, *The Cosmographical Glasse, Conteinyng the Pleasant Principles of Cosmographie, Geographie, Hydrographie, or Nauigation* (London, 1559), A6v and p. 6.
32. Peter Heylyn, *Microcosmus, Or, A Little Description of the Great World. A Treatise Historicall, Geographicall, Politicall, Theologicall* (Oxford, 1621), p. 1. A marginal note suggests Heylyn was drawing on Keckermann here.
33. Gordon, *Geography Anatomiz'd*, p. 3.
34. Sitwell, *Four Centuries of Special Geography*, p. 4. See also Paul Carter, *The Road to Botany Bay: an Essay in Spatial History* (London: Faber and Faber, 1987), pp. 69–75 for this, Carter's interest being in the genre of the travel narrative rather than the geography book.
35. Guthrie, *Geographical Grammar*, pp. iv–v.
36. James Barclay, *A Treatise on Education* (Edinburgh, 1743), p. 206.
37. See Mona Domosh, 'Toward a Feminist Historiography of Geography', *Transactions of the Institute of British Geographers*, N.S. 16 (1991) pp. 95–104; and Denis Cosgrove and Mona Domosh, 'Writing and Authority: Writing the New Cultural Geography', in James Duncan and David Ley, eds, *Place/Culture/Representation* (London: Routledge, 1993), pp. 25–38.
38. Skinner, in Tully, ed., *Meaning and Context*, p. 63.
39. See Champion, *Pillars of Priestcraft Shaken*, pp. 25–7.
40. Heylyn here expanded on what he had already said on this subject in *Microcosmus*, pp. 10–11.
41. Peter Heylyn, *Cosmographie in Four Books, Containing the Chorographie and Historie of the Whole World*, 2nd edn. (London, 1657), p. 19.
42. This commonplace starts with Ortelius, and no doubt owes something to Ptolemy's previously-discussed imagery. For conceptions of the nature of historical enquiry in early-modern England, see Philip Hicks, *Neoclassical History and English Culture: From Clarendon to Hume* (London: Macmillan, 1996), esp. pp. 7–22.
43. Heylyn, *Cosmographie*, p. 19.
44. Edward Wells, *A Treatise of Antient and Present Geography* (Oxford, 1701), A4r.
45. Charles Rollin, *New Thoughts Concerning Education* (London, 1735), p. 21.
46. Barrow, *Essay on Education*, II 58.
47. Vicesimus Knox, *Liberal Education: Or, a Practical Treatise on the Methods of Acquiring Useful and Polite Learning* (London, 1781), p. 162. The notion of gentlemanly learning was particularly important in the eighteenth century, but Knox is unusual in linking geography to this discourse. For the importance of gentlemanly learning as a discourse to science, see Steven Shapin,

A Social History of Truth: Civility and Science in Seventeenth-Century England (Chicago: Chicago University Press, 1994) and John Gascoigne, *Joseph Banks and the English Enlightenment: Useful Knowledge and Polite Culture* (Cambridge: Cambridge University Press, 1994).

48. See P. Cunish *et al.*, *A History of Magdalene College Cambridge, 1428–1988* (Cambridge: Magdalene College Press, 1994), pp. 160–2.
49. Daniel Waterland, *Advice to a Young Student* [Cambridge, 1730] (Oxford, 1755), p. 14.
50. Jonathan Smith, 'State Formation, Geography, and a Gentleman's Education', *Geographical Review*, 86 (1996), pp. 91–100.
51. Heylyn, *Cosmographie*, p. 19.
52. Thomas Watts, *An Essay on the Proper Method for Forming the Man of Business* [1716], Arthur H. Cole, ed. (Boston: Harvard Graduate School, 1946), p. 24.
53. Martin Clare, *Youth's Introduction to Trade and Business*, 10th edn. (London, 1769), p. vi.
54. Watts, *Essay*, p. 25.
55. J. Bettesworth and H. Fox, *Observations on Education; but Particularly on Naval Education* (London, 1782), pp. 13–14.
56. For more on this subject, see Charles Withers, 'Towards a History of Geography in the Public Sphere', *History of Science*, 37 (1999), pp. 47–78, at pp. 52–63.
57. Richard Sorrenson, 'Towards a History of the Royal Society in the Eighteenth Century', *Notes and Records of the Royal Society of London*, 50 (1996), pp. 29–46 at p. 30. See also Lorraine Daston, *Classical Probability in the Enlightenment* (Princeton: Princeton University Press, 1988), pp. 53–6.
58. Larry Stewart, *The Rise of Public Science: Rhetoric, Technology and Natural Philosophy in Newtonian Britain, 1660–1750* (Cambridge: Cambridge University Press, 1992); and Simon Schaffer, 'The Consuming Flame: Electrical Showmen and Tory Mystics in the World of Goods', in John Brewer and Roy Porter, eds, *Consumption and the World of Goods* (London: Routledge, 1993), pp. 489–526.
59. Benjamin Martin, *A Course of Lectures in Natural and Experimental Philosophy, Geography and Astronomy* (Reading, 1743), Preface.
60. Baker, *History of Geography*, p. 89; and Martin and James, *All Possible Worlds*, pp. 108–9.
61. Anton Büsching, *A New System of Geography* (trans., London, 1762), I 1.
62. Strabo, *Geography*, 1.1.22.
63. For more on this, see Robert Mayhew, 'Geography in Eighteenth Century British Education', *Paedagogica Historica*, N.S. 34 (1998), pp. 731–69.
64. Sitwell, *Four Centuries of Special Geography*, p. 24.
65. Pat Rogers, *Grubstreet: Studies in a Subculture* (London: Methuen, 1972), p. 281. For geography and the dunces at an earlier time, see Smith, 'State Formation and Gentleman's Education', pp. 98–9.
66. James Boswell, *Life of Johnson*, G.B. Hill, ed., L.F. Powell, rev. (Oxford: Clarendon Press, 1934–50), II 52.
67. Alexander Pope, *The Dunciad* [1742 edn.] IV 193–4, in *Poems*, ed. John Butt (London: Methuen, 1963).
68. Thomas Hearne, quoted in Robin Butlin, 'Ideological Contexts and the Reconstruction of Biblical Landscapes in the Seventeenth and Early

Eighteenth Centuries: Dr Edward Wells and the Historical Geography of the Holy Land', in Alan Baker and Gideon Biger, eds, *Ideology and Landscape in Historical Perspective: Essays on the Meanings of Some Places in the Past* (Cambridge: Cambridge University Press, 1992), pp. 31–62 at p. 33.

69. For a detailed study of this process in operation in Jedidiah Morse's *American Geography*, see J.K. Wright, 'Some British "Grandfathers" of American Geography', in R. Miller and J. Wreford Watson, eds. *Geographical Essays in Memory of Alan G. Ogilvie* (London: Nelson, 1959), pp. 144–65.
70. See Adrian Johns, *The Nature of the Book: Print and Knowledge in the Making* (Chicago: Chicago University Press, 1998), Chapters 2 and 3; and Rose, *Owners and Authors*.
71. Rose, *Owners and Authors*, pp. 136–7.
72. Foucault, 'What is an Author?', p. 109.
73. Eisenstein, *Printing Press as an Agent of Change*, pp. 153 and 155. See also Alvin Kernan, *Samuel Johnson and the Age of Print*, new edn. (Princeton: Princeton University Press, 1989).
74. Sitwell, *Four Centuries of Special Geography*, p. 301.
75. Ibid., p. 500. The editions of Salmon are chronicled pp. 500–5.
76. Ibid., p. 284.
77. Clark, *English Society 1688–1832* emphasises the centrality of religion to politics, where Bradley, *Religion, Revolution and English Radicalism* emphasises the centrality of political aims in mobilising the dissenting community.
78. Sitwell, *Four Centuries of Special Geography*, p. 95.

3 'Geography is twinned with divinity': the Laudian Geography of Peter Heylyn, 1621–57

1. Strabo, *Geography*, Book 1.1.
2. Cunningham, *Cosmographical Glasse*, Aiiiir.
3. *Camden's Britannia*, Edmund Gibson, ed. (London, 1695), p. iv. Gibson's translation is faithful to the original where England 'est enim naturae gaudentis opus, quam quasi alterum orbem': William Camden, *Britannia: sive Florentissimorum Regnorum, Angliæ, Scotiae, Hiberniae, et Insularum Adiacentum ex Intima Antiquitate Chorographica Descriptio* (London, 1586), p. 4. The image of Britain as cut off from the rest of the world had its roots in Virgil's description, *Eclogues*, I 36.
4. William Camden, *Britain, Or A Chorographical Description of the Most flourishing Kingdomes, England, Scotland, and Ireland*, Philemon Holland, trans. (London, 1610), p. 163. Holland's translation was overseen by Camden.
5. Ibid, pp. 67–8.
6. Ibid, ¶5r.
7. For longer discussions of Camden's politics, see F.J. Levy, 'The Making of Camden's *Britannia*', in *Bibliothèque D'Humanisme et Renaissance*, 26 (1964) pp. 70–97; Kevin Sharpe, *Sir Robert Cotton 1586–1631: History and Politics in Early Modern England* (Oxford: Oxford University Press, 1979); Richard Helgerson, *Forms of Nationhood: the Elizabethan Writing of England* (Chicago: Chicago University Press, 1992), Ch. 3; Stuart Piggott, 'William Camden and the *Britannia*', *Proceedings of the British Academy*, 37 (1951) pp. 199–217;

and William Rockett, 'Historical Topography and British History in Camden's *Britannia*', *Renaissance and Reformation*, 26 (1990) pp. 71–80.

8. See Walter Ullmann, 'This Realm of England is an Empire', *Journal of Ecclesiastical History*, 30 (1979) pp. 175–203; and David Armitage, 'Literature and Empire' in Nicholas Canny, ed. *The Oxford History of the British Empire: Volume 1: The Origins of Empire* (Oxford: Oxford University Press, 1998), pp. 99–123 at pp. 103–4.
9. George Abbot, *A Briefe Description of the Whole Worlde*, 3rd edn. (London, 1605), N1v–N2r. This section was not in the 1599 edition.
10. It should be noted, however, that the first edition of *Britannia* was itself only an octavo, unlike the massive folio versions to which both the eighteenth century and the present century are accustomed. *Britannia* is a rare case where the length and monumentality of a geographical work increased with its popularity.
11. See Nicholas Tyacke, 'Science and Religion at Oxford before the Civil War' in Donald Pennington and Keith Thomas, eds, *Puritans and Revolutionaries: Essays in Seventeenth Century History Presented to Christopher Hill* (Oxford: Clarendon Press, 1978), pp. 73–93, esp. pp. 81–4 and pp. 92–3; and Nicholas Tyacke, *Anti-Calvinists: the Rise of English Arminianism, c.1590–1640* (Oxford: Clarendon Press, 1987), pp. 66 and 120.
12. The details of Heylyn's life are contained in: Peter Heylyn, 'Heylyn's own Memoranda' in *Memorial of Bishop Waynflete*, ed. John R. Bloxam (London: Caxton Society, 1851), pp. x–xxiv; the 'Life of Heylyn' prefixed to *Keimhaia Ecclesiastica. The Historical and Miscellaneous Tracts of the Reverend and Learned Peter Heylyn* (London, 1681); George Vernon, *The Life of the Learned and Reverend Dr Peter Heylyn* (London, 1682); and John Barnard, *Theologo-Historicus, or the True Life of the Most Reverend Divine, and Excellent Historian Peter Heylyn* (London, 1683). Vernon claims to correct the life in Heylyn's miscellaneous tracts, while Barnard claims to correct Vernon and accuses him of plagiarising his (Barnard's) notes. In fact, the three lives clearly derive from one basic source, whoever wrote it, and the information they give is substantially the same.
13. Peter Heylyn, *Microcosmus, or A Little Description of the Great World. A Treatise Historicall, Geographicall, Politicall, Theologicall* (Oxford, 1621); and *idem*, *Cosmographie in Four Books. Containing the Chorographie and Historie of the Whole World*, 2nd edn. (London, 1657). These editions are referred to throughout, further references in the text to *M* and *C* respectively. It should be noted that *Microcosmus* appeared in a substantially enlarged edition in 1625, much of the material from which was reprinted in unaltered form in *Cosmographie*.
14. Peter Heylyn, *A Full Relation of Two Journeys. The one into the Main-Land of France. The Other into some of the Adjacent Ilands* (London, 1656), p. 118.
15. Helgerson, *Forms of Nationhood*.
16. Heylyn, 'Memoranda,' p. xvi.
17. For Heylyn's other efforts to impress Laud, see Kenneth Fincham, 'Oxford and the Early Stuart Polity' in Nicholas Tyacke, ed. *The History of the University of Oxford: Volume IV: Seventeenth-Century Oxford* (Oxford: Oxford University Press, 1998), pp. 179–210 at pp. 180–1. It should, however, be emphasized that Heylyn had not met Laud before the publication of

Microcosmus (in its 1621 or 1625 editions): see Heylyn, 'Memoranda,' p. xvii.

18. See Tyacke, *Anti-Calvinists*, p. 81; and Anthony Milton, *Catholic and Reformed: the Roman and Protestant Churches in English Protestant Thought, 1600–1640* (Cambridge: Cambridge University Press, 1995), Part II.
19. Kenneth Hylson-Smith, *High Churchmanship in the Church of England: From the Sixteenth Century to the Late Twentieth Century* (Edinburgh: T&T Clark, 1993), pp. 35–6; the best analysis of Laudian doctrine is in Milton, *Catholic and Reformed*.
20. Ibid., p. 38.
21. See Colin Kidd, *British Identities before Nationalism: Ethnicity and Nationhood in the Atlantic World, 1600–1800* (Cambridge: Cambridge University Press, 1999); and Clark, *Language of Liberty*, pp. 46–62.
22. Heylyn, *Full Relation*, p. 248. For absolutism in English political thought, see J.P. Sommerville, *Politics and Ideology in England 1603–1640* (London: Longman, 1986), Ch.1.
23. Heylyn, *Historical and Miscellaneous Tracts*, p. 715.
24. See Glenn Burgess, *The Politics of the Ancient Constitution: an Introduction to English Political Thought, 1603–1642* (London: Macmillan, 1992), Part II, for the way in which absolutism and common law were normally complementary languages in the period, even for Laudian divines like Heylyn (pp. 181–8).
25. Pocock, *Ancient Constitution and Feudal Law*, pp. 52–3.
26. Heylyn, *Full Relation*, p. 8. See Mark Goldie, 'John Locke and Anglican Royalism', *Political Studies*, 31 (1983) pp. 61–83, esp. pp. 69–70.
27. See in particular Peter Heylyn, *Examen Historicum: Or a Discovery of the Mistakes, Falsities, and Defects in some Modern Historians* (London, 1659), pp. 125 and 201.
28. Heylyn, *Historical and Miscellaneous Tracts*, p. 716. On English sovereignty as more absolute than any in Europe, see Clark, *Language of Liberty*, pp. 66–7 and 75–7.
29. Milton, *Catholic and Reformed*, pp. 272–7.
30. Heylyn's views are spelled out at length in *Ecclesia Restaurata: Or, the History of the Reformation of the Church of England* (London, 1661). Laud's paradigmatic position is clear from Heylyn's hagiographical *Cyprianus Anglicus: Or, The History of the Life and Death, of the most Reverend and Renowned Prelate William by Divine Providence, Lord Archbishop of Canterbury* (London, 1668).
31. It is noticeable that Hylson-Smith's *High Churchmanship in the Church of England*, p. 25 comes up with exactly the same names (with the addition of John Whitgift). Heylyn's Reformation is in fact but one branch of the Reformation. Heylyn does mention Wycliffe, and thereby the 'radical' tradition of medieval heresies which Calvinists emphasised in genealogies of Protestantism, but his view was ambivalent: '[Wycliffe] so valiantly opposed the power and Errors of the Church of *Rome*; though he vented so many of his own' (*C.*, p. 303). Laudians by and large downplayed the historical continuities between medieval heresies and modern Protestantism: see Milton, *Catholic and Reformed*, pp. 301–6.
32. See also Heylyn, *Full Relation*, pp. 242–4 where his animosity to the Huguenots is detailed at greater length.

33. Heylyn saw multiple independent but Catholic churches as the doctrine of Cyprian, and Laud's use of the doctrine in England explains why Heylyn's biography of Laud was called *Cyprianus Anglicanus*: see Champion, *Pillars of Priestcraft Shaken*, pp. 67–8. It was also because of this conception of Catholicity that Heylyn said that he was a Catholic when asked in France, see *Full Relation*, p. 163. For Laudian praise of the Gallican church, see Milton, *Catholic and Reformed*, pp. 263–9. High Church Anglican admiration of the Gallican conception of Catholicism continued into the eighteenth century: see B.W. Young, 'A History of Variations: the Identity of the Eighteenth-Century Church of England' in Tony Claydon and Ian McBride, eds, *Protestantism and National Identity: Britain and Ireland, c.1650–c.1850* (Cambridge: Cambridge University Press, 1998), pp. 105–28 at pp. 108–9; and Kidd, *British Identities before Nationalism*, pp. 236–9.
34. Vernon, *Life of Heylyn*, pp. 172 and 142.
35. E.W. Gilbert, *British Pioneers in Geography* (Newton Abbot: David and Charles, 1972), pp. 46 and 50.
36. Champion, *Pillars of Priestcraft Shaken*, p. 34, a point reinforced for Heylyn in particular, pp. 64–77.
37. Heylyn, *Full Relation*, pp. 205 and 246.
38. Ibid., p. 243.
39. As Kidd, *British Identities before Nationalism*, p. 287 summarises it: 'provenance was the keystone of legitimacy' in early-modern political argument.
40. In the edition I have consulted, p. 1105 is mispaginated as a second p. 1095.
41. 'Life of Heylyn,' prefixed to *Historical and Miscellaneous Tracts*, p. xviii.

4 John Ogilby and the Iconographic Roads to a Restored Royalist Geography, c.1660–75

1. John Ogilby, *Africa* (London, 1670), c.r.
2. Robert Bosher, *The Making of the Restoration Settlement: the Influence of the Laudians 1649–62* (London: Dacre Press, 1951), pp. 36 and 37–8.
3. The politics of Ogilby's Aesopian fables is discussed in Mark Loveridge, *A History of Augustan Fable* (Cambridge: Cambridge University Press, 1998), pp. 102–41; Jayne Elizabeth Lewis, *The English Fable: Aesop and Literary Culture, 1651–1740* (Cambridge: Cambridge University Press, 1996), pp. 14–23; and Annabel Patterson, *Fables of Power: Aesopian Writing and Political History* (Durham, NC: Duke University Press, 1991), pp. 85–94.
4. John Ogilby, *The Fables of Aesop*, 3rd edn. (London, 1673), p. 39.
5. Cited in Katherine Van Eerde, *John Ogilby and the Taste of his Times* (Folkestone: Dawson Press, 1976), p. 27.
6. Ogilby, *Fables*, pp. 114 and 120.
7. Loveridge, *History*, p. 118; see pp. 105–6 and pp. 110–11 for Parliamentarian readings of the *Fables*.
8. I.M. Green, *The Re-Establishment of the Church of England, 1660–1663* (Oxford: Oxford University Press, 1978), p. 91.
9. Van Eerde, *Ogilby and the Taste of his Times*, p. 71.
10. See Sir Edward Walker, *A Circumstantial Account of the Preparations for the Coronation of His Majesty King Charles the Second* (London, 1820), p. 85.

11. Lawrence Manley, *Literature and Culture in Early Modern London* (Cambridge: Cambridge University Press, 1995), p. 214. See also Roy Strong, *Art and Power: Renaissance Festivals, 1450–1650* (Woodbridge: Boydell and Brewer, 1984); Stephen Orgel, *The Illusion of Power: Political Theatre in the English Renaissance* (Berkeley: University of California Press, 1975); and Vaughan Hart, *Art and Magic in the Court of the Stuarts* (London: Routledge, 1994), Chapter 7.
12. Edward Hyde, *The Life of Edward Earl of Clarendon* (Oxford, 1759), II 186.
13. Samuel Pepys, *Diary*, transcribed by Robert Latham and William Matthews (London: G. Bell and Sons, 1970), II 81.
14. Sir Richard Baker [Continued by Edward Philips?], *A Chronicle of the Kings of England* (London, 1674), p. 759.
15. See also Charles Withers, 'Geography, Royalty and Empire: Scotland and the Making of Great Britain, 1603–1661', *Scottish Geographical Magazine*, 113 (1997), pp. 22–32.
16. John Ogilby, *The Relation of His Majesties Entertainment Passing through the City of London, to his Coronation: With a Description of the Triumphal Arches, and Solemnity*, 2nd edn. (London, 1661), facing the title page. Ogilby also published a longer account of the procession which elaborated on the classical precedents for the procession, under the title of *The Entertainment of His Most Excellent Majestie Charles II, in his Passage through the City of London to his Coronation* (London, 1662). These works are hereafter referred to in the text as *Relation* and *Entertainment* respectively.
17. See John Keay, *The Honourable Company: a History of the English East India Company* (London: Harper Collins, 1991), Chs.1–6.
18. John Ogilby, *America: Being the Latest, and Most Accurate Description of the New World* (London, 1671), p. 169.
19. See David Berkowitz, *John Selden's Formative Years: Politics and Society in Early Seventeenth-Century England* (London: Associated University Press, 1988), pp. 51–4.
20. Steven Pincus, *Protestantism and Patriotism: Ideologies and the Making of English Foreign Policy, 1650–1668* (Cambridge: Cambridge University Press, 1996).
21. A distinction was drawn in seventeenth-century political thought between pernicious 'universal monarchy' and England's beneficent commercial empire: but there is no sign that Ogilby's procession draws such a distinction, simply claiming instead universal monarchy for England. For the distinction: see John Robertson, 'Gibbon's Roman Empire as a Universal Monarchy: the *Decline and Fall* and the Imperial Idea in Early Modern Europe', in Rosamond McKitterick and Roland Quinault, eds. *Edward Gibbon and Empire* (Cambridge: Cambridge University Press, 1997), pp. 247–70, at pp. 254–5. See also Steven Pincus, 'The English Debate over Universal Monarchy' in John Robertson, ed., *A Union for Empire: Political Thought and the British Union of 1707* (Cambridge: Cambridge University Press, 1995), pp. 37–62.
22. See Richard Tuck, 'Grotius and Selden' in J.H. Burns with Mark Goldie, eds, *The Cambridge History of Political Thought, 1450–1700* (Cambridge: Cambridge University Press, 1991), pp. 499–529, at pp. 527–8.
23. George Downing, cited in Pincus, *Protestantism and Patriotism*, pp. 267–8.

24. See Hart, *Art and Magic*, pp. 171–2.
25. See Derek Hirst, 'The English Republic and the Meaning of Britain', in Brendan Bradshaw and John Morrill, eds, *The British Problem, c.1534–1707: State Formation in the Atlantic Archipelago* (London: Macmillan, 1996), pp. 192–219, at pp. 192–4.
26. Manley, *Literature and Culture*, p. 218.
27. See Bosher, *Making of the Restoration Settlement*; and Green, *Re-Establishment of the Church of England*.
28. See the balanced discussion in Ronald Hutton, *The Restoration: a Political and Religious History of England and Wales 1658–1667* (Oxford: Clarendon Press, 1985), pp. 181 and 288–9.
29. Paul Seaward, *The Cavalier Parliament and the Reconstruction of the Old Regime, 1661–1667* (Cambridge: Cambridge University Press, 1989), Ch. 3.
30. I do not include the series of atlases Ogilby produced in the 1670s, as they were translations which only appeared under his name, Ogilby having little to do with their textual production: see Van Eerde, *Ogilby and the Taste of his Times*, p. 96.
31. See Hutton, *Restoration*, p. 185; and Seaward, *Cavalier Parliament*, pp. 44–55.
32. Those who see elements of absolutism include: J.R. Jones, *Charles II: Royal Politician* (London: Allen and Unwin, 1987), Ch. 5; and Maurice Lee, *The Cabal* (Urbana: University of Illinois Press, 1965). Those who oppose this view include Ronald Hutton, *Charles the Second: King of England, Scotland and Ireland* (Oxford: Clarendon Press, 1989), pp. 318–19; and John Miller, 'The Potential for "Absolutism" in Later Stuart England', *History*, 69 (1984) pp. 187–207.
33. Pincus, *Protestantism and Patriotism*, p. 440.
34. John Ogilby, *Britannia, Volume the First: Or, An Illustration of the Kingdom of England and Dominion of Wales: By a Geographical and Historical Description of the Principal Roads Thereof* (London, 1675), Ar. At Ogilby's death in 1676, *Britannia* was projected to have two further volumes, one a description of 25 cities and the other a topographical description of Britain (Br-v). This work is hereafter referenced in the text as *Britannia*.
35. Ogilby, *Africa*, br-v. (Original in italics.) See also Cynthia Wall, *The Literary and Cultural Spaces of Restoration London* (Cambridge: Cambridge University Press, 1998).
36. John Ogilby and William Morgan, *London Survey'd: Or, An Explanation of the Large Map of London* (London, 1677), av.
37. Ibid., A2r. Almost the same wording is used in *Britannia*, Dv.
38. Ogilby, *Fables, Part II* (new pagination), pp. 13–5, 38 and 43.
39. Van Eerde, *Ogilby and the Taste of his Times*, p. 151; see also Herbert Fordham, *Notes on British and Irish Itineraries and Road Books* (Hertford: Stephen Austin Press, 1912), pp. 7–8.
40. Emanuel Bowen, *Britannia Depicta or Ogilby Improv'd* (London, 1720). This work went through a series of editions between 1720 and 1764.
41. Helgerson, *Forms of Nationhood*, Ch. 3.
42. R. & J. Dodsley [printers], *A New and Accurate Description of the Present Great Roads and the Principal Cross Roads of England and Wales* (London, 1756). Roman roads are treated after modern ones, in a separate pagination, pp. i–xxxviii.

43. John Potter, *The Traveller's Pocket-Book; or, Ogilby and Morgan's Book of the Roads Improved and Amended*, 20th edn. (London, 1780), p. 51. The persistent influence of Ogilby is clear from the many ventures still deriving their origin from his work and/or invoking him in their title.
44. John Cary, *Cary's New Itinerary; or, An Accurate Delineation of the Great Roads, both Direct and Cross, throughout England and Wales* (London, 1798), pp. 743–96.

5 The Political and Geographical Appropriations of Edmund Bohun, 1684–1710

1. Edmund Bohun, *The Diary and Correspondence of Edmund Bohun*, S. Wilton Rix, ed. (Beccles: Privately Printed, 1853), pp. 76 and 78.
2. Ibid., pp. 78–9.
3. Robert Filmer, *Patriarcha: Or the Natural Power of Kings*, ed. with Preface and Postscript by Edmund Bohun, 2nd edn. (London, 1685), 'The copy of a letter written by the late learned Dr Peter Heylyn to Sir Edward Filmar [*sic*], son of the worthy author'.
4. Bohun Collection, Cambridge University Library, Add.4521.237.
5. Mark Goldie has illuminated Bohun's politics and its context in a series of outstanding essays to which I am indebted: 'Edmund Bohun and *Jus Gentium* in the Revolution Debate, 1689–93', *Historical Journal*, 20 (1977) pp. 569–86; 'The Revolution of 1689 and the Structure of Political Debate', *Bulletin of Research in the Humanities*, 83 (1980) pp. 473–564; 'John Locke and Anglican Royalism', *Political Studies*, 31 (1983) pp. 61–85; and 'The Political Thought of the Anglican Revolution', in Robert Beddard, ed., *The Revolutions of 1688: the Andrew Browning Lectures 1988* (Oxford: Oxford University Press, 1991), pp. 102–36.
6. [Edmund Bohun], *A Defence of Sir Robert Filmer, against the Mistakes and Misrepresentations of Algernon Sidney, Esq.* (London, 1684), p. 11. See also *Patriarcha*, (e4r)–(e4v).
7. Ibid., p. 5.
8. Ibid., p. 15.
9. Bohun, *Diary*, pp. 45–6.
10. See E[dmund] B[ohun], *An Apologie for the Church of England, Against the Clamours of the Men of No-Conscience* (London, 1685).
11. Edmund Bohun, *A Geographical Dictionary* (London, 1688), A7v. For this study I have consulted the first (1688), second ('1689' [actually 1690]) and fourth (1710) editions. I will hereafter cite them in the text as *GD1*, *GD2* and *GD4*.
12. The political positions on 1688 are comprehensively treated in Goldie, '1688 and Structure of Political Argument'.
13. For the doctrine of non-assistance, see Goldie, 'Political Thought of the Anglican Revolution', pp. 113–17.
14. For conquest, see Goldie, 'Bohun and *Jus Gentium*'.
15. Edmund Bohun, *The Doctrine of Non-Resistance or Passive Obedience No Way Concerned in the Controversies now Depending between the Williamites and the Jacobites* (London, 1689), p. 7.

16. See Goldie, 'Bohun and *Jus Gentium*', pp. 582–6; and Goldie, '1688 and Structure of Political Argument', pp. 516–17.
17. Peter Heylyn, *Cosmographie in Four Books, Improv'd with an Historical Continuation to the Present Times by Edmund Bohun, Esq.*, 7th edn. (London, 1703), p. 286. Hereafter cited in the text as *EBC*.
18. [Edmund Bohun], *The History of the Desertion, or an Account of all the Publick Affairs of England, from the Beginning of September 1688 to the Twelfth of February following* (London, 1689).
19. Goldie, 'Bohun and *Jus Gentium*', p. 575.
20. Edmund Bohun, *Three Charges Delivered at the General Quarter Sessions* (London, 1693), A2r–A3v.
21. Laurence Echard, *A Most Compleat Compendium of Geography*, 3rd edn. (London, 1693), A7v–A8r. This first appeared in the second edition of 1691.
22. See, for example, Erskine-Hill, *Poetry of Opposition and Revolution*.
23. Paul Monod, *Jacobitism and the English People, 1688–1788* (Cambridge: Cambridge University Press, 1989), p. 5.
24. Ibid., p. 6.
25. The parallel of Cromwell and William III was a commonplace in Jacobite rhetoric: Ibid., pp. 51–2.
26. Bohun, 'Letter to Mr Echard', A8r.
27. For which, see [Charles Blount], *Reasons Humbly Offered for the Liberty of Licens'd Printing. To which is subjoin'd, the Just and True Character of Edmund Bohun, the Licenser of the Press* (London, 1693), pp. 10–32.
28. C.U.L. Sel.3.237.357 Arthur Charlett to Edmund Bohun, ? 9th 1692/3.
29. On Barnard and his circle, see Robert Beddard, 'James II and the Catholic Challenge' in Tyacke, ed. *Seventeenth-Century Oxford*, pp. 907–54 at pp. 917–29, p. 950 n. 303 and p. 953 n. 328.
30. Herman Moll, *A System of Geography* (London, 1701), p. 268 (Modena), pp. 83–4 (St Germain en Laye) and p. 55 (Boyne).

6 Edmund Gibson's Editions of *Britannia*: Loyalist Chorography and the Particularist Politics of Precedent, 1695–1722

1. See Edmund Gibson, *New Proposals for Printing by Subscription, Cambden's [sic] Britannia, English* (London, 1693).
2. Edmund Gibson, ed., *Camden's Britannia, Newly Translated into English: With Large Additions and Improvements* (London, 1695), sig.A2v-a.r. Hereafter cited in the text as *1695*.
3. Parry, *Trophies of Time*, p. 331.
4. Even if Gibson's notes were often reliant on published county histories such as Dugdale's for Warwickshire and Plot's for Oxfordshire and Staffordshire: see *1695*, sig.a.r-v.
5. David Douglas, *English Scholars* (London: Jonathan Cape, 1939), p. 266.
6. Ibid., p. 17. See also Pocock, *Ancient Constitution and Feudal Law*; Hicks, *Neoclassical History and English Culture*; and Laird Okie, *Augustan Historical Writing: Histories of England in the English Enlightenment* (Lancham, Maryland: University Press of America, 1991).

7. G.V. Bennett, *White Kennett 1660–1728. Bishop of Peterborough. A Study in the Political and Ecclesiastical History of the Early Eighteenth Century* (London: SPCK, 1957), p. 165; see also Norman Sykes, *Edmund Gibson, Bishop of London 1669–1748: a Study in Politics and Religion in the Eighteenth Century* (London: Oxford University Press, 1926), p. 16.
8. Douglas, *English Scholars*, pp. 335–6; see also Parry, *Trophies of Time*, Chapter 12. Gwyn Walters and Frank Emery, 'Edward Lhuyd, Edmund Gibson, and the Printing of Camden's *Britannia*, 1695', *The Library*, 5th series 32 (1977) pp. 109–37, detail processes of textual construction, but do not look at the politics of these processes.
9. Moses Pitt, ed., *The English Atlas* (Oxford, 1680–83), 'Proposals for Printing the English Atlas', I n.p.
10. For Nicolson's authorship, see his letter in Ralph Thoresby, *Letters of Eminent Men, Addressed to Ralph Thoresby* (London, 1832), I 122.
11. See Sykes, *Edmund Gibson*, p. 18; and Bennett, *White Kennett*, pp. 4–5 and 12.
12. William Drummond, *Polemo-Middinia*, Edmund Gibson, ed. (Oxford, 1691); Edmund Gibson, ed., *Chronicon Saxonicum* (Oxford, 1692), see Gibson's textual appendix, 'Nominum Locorum Quae in Chronico Saxonico memorantur, Explicatio'; for Somner, see Bennett, *White Kennett*, p. 160.
13. Gibson, *Proposals*, n.p.
14. William Nicolson, *The English Historical Library* (London, 1696–99), I 10 and I 13.
15. Somers had been the counsel for the seven Anglican bishops imprisoned by James II for opposing his declaration of liberty of religious conscience.
16. For further discussion of humanism and geographical textual editing see Robert Mayhew, 'Was William Shakespeare an Eighteenth-Century Geographer? Constructing Histories of Geographical Knowledge', *Transactions of the Institute of British Geographers*, N.S. 23 (1998) pp. 21–37. More general information about humanist editing can be gleaned from: Anthony Grafton, *Defenders of the Text: the Traditions of Scholarship in an Age of Science, 1450–1800* (Cambridge, Mass.: Harvard University Press, 1991); Anthony Grafton and Lisa Jardine, *From Humanism to the Humanities: Education and the Liberal Arts in Fifteenth- and Sixteenth-Century Europe* (London: Duckworth, 1986); and Joseph Levine, *Humanism and History: Origins of Modern English Historiography* (Ithaca: Cornell University Press, 1987).
17. Gibson *et al.* shared a vehement anti-Catholicism: see Sykes, *Edmund Gibson*, pp. 295–6; and Bennett, *White Kennett*, p. 85.
18. J.P. Kenyon, *Revolution Principles: the Politics of Party, 1689–1720* (Cambridge: Cambridge University Press, 1977), p. 35; see also Chs. 4 and 5.
19. Robert Beddard, 'The Unexpected Whig Revolution of 1688', in Beddard, ed., *The Revolutions of 1688*, pp. 11–101 at pp. 67–8.
20. Tony Claydon, *William III and the Godly Revolution* (Cambridge: Cambridge University Press, 1996), p. 4 and *passim*; see also Kenyon, *Revolution Principles*, p. 24.
21. Geoffrey Holmes, *Politics in the Age of Queen Anne* (London: Macmillan, 1967), p. 20. See also Henry Horwitz, 'The Structure of Parliamentary Politics' in Geoffrey Holmes, ed., *Britain After the Glorious Revolution, 1689–1714* (London: Macmillan, 1969), pp. 96–114.

22. See Geoffrey Holmes, 'Religion and Party in Late Stuart England' in his *Politics, Religion and Society in England 1679–1742* (London: Hambledon Press, 1986), pp. 181–215.
23. See Henry Horwitz, *Parliament, Policy and Politics in the Reign of William III* (Manchester: Manchester University Press, 1977), p. 312.
24. Francis Atterbury, *The Rights, Powers, and Privileges of an English Convocation*, 2nd edn. (London, 1701), p. 31n.
25. Ibid., p. 67. Atterbury accurately transcribes from Camden's 1607 Latin edition, p. 131.
26. These disputes are fully covered in Francis James, *North Country Bishop: a Biography of William Nicolson* (New Haven: Yale University Press, 1956); and G.V. Bennett, *The Tory Crisis in Church and State 1688–1730: the Career of Francis Atterbury, Bishop of Rochester* (Oxford: Clarendon Press, 1975).
27. John Nichols, ed., *Letters on Various Subjects, Literary, Political, and Ecclesiastical, to and from William Nicolson* (London, 1809), I 239.
28. White Kennett, *Ecclesiastical Synods and Parliamentary Convocations in the Church of England Historically Stated and Justly Vindicated from the Misrepresentations of Mr Atterbury* (London, 1701), p. 215.
29. Edmund Gibson, *Britannia: Or a Chorographical Description of Great Britain and Ireland*, 2nd edn. (London, 1722), col.1093. I have used the 1728 reprint, which is hereafter cited in the text as *1722*.
30. [Edmund Gibson], *Synodus Anglicana: Or, the Constitution and Proceedings of an English Convocation* (London, 1702), pp. ii–iii.
31. [Edmund Gibson], *A Short State of Some Present Questions in Convocation* (London, 1703), p. 39. White Kennett also attacked Atterbury as an historian: see *Ecclesiastical Synods*, pp. i–ii and 34.
32. Kennett, *Ecclesiastical Synods*, pp. x and 16.
33. Ibid., p. 217.
34. Gibson, *Short State*, p. 60; see also p. 11 and Gibson, *Synodus Anglicana*, p. 12.
35. Edmund Gibson, *Five Pastoral Letters* [4th Letter, 1739] (London, 1760), p. 279.
36. Edmund Gibson, *Codex Juris Ecclesiastici Anglicani* (London, 1713) I xxvii.
37. Kennett, *Ecclesiastical Synods*, p. 233.
38. Thoresby, *Letters*, I 300. The letter was written in 1697.
39. See Pocock, *Ancient Constitution and Feudal Law*, Chs 2 and 3; Burgess, *Politics of the Ancient Constitution*; Sykes, *Edmund Gibson*, p. 332; and Bennett, *White Kennett*, p. 248.
40. Jennifer Carter, 'The Revolution and the Constitution' in Holmes, ed., *Britain after the Glorious Revolution*, pp. 39–58 at p. 53.
41. *Pace* J.H. Plumb, *The Growth of Political Stability in England, 1657–1725* (London: Macmillan, 1967).
42. Edmund Gibson, *The Deliverances and Murmurings, of the Israelites, and these Nations, Compar'd* (London, 1716), p. 12; see also Edmund Gibson, *Religion, the Best Security to Church and State* (London, 1715), p. 25.
43. White Kennett, *A Thanksgiving Sermon for the Blessing of God, in Suppressing the Late Unnatural Rebellion* (London, 1716), pp. 12 and 15.
44. [Edmund Gibson], *The Causes of the Discontents in Relation to the Plague* (London, 1721), p. 8.

45. Kennett, *Thanksgiving Sermon*, p. 23; and Gibson, *Deliverances and Murmurings, passim.*
46. On the conservatism of Whigs, see Kenyon, *Revolution Principles*, Chs. 10–11. On this in Gibson *et al.*, see Sykes, *Edmund Gibson*, pp. 182 and 391; Bennett, *White Kennett*, pp. 22 and 166; and James, *North Country Bishop*, p. 146.
47. White Kennett, *A Complete History of England, Volume 3* (London, 1706), p. 86.
48. White Kennett, *The Happiness of this Church and Nation* (London, 1705), pp. 6 and 8.
49. Ibid., p. 7.
50. Kidd, *British Identities before Nationalism*, Ch. 4.
51. See Camden, *Britannia* (1586 Latin edition), p. 44.
52. Gibson, *Deliverances and Murmurings*, p. 14.
53. Gibson also dedicated his *Codex* to Tenison.
54. Similar praise of the Prince of Wales was inserted at *1722*, col. 832.
55. Thoresby, *Letters*, II 215.
56. G.M. Townend, 'Religious Radicalism and Conservatism in the Whig Party under George I: The Repeal of the Occasional Conformity and Schism Acts', *Parliamentary History*, 7 (1988), pp. 24–44 at p. 40.
57. Richard Gough, *Anecdotes of British Topography* (London, 1768), p. 26.
58. Nathaniel Salmon, *A New Survey of England* (London, 1731), Title page and I 3.
59. Gough, *Anecdotes*, p. xviii.
60. Richard Gough, *Britannia: Or, A Chorographical Description of the Flourishing Kingdoms of England, Scotland, and Ireland* (London, 1789), I xiii–xiv (new pagination) and I viii (first pagination).

7 Varieties of Orthodox Geography, 1700–50: Three Vignettes: Echard, Wells and Salmon

1. Laurence Echard, *A Most Compleat Compendium of Geography*, 4th edn. (London, 1697), A5v. I have used this edition in this section. Further references in the text as *CC*.
2. See John Oldmixon, *The Critical History of England, Ecclesiastical and Civil* (London, 1724); Edmund Calamy, *A Letter to Mr Archdeacon Echard, Upon Occasion of his History of England* (London, 1718); and Henry Fielding, *Joseph Andrews*, Douglas Brooks-Davies, ed. (Oxford: Oxford University Press, 1966), Bk III, Ch. 1. For a beautiful parody of Echard and others, see Henry Fielding, *The Covent Garden Journal*, Bertrand Goldgar, ed. (Oxford: Clarendon Press, 1988), No. 17, 29 Feb. 1752.
3. Deborah Stephan, 'Laurence Echard – Whig Historian', *Historical Journal*, 32 (1989), pp. 843–66; and Hicks, *Neoclassical History and English Culture*, pp. 102–9.
4. Laurence Echard, *The History of England* (London, 1707–18) II v. Hereafter referenced in the text as *HE*.
5. Stephan, 'Laurence Echard', p. 844.
6. See John Marshall, 'The Ecclesiology of the Latitude-Men 1660–1689: Stillingfleet, Tillotson and "Hobbism"', *Journal of Ecclesiastical History*, 26 (1985), pp. 407–27.

7. Laurence Echard, *The History of the Revolution, and the Establishment of England in the Year 1688* (London, 1725), pp. 238–9.
8. Oldmixon, *Critical History*, p. 224.
9. Although significant qualifications need to be entered about the historical accuracy of these stereotypes: see B.W. Young, *Religion and Enlightenment in Eighteenth-Century England: Theological Debate from Locke to Burke* (Oxford: Clarendon Press, 1998).
10. See John Gascoigne, *Cambridge in the Age of the Enlightenment: Science, Religion and Politics from the Restoration to the French Revolution* (Cambridge: Cambridge University Press, 1989).
11. See Clark, *English Society: 1688–1832*, p. 51, n. 27; G.V. Bennett, 'The Era of Party Zeal 1702–14', in L.S. Sutherland and L.G. Mitchell, eds, *The History of the University of Oxford: Volume V: The Eighteenth Century* (Oxford: Oxford University Press, 1986), pp. 61–98.
12. Edward Wells, *Dr Wells's Controversial Tracts Against the Dissenters*, 6th edn (London, 1716), p. 10; see also p. 82. This book collects all of Wells's important tracts, and is hereafter cited in the text as *Tracts*.
13. Robert Cornwall, *Visible and Apostolic: the Constitution of the Church in High Church Anglican and Non-Juror Thought* (Newark: University of Delaware Press, 1993).
14. On the rhetoric of the *via media*, see J.G.A. Pocock, 'Within the Margins: the Definitions of Orthodoxy', in R.D. Lund, ed., *The Margins of Orthodoxy: Heterodox Writing and Cultural Response, 1660–1750* (Cambridge: Cambridge University Press, 1995), pp. 33–53.
15. Holmes, *Politics in the Age of Queen Anne*, pp. 17–8.
16. Edward Wells, *Remarks on Dr Clarke's Introduction to his Scripture-Doctrin of the Trinity* (Oxford, 1713), p. 53.
17. Ibid., pp. 18 and 21.
18. Butlin, 'Edward Wells', pp. 50–8.
19. Edward Wells, *An Historical Geography of the New Testament* (London, 1708), A2v and A3v–A4r. Further references to *NT* are in the text.
20. Edward Wells, *An Historical Geography of the Old Testament* (London, 1711–12), I 281. Hereafter referred to in the text as *OT*.
21. Butlin, 'Edward Wells', p. 31.
22. On this and other versions of England's primitive ecclesiastical identity, see Kidd, *British Identities before Nationalism*, Ch. 5.
23. See Thomas Salmon, *An Impartial Examination of Bishop Burnet's History of His Own Times* (London, 1724), pp. 408 and 1064–5. The pagination in the edition I have read inexplicably starts at p. 291 and is then continuously and correctly paginated throughout.
24. Thomas Salmon, *A Review of the History of England* (London, 1722–4), II, pp. vi–vii.
25. See Linda Colley, *In Defiance of Oligarchy: the Tory Party, 1714–60* (Cambridge: Cambridge University Press, 1982).
26. See Salmon, *Review*, II 210. On the division between Tories and Jacobites in the period, see Colley, *In Defiance of Oligarchy*, p. 30ff.
27. Thomas Salmon, *The History of the Life and Reign of her Late Majesty Queen Anne* (London, 1740), pp. 249–50. For the Treaty of Utrecht, see pp. 483–4.

28. See, for example, Pocock, *Machiavellian Moment*, Chs. XII–XIV; J.G.A. Pocock, *Virtue, Commerce and History: Essays on Political Thought and History, Chiefly in the Eighteenth Century* (Cambridge: Cambridge University Press, 1985), esp. pp. 230–53; and Christine Gerrard, *The Patriot Opposition to Walpole: Politics, Poetry and National Myth, 1725–42* (Oxford: Oxford University Press, 1994).
29. Colley, *In Defiance of Oligarchy*, p. 115.
30. See Salmon, *Review*, I xi and 464 against contract; and II 64–5 against conquest (a doctrine Salmon associated with the Whig Burnet rather than Bohun – for which see Goldie, 'Bohun and *Jus Gentium*', pp. 582–3).
31. Thomas Salmon, *Considerations on the Bill for a General Naturalization* (London, 1748), pp. 54–5 and 56–7.
32. Ibid., p. 27.
33. See Gerrard, *Patriot Opposition to Walpole*, p. 9; and Colley, *In Defiance of Oligarchy*, pp. 86–9.
34. Salmon, *Review*, II 260–1.
35. Salmon, *Queen Anne*, pp. 424–5.
36. Ibid., pp. 484 and 505.
37. Thomas Salmon, *A New Geographical and Historical Grammar* (London, 1749), pp. 192–3 (climate) and p. 195 (people). Further references to *GHG* are in the text.
38. The best available study is Frank O'Gorman, *Voters, Patrons and Parties: the Unreformed Electorate in Hanoverian England, 1734–1832* (Oxford: Oxford University Press, 1989).
39. Gerrard, *Patriot Opposition to Walpole*, pp. 24–7.
40. See Monod, *Jacobitism and the English People*, Maps 4–6.
41. See Robert Harris, *A Patriot Press: National Politics and the London Press in the 1740s* (Oxford: Oxford University Press, 1993), pp. 60–1 and 75–6.
42. Salmon died in 1767.
43. Thomas Salmon, *Salmon's Geographical and Astronomical Grammar*, 13th edn. (London 1785), pp. 255 and 245.

8 The Denominational Politics of Travel-Writing: the Case of Tory Anglicans in the 1770s

1. Daniel Defoe, *A Tour Thro' the Whole Island of Great Britain*, G.D.H. Cole, ed. (London: Frank Cass, 1968), p. 78; see also p. 587. For the relation of geography and travel writing, see also Percy G. Adams, *Travel Literature and the Evolution of the Novel* (Lexington: University Press Of Kentucky, 1983), p. 77; and especially Charles Batten, *Pleasurable Instruction: Form and Convention in Eighteenth-Century Travel Literature* (Berkeley: University of California Press, 1978), p. 32.
2. See Batten, *Pleasurable Instruction*, p. 88.
3. See Pocock, *Virtue, Commerce and History*, Chapter 9 for Tucker; and Jack P. Greene, *Peripheries and Center: Constitutional Development in the Extended Polities of the British Empire and the United States, 1607–1788* (Athens, Ga.: University of Georgia Press, 1986) for the geographical angle on the constitutional issues at stake.

4. Josiah Tucker, *Instructions for Travellers* (Dublin, 1758), p. 4.
5. Ibid., pp. 14–15.
6. Tobias Smollett, *Travels through France and Italy*, Frank Felsenstein, ed. (Oxford: Oxford University Press, 1981), p. 296.
7. Tucker, *Instructions*, p. 5.
8. Ibid., p. 7.
9. For a good summary on these themes, see Jeremy Black, *The British Abroad: the Grand Tour in the Eighteenth Century* (Stroud: Alan Sutton, 1992), Chapters 10 and 11.
10. Thomas Pennant, *Some Account of London*, 2nd edn. (London, 1791), p. 216.
11. Thomas Pennant, *The Literary Life of the Late Thomas Pennant* (London, 1793), pp. 135–44.
12. Ibid., p. 32; see also [Thomas Pennant], *American Annals; or, Hints and Queries for Parlement [sic] Men* (Darlington, 1778).
13. Ibid., p. 13.
14. Thomas Pennant, *A Journey from London to the Isle of Wight* (London, 1801), I v.
15. Thomas Pennant, *A Tour in Wales* (London, 1781–84), II 92.
16. Ibid., I 1.
17. Thomas Pennant, *The Journey from Chester to London* (Dublin, 1783), p. 250.
18. Pennant, *Tour in Wales*, I 225. See E.P. Thompson, *Whigs and Hunters: the Origin of the Black Acts* (Harmondsworth: Penguin, 1975).
19. Pennant, *Journey from London to Isle of Wight*, I 44.
20. Pennant, *Tour in Wales*, II 48–9. See also Pennant, *Some Account of London*, p. 190.
21. Pennant, *Some Account of London*, pp. 80, 115 and 80.
22. Pennant, *Tour in Wales*, I 402. Pennant calls this the 'Glisseg valley'.
23. Thomas Pennant, *A Tour in Scotland, and Voyage to the Hebrides; MDCCLXXII* (Chester, 1774) II 279–80. On this theme, see Nigel Everett, *The Tory View of Landscape* (New Haven: Yale University Press, 1994).
24. Pennant, *Literary Life*, p. 11.
25. Thomas Pennant, *A Tour in Scotland; MDCCLXIX*, 3rd edn. (Warrington, 1774), p. 75; see also p. 158; and Pennant, *Tour in Scotland; MDCCLXXII*, I 211–12.
26. Pennant, *Tour in Scotland; MDCCLXIX*, p. 42.
27. Pennant, *Tour in Scotland; MDCCLXXII*, I 294.
28. Ibid., I 368.
29. Ibid., I 294.
30. Pennant, *Literary Life*, p. 21; and Thomas Pennant, *The View of Hindoostan* (London, 1798–1800), II ix.
31. Pennant, *Tour in Wales*, I 282; see also Pennant, *Journey from Chester*, p. 90.
32. Thomas Pennant, *Tour on the Continent, 1765*, G.R. de Beer, ed. (London: Ray Society, 1948).
33. Pennant, *Tour in Scotland; MDCCLXIX*, p. 47; and Pennant, *Tour in Scotland; MDCCLXXII*, II 78.
34. Pennant, *Tour on the Continent*, pp. 5–6. See also Pennant, *Tour in Scotland; MDCCLXXII*, II 129–30.
35. Pennant, *Tour in Scotland; MDCCLXIX*, p. 169n.
36. Ibid., pp. 155–6.

37. Pennant, *Tour in Scotland; MDCCLXXII*, II 373.
38. Ibid., I 255–6. For this view, see Kidd, *British Identities before Nationalism*, pp. 129–31.
39. John Wesley, *Letters*, J. Telford, ed. (London: Epworth Press, 1931), VII 305.
40. John Wesley, *Journal*, Nehemiah Curnock, ed. (London: Epworth Press Reprint, 1938), VI 109. Hereafter cited in the text as *Journal*.
41. On the persistence of belief in witches and ghosts, see R.D. Stock, *The Holy and the Daemonic from Sir Thomas Browne to William Blake* (Princeton: Princeton University Press, 1982).
42. Wesley, *Letters*, VII 28.
43. In this Wesley was by no means alone: see Adams, *Travel and the Evolution of the Novel*, p. 80; and Batten, *Pleasurable Instruction*, p. 1.
44. See *Journal*, VI 186–7 and VII 241 for Raynal; and John Wesley, *Sermons, Volume II*, Albert C. Outler, ed. (Nashville: Abingdon Press, 1985), pp. 486–7 for Wortley Montagu.
45. See also Wesley's other comments: *Journal*, VII 464, VII 521–2 and VIII 29; and Wesley, *Sermons*, p. 486. See also Bernard Smith, *European Vision and the South Pacific*, 2nd edn. (New Haven: Yale University Press, 1985), p. 44.
46. Wesley, *Sermons*, pp. 488, 491 and 490.
47. Ibid., p. 390.
48. Henry Rack, *Reasonable Enthusiast: John Wesley and the Rise of Methodism*, 2nd edn. (London: Epworth Press, 1992), p. 113.
49. For other examples, see: *Journal* II 425–6, III 486–7 and IV 113. This response to weather changes as the operation of particular providence was one Wesley shared with his fellow Evangelist, George Whitefield: see George Whitefield, *Journals* (Edinburgh: Banner of Truth Trust, 1960), pp. 263–5, 323 and 497–8.
50. See Wesley, *Sermons*, number 67, 'On divine providence' (pp. 534–50) and number 69 'The imperfection of human knowledge' (pp. 567–86).
51. Wesley, *Letters*, VIII 125.
52. This was a more general trend in Wesley's writing: see Rack, *Reasonable Enthusiast*, p. 111.
53. Although Wesley also argued for the essential similarity of bishops and presbyters: see Rack, *Reasonable Enthusiast*, p. 295. This was a position which was anathema to Anglicans of a more episcopal bent, as we have already seen in a geographical context in the work of Heylyn and Wells.
54. Johnson, *Dictionary*, *sub.* tory and whig.
55. James Boswell, *Life of Johnson*, G.B. Hill, ed., L.F. Powell, rev. (Oxford: Clarendon Press, 1934–50), I 431 and III 326. Hereafter cited in the text as *Life*.
56. As well as Boswell, see Sir John Hawkins, *The Life of Samuel Johnson*, Bertram Davis, ed. (London: Jonathan Cape, 1962), pp. 223–4; and Hester Piozzi, *Anecdotes of the Late Samuel Johnson*, in *Johnsonian Miscellanies*, G.B. Hill, ed. (Oxford: Clarendon Press, 1898), I 172.
57. See Donald Greene, *The Politics of Samuel Johnson*, 2nd edn. (Athens, Ga.: University of Georgia Press, 1990); and J.C.D. Clark, *Samuel Johnson: Literature, Religion and English Cultural Politics from the Restoration to Romanticism* (Cambridge: Cambridge University Press, 1994): these two writers have very different definitions of 'Tory', but neither have any hesita-

tion in labelling Johnson as one. The protracted debates in *The Age of Johnson*, volumes 7 and 8 (1996–7), and in *English Literary History*, 64.4, 1997 on Johnson's Jacobitism have not questioned this ascription.

58. Piozzi, *Anecdotes*, I 210; and Samuel Johnson, *Sermons*, Jean Hagstrum and James Gray, eds (New Haven: Yale University Press, 1978), p. xliv. On Johnson's Anglicanism, see, *inter alia*: Maurice Quinlan, *Samuel Johnson: a Layman's Religion* (Madison: Wisconsin University Press, 1964); Chester Chapin, *The Religious Thought of Samuel Johnson* (Ann Arbor: University of Michigan Press, 1968); James Gray, *Johnson's Sermons* (Oxford: Clarendon Press, 1972); Charles Pierce, *The Religious Life of Samuel Johnson* (London: Athlone Press, 1983); and Michael Suarez, 'Johnson's Christian Thought', in Greg Clingham, ed., *The Cambridge Companion to Samuel Johnson* (Cambridge: Cambridge University Press, 1997), pp. 192–208.
59. Wesley, *Journal*, VI 66–7n.
60. Pennant, *Some Account of London*, p. 216.
61. Samuel Johnson, *A Journey to the Western Islands of Scotland*, J.D. Fleeman, ed. (Oxford: Clarendon Press, 1985): this edition is referred to in the text as *Journey*.
62. Barbara Stafford, *Voyage into Substance: Art, Science and the Illustrated Travel Account, 1760–1840* (Cambridge, Mass.: MIT Press, 1984), p. 1.
63. Johnson was recorded by Mrs Piozzi (in *Thraliana: the Diary of Mrs Hester Lynch Thrale (later Mrs Piozzi), 1776–1809*, Katharine C. Balderston, ed. [Oxford: Clarendon Press, 1951]) as being sceptical of all exceptional natural events, Johnson himself admitting 'I did not give Credit a long time to the Earthquake at Lisbon' (p. 468).
64. Ibid, p. 716.
65. See Robert Mayhew, 'William Gilpin's Latitudinarian Picturesque', *Eighteenth-Century Studies* 33 (2000), pp. 349–66.
66. William Gilpin, *Observations, Relative Chiefly to Picturesque Beauty, made in the Year 1776, on Several Parts of Great Britain; Particularly the High-Lands of Scotland*, 2nd edn. (London, 1792), II 120–1.
67. Samuel Johnson, 'Review of Soame Jenyns's *Free Enquiry*', *Literary Magazine, or Universal Review* (1756–57), XIII, pp. 171–5, XIV pp. 251–3 and XV pp. 301–6, at XV, p. 306.
68. See also Johnson's comments at Arbroath (*Journey*, pp. 7–8).
69. I here use a similar method to develop a characterisation of Johnson's politics as a traveller to that deployed for Johnson's politics in general by J.C.D. Clark, 'Religious Affiliation and Dynastic Allegiance in Eighteenth-Century England: Edmund Burke, Thomas Paine and Samuel Johnson', *English Literary History*, 64 (1997), pp. 1029–67.
70. Hawkins, *Life of Samuel Johnson*, p. 45. See also pp. 116 and 243 for Johnson's debt to seventeenth-century Anglican thought.
71. This argument is rehearsed in greater detail in Robert Mayhew, 'Samuel Johnson's Intellectual Character as a Traveler: A Reassessment', *The Age of Johnson*, 10 (1999), pp. 35–67. For eighteenth-century High Churchmanship, see Cornwall, *Visible and Apostolic*, pp. 12, 86, 97 and p. 105ff on the concern for the visible church; and Peter B. Nockles, *The Oxford Movement in Context: Anglican High Churchmanship, 1760–1857* (Cambridge: Cambridge University Press, 1994), pp. 146 and 153 on episco-

pal order, pp. 184 and 208–13 on the importance of external forms and pp. 203–4 on the limits of rational proofs of God.

72. See Albert Lyles, *Methodism Mocked: the Satiric Reaction to Methodism in the Eighteenth Century* (London: Epworth Press, 1960).
73. Clarence Tracy, *A Portrait of Richard Graves* (Toronto: Toronto University Press, 1987), pp. 43–5.
74. Richard Graves, *The Spiritual Quixote: Or the Summer's Ramble of Mr Geoffry Wildgoose: a Comic Romance*, Clarence Tracy, ed. (London: Oxford University Press, 1967), p. 19. Hereafter cited as *Quixote* in the text.
75. *The Gentleman's Magazine*, 45, 1775, pp. 35–8 and 83–6, at pp. 36 and 85. See also Clark, *Samuel Johnson*, pp. 219–25 for the *Journey* as Johnson's Jacobite expedition and expiation.
76. Donald McNicol, *Remarks on Dr Samuel Johnson's Journey to the Hebrides* (London, 1779), p. 40. Further references to the *Remarks* are in the text.
77. There was a more general unease with positive responses to Catholic ruins: see Ian Ousby, *The Englishman's England: Taste, Travel and the Rise of Tourism* (Cambridge: Cambridge University Press, 1990), pp. 104 and 126. For joy at Catholicism being visibly in ruins as a traveller's trope, see Michael Charlesworth, 'The Ruined Abbey: Picturesque and Gothic Values', in Stephen Copley and Peter Garside, eds, *The Politics of the Picturesque: Literature, Landscape and Aesthetics since 1770* (Cambridge: Cambridge University Press, 1994), pp. 62–80.
78. *The Critical Review, Or Annals of Literature*, 49, 1779, pp. 241–8 at pp. 247–8.
79. Black, *The British Abroad*, p. 222.
80. See Clark, *Language of Liberty*, for the American war; and Young, *Religion and Enlightenment in Eighteenth-Century*, Ch. 2 for the Subscription controversy.

9 The Scottish Enlightenment and British Geography (I): Guthrie and Pinkerton, c.1770–1802

1. A version of this section appeared in the *Scottish Geographical Journal*, 115 (1999), pp. 19–34.
2. Full details are given by Harris, *Patriot Press*.
3. For discussions of this, see Duncan Forbes, *Hume's Philosophical Politics* (Cambridge: Cambridge University Press, 1975), Part III; Colin Kidd, *Subverting Scotland's Past: Scottish Whig Historians and the Creation of an Anglo-British Identity, 1689–c.1830* (Cambridge: Cambridge University Press, 1993); and many of the essays in Stewart Brown, ed., *William Robertson and the Expansion of Empire* (Cambridge: Cambridge University Press, 1997).
4. See Richard Sher, *Church and University in the Scottish Enlightenment: the Moderate Literati of Edinburgh* (Edinburgh: Edinburgh University Press, 1985), pp. 9–11; and Forbes, *Hume's Philosophical Politics*, p. 253, n. 3.
5. William Guthrie, *A General History of Scotland* (London, 1767–68) II 25 and II 265–7.
6. Kidd, *Subverting Scotland's Past*, pp. 86–90.
7. Guthrie, *History of Scotland*, X 269.
8. Ibid., VI 49–50 and 63 for iconoclasm and X 7–8 on Covenanters.
9. Ibid., X 303.

10. Ibid., X 354–76.
11. William Guthrie, *A New Geographical, Historical and Commercial Grammar*, 3rd edn. (London 1771), p. 235. I have used this and the second Dublin edition, which was an exact reprint of the first edition. They are substantially the same, although some textual alterations will be noted in my discussion. They are referred to hereafter in the text as *Grammar 2* and *3*.
12. It should be noted, however, that by this time English Tories had a long history of reverting to arguments about the Ancient Constitution whilst many English Whigs stressed that 1688 was a break in constitutional history; see Pocock, *Ancient Constitution and Feudal Law*, pp. 335–87. But the point that Guthrie could move between the two images to stress his impartiality still holds.
13. See Forbes, *Hume's Philosophical Politics*, pp. 253–8 for a discussion of this work. It was praised at the time by Henry Fielding in *The Jacobite's Journal and Related Writings*, W.B. Colley, ed. (Oxford: Clarendon Press, 1974), pp. 177 and 196.
14. On military valour and modern luxury in Scottish Enlightenment thought, see John Robertson, *The Scottish Enlightenment and the Militia Issue* (Edinburgh: Donald, 1985); and *idem*, 'The Scottish Enlightenment and the Limits of the Civic Tradition', in Istvan Hont and Michael Ignatieff, eds, *Wealth and Virtue: the Shaping of Political Economy in the Scottish Enlightenment* (Cambridge: Cambridge University Press, 1983), pp. 137–78.
15. For Scottish Enlightenment fears of Empire, see Sher, *Church and University*, pp. 271–2. The more widespread fears about empire intruding into political discussion are chronicled in Kathleen Wilson, *The Sense of the People: Politics, Culture and Imperialism in England, 1715–85* (Cambridge: Cambridge University Press, 1995), pp. 237, 246–8 and 277; and Peter Miller, *Defining the Common Good: Empire, Religion and Philosophy in Eighteenth-Century Britain* (Cambridge: Cambridge University Press, 1994), Chs. 3 and 4.
16. I discuss the sources Guthrie used below. Commentaries on stadial theory include: Knud Haakonssen, *The Science of the Legislator: the Natural Jurisprudence of David Hume and Adam Smith* (Cambridge: Cambridge University Press, 1981), Ch. 7; H.M. Höpfl, 'From Savage to Scotsman: Conjectural History in the Scottish Enlightenment', *Journal of British Studies*, 17 (1978), pp. 19–40; and Michael Ignatieff, 'John Millar and Individualism', in Hont and Ignatieff, *Wealth and Virtue*, pp. 317–43.
17. On the impact of Robertson's *History of Charles V*, see Karen O'Brien, 'Robertson's Place in the Development of Eighteenth-Century Narrative History'; and Richard Sher, '*Charles V* and the Book Trade: an Episode in Enlightenment Print Culture', both in Brown, ed., *Robertson and Expansion of Empire*, pp. 74–91 and 164–95 respectively.
18. See Adam Ferguson, *An Essay on the History of Civil Society*, Fania Oz-Salzberger, ed. (Cambridge: Cambridge University Press, 1995), pp. 116–18.
19. David Hume, 'Of the Rise and Progress of Arts and Sciences', in his *Essays, Moral, Political and Literary*, Eugene Miller, ed., 2nd edn. (Indianapolis: Liberty Press, 1987), pp. 111–37 at p. 121.
20. See Nicholas Phillipson, 'Providence and Progress: an Introduction to the Historical Thought of William Robertson', in Brown, ed., *Robertson and the Expansion of Empire*, pp. 55–73; and David Allan, *Virtue, Learning and*

the Scottish Enlightenment: Ideas of Scholarship in Early Modern History (Edinburgh: Edinburgh University Press, 1993), Ch. 4.

21. See Ferguson, *Essay*, p. 118.
22. See P.J. Marshall and Glyndwr Williams, *The Great Map of Mankind: British Perceptions of the World in the Age of Enlightenment* (London: Dent, 1982).
23. See in particular Hugh Blair, *A Critical Dissertation on the Poems of Ossian* (London, 1763). Ferguson also lavished praise on the poems. As Kidd points out, Macpherson was well aware of Scottish stadial theory, such that the way he tailored the oral sources at his disposal reflected prevailing theories about primitive societies: Kidd, *Subverting Scotland's Past*, pp. 221–31. The circular relationship between evidence and theory which the construction and reception of Ossian demonstrates was a recurrent pattern with stadial theorists: evidence about a society which did not fit its designated stage was often omitted: see Bruce Lenman, '"From Savage to Scot" via the French and the Spaniards: Principal Robertson's Spanish sources', in Brown, ed., *Robertson and the Expansion of Empire*, pp. 196–209.
24. W. Gordon East, 'An Eighteenth-Century Geographer. William Guthrie of Brechin', *Scottish Geographical Magazine*, 72 (1956), pp. 32–7; summary assessment on p. 37.
25. On this intellectual stance more generally, see Michael Kugler, 'Provincial Intellectuals: Identity, Patriotism, and Enlightened Peripheries', *The Eighteenth Century: Theory and Interpretation*, 37 (1996), pp. 156–73.
26. Both phrases are used by Sher, who provides a good summary of the politics of this position in *Church and University*, pp. 262–72. England also saw similar types of conservatism, often influenced by the Scots: see Pocock, *Virtue, Commerce and History*, pp. 157–91.
27. Colin Kidd, 'North Britishness and the Nature of Eighteenth-Century British Patriotisms', *Historical Journal*, 39 (1996), pp. 361–82.
28. For this shift, see John Dunn, 'From Applied Theology to Social Analysis: the Break between John Locke and the Scottish Enlightenment', in Hont and Ignatieff, eds, *Wealth and Virtue*, pp. 119–35.
29. This continuity in political thought which is reflected in the politics of geography is discussed by J.G.A. Pocock, 'Political Thought in the English-Speaking Atlantic, 1760–1790: (ii) Empire, Revolution and the End of Early Modernity', in J.G.A. Pocock *et al.*, eds, *The Varieties of British Political Thought, 1500–1800* (Cambridge: Cambridge University Press, 1993), pp. 283–317.
30. Joseph Priestley, *Lectures on History and General Policy* (Birmingham, 1788), p. 106; and *idem*, 'An Essay on a Course of Liberal Education for Civil and Active Life', in *Miscellaneous Observations Relating to Education* (Bath, 1778), pp. 260–1.
31. Priestley, *Lectures on History*, p. 528.
32. See Alan Tapper, 'Priestley on Politics, Progress and Moral Theology', in Haakonssen, ed. *Enlightenment and Religion*, pp. 272–86.
33. On this tradition see Anne Secord, 'Artisan Botany', in Jardine *et al.*, eds. *Cultures of Natural History*, pp. 378–93.
34. John Walker, *Elements of Geography, with the Principles of Natural Philosophy, and Sketches of General History* (Dublin, 1788), Postscript to the Advertisement, n.p.

35. Ibid.
36. Ibid., pp. 134–7 and 110–22.
37. Ibid., p. 3.
38. Ibid., A3v.
39. Ibid., Book II, p. 4.
40. The role of Dissent in radicalism is discussed in Haakonssen, ed. *Enlightenment and Religion*; Bradley, *Religion, Revolution and English Radicalism*; and Clark, *English Society 1688–1832*, Ch. 5.
41. 'Robert Heron' [i.e. John Pinkerton], *Letters of Literature* (London, 1785), pp. 340 and 341.
42. See 'Philostor' [i.e. John Pinkerton], 'Letters to the People of Great Britain on the Cultivation of their National History', *Gentleman's Magazine*, 58 (1788), February–December.
43. For Pinkerton as a historian, see Kidd, *Subverting Scotland's Past*, pp. 251–3.
44. John Pinkerton, *The History of Scotland from the Accession of the House of Stuart to that of Mary* (London, 1797), I 355–6.
45. See Sher, *Church and University*, for the Scottish Enlightenment; Pocock, *Virtue, Commerce and History*, pp. 274–94, for English Whig political language; and Frank O'Gorman *The Whig Party and the French Revolution* (London: Macmillan, 1967) for the parliamentary Whigs.
46. [Pinkerton], *Letters of Literature*, pp. 351–2.
47. The terms new and old Whig, while applying to the same groupings, took on new meanings after the French Revolution, the new Foxite Whigs now being associated with radicalism rather than corruption; the recasting of terms was effected in Burke's *An Appeal from the New to the Old Whigs* (1791). See Pocock, *Virtue, Commerce and History*, pp. 279–87.
48. John Pinkerton, *Literary Correspondence* (London, 1830), I 198–208.
49. [Pinkerton], 'Letters on History', p. 197.
50. William Doyle, *The Oxford History of the French Revolution* (Oxford: Oxford University Press, 1989), p. 161.
51. Pinkerton, *Correspondence*, I 354–5.
52. Pinkerton, *History of Scotland*, I x–xi.
53. Ibid., I 2.
54. Ibid., I 340.
55. John Pinkerton, *Modern Geography* (London, 1802) I iv. Further references in the text to *MG*.
56. See Linda Colley, *Britons: Forging the Nation 1707–1837* (New Haven: Yale University Press, 1992), pp. 33–5 and 88–90.
57. John Pinkerton, *Recollections of Paris, in the Years 1802–3–4–5* (London, 1806) I viii, I 138 and I 140.
58. Ibid., I 288–9. There was a more general backlash against concepts of the 'noble savage' after Cook's death in 1779: see Kathleen Wilson, 'The Island Race: Captain Cook, Protestant Evangelism and the Construction of English National Identity, 1760–1800' in Claydon and McBride, eds, *Protestantism and National Identity*, pp. 265–90 at pp. 279–85.
59. John Pinkerton, ed., *A General Collection of the Best and Most Interesting Voyages and Travels* (London, 1808–14).
60. See A.A. Wilcock, '"The English Strabo": the Geographical Publications of John Pinkerton', *Transactions of the Institute of British Geographers*, 61 (1974), pp. 35–45, at pp. 36–8; and Pinkerton, *Correspondence*, II 229–30.

61. A similar trajectory can be traced in the political thought of Edward Gibbon: see David Womersley, 'Gibbon's Unfinished History: the French Revolution and English Political Vocabularies', *Historical Journal*, 35 (1992), pp. 63–89.
62. Pocock, *Virtue, Commerce and History*, p. 280.

10 On the Cusp of Modern Geography: Fieldwork and Textuality in the Career of James Rennell, 1764–1830

1. Clements Markham, *Major James Rennell and the Rise of Modern English Geography* (London: Cassell, 1895), p. 105. Such positions are fully chronicled in O'Gorman, *Whig Party and French Revolution.*
2. 'An Old Englishman' [James Rennell], *War with France the Only Security of Britain* (London, 1794), p. 3.
3. See Michael Bravo, 'Precision and Curiosity in Scientific Travel: James Rennell and the Orientalist Geography of the New Imperial Age (1760–1830)' in Jas Elsner and Joan-Pau Rubiés, eds, *Voyages and Visions: Towards a Cultural History of Travel* (London: Reaktion, 1999), pp. 162–83. See also Brian Hudson, 'The New Geography and the New Imperialism: 1870–1918', *Antipode*, 9.2 (1977), pp. 12–19.
4. Richard Drayton, 'Knowledge and Empire' in P.J. Marshall, ed. *The Oxford History of the British Empire: Volume 2: The Eighteenth Century* (Oxford: Oxford University Press, 1998), pp. 231–52 at p. 244.
5. See James Rennell, *Journals*, T.H.D. La Touche, ed., in *Memoirs of the Asiatic Society of Bengal*, 3 (1910–14) (Calcutta, 1914), pp. 95–248.
6. These claims can be verified by looking at Rennell's *Memoirs of a Map of Hindoostan* (London, 1783), which detailed the collations and weighing of evidence by which he had constructed a map of India. The rapid expansion of the work in its second (1788) and third (1793) editions shows Rennell incorporating new data to improve cartographic accuracy. Rennell's work for the Africa Association followed the same pattern, using the information from Mungo Park and Frederick Horneman to fill in details about the geography of the African interior: see Mungo Park, *Travels in the Interior of Africa* (London, 1799), 'Appendix: Geographical Illustrations of Africa', by Rennell; and *African Researches* (London, 1799?), 'Geographical Illustrations of Mr Horneman's Route', pp. 143–208.
7. Rennell saw his life's work as leading to a comparative geography of Asia, a project which he sketched, but at his death he left only a fragment of it, which was published posthumously as *A Treatise on the Comparative Geography of Western Asia* (London, 1831).
8. See James Rennell, *An Investigation of the Currents of the Atlantic Ocean* (London, 1832); *The Geographical System of Herodotus Examined* (London, 1800), Section XVIII; and *Map of Hindoostan*, 1793 ed., 'An Account of the Ganges and Burrumpooter Rivers'.
9. Rennell, *Geographical System of Herodotus*, Section XXVI. Rennell's method was very similar to that of the French 'rational geographers', centred on the Académie des Inscriptions and led by D'Anville: see Guido Abbattista, 'Establishing the "Order of Time and Place": "Rational Geography", French Erudition and the Emplacement of History in Gibbon's Mind' in David

Womersley, ed., *Edward Gibbon: Bicentenary Essays*, Studies on Voltaire and the Eighteenth Century, 335 (Oxford: Voltaire Foundation, 1997), pp. 45–72, esp. pp. 54–65.

10. See Anne Goldgar, *Impolite Learning: Conduct and Community in the Republic of Letters, 1680–1750* (New Haven: Yale University Press, 1995).
11. Cambridge University Library, *Minutes of the African Association*: Add. 7086, fol.4r-v.
12. Ibid., Add. 7087, fol.6v.
13. See also Anne Godlewska, 'Napoleon's Geographers (1797–1815): Imperialists and Soldiers of Modernity', in Godlewska and Smith, eds, *Geography and Empire*, pp. 31–53, at pp. 32–9.
14. For a similar nexus of assumptions at the same time, see William Ravenhill, 'The honourable Robert Edward Clifford, 1767–1817: a Cartographer's Response to Napoleon', *Geographical Journal*, 160 (1994), pp. 159–72. See also Bravo, 'Precision and Curiosity', pp. 175–6.
15. Rennell, *Comparative Geography of Asia*, I 340.
16. Rennell, *Map of Hindoostan*, 1783 ed., p. 26. See also Bravo, 'Precision and Curiosity', p. 179.
17. See P.J. Marshall, *East India Fortunes: the British in Bengal in the Eighteenth Century* (Oxford: Oxford University Press, 1976), pp. 8–9.
18. Rennell, *Map of Hindoostan*, 1793 ed., pp. xxii and xxviii.
19. Park, *Travels in Africa*, Appendix, p. xcii.
20. Rennell, *Currents of Atlantic Ocean*, p. 6.
21. Quoted in Markham, *Rennell and Rise of Geography*, p. 148.
22. Rennell, *Geographical System of Herodotus*, p. 7.
23. Ibid., p. 525.
24. See Colley, *Britons*, pp. 367–70.
25. Rennell, *Map of Hindoostan*, 1793 ed., p. cv.
26. Ibid., p. lxxviii.
27. Ibid., pp. cv–cvi.
28. On limited empire, see Marshall and Williams, *Great Map of Mankind*, pp. 156–7; and Geoffrey Carnall, 'Robertson and Contemporary Images of India', in Brown, ed., *Robertson and Expansion of Empire*, pp. 210–30, at p. 229.
29. Matthew Edney, *Mapping an Empire: the Geographical Construction of British India, 1765–1843* (Chicago: Chicago University Press, 1997), pp. 100 and 104; see also pp. 29–30, 97–103 and 133–6.
30. Rennell, *Geographical System of Herodotus*, Section XXIV.
31. See Grafton, *Defenders of the Text*; and Abbattista, 'Establishing the "Order of Time and Place"'.
32. Rennell, *Observations on the Topography of the Plain of Troy* (London, 1814), pp. 103–4.
33. See, for example, Rennell, *Map of Hindoostan*, 1783 ed., p. 6; and also James Rennell, *The Marches of the British Armies in the Peninsula of India, during the Campaigns of 1790 and 1791* (London, 1792), esp. pp. 52–9.
34. James Rennell, *Illustrations (Chiefly Geographical) of the History of the Expedition of Cyrus* (London, 1816) pp. vii–viii. See Anne Godlewska, 'Map, Text and Image. The Mentality of Enlightened Conquerors: a New Look at the *Description de l'Egypte*', *Transactions of the Institute of British Geographers*, N.S. 20 (1995), pp. 5–28.

35. Ibid., p. 286.
36. Ibid., p. 287.
37. Rennell, *Comparative Geography of Western Asia*, I lv.
38. Quoted in Baker, *History of Geography*, p. 155.
39. Markham, *Rennell and Rise of Modern Geography*, pp. 10–11.
40. Ibid., p. 93.
41. James Rennell, 'On the Topography of Ancient Babylon,' in Claudius Rich, *Narrative of a Journey to the Site of Babylon* [1815] (London, 1839).
42. Baker, *History of Geography*, p. 156.

11 The Scottish Enlightenment and British Geography (II): James Bell and J.R. McCulloch, 1830–50

1. See Richard Sher and Andrew Hook, 'Introduction: Glasgow and the Enlightenment', Andrew Hook and Richard Sher, eds, *The Glasgow Enlightenment* (East Linton: Tuckwell Press, 1995), pp. 1–17.
2. James Bell, *A System of Geography, Popular and Scientific. Or, A Physical, Political, and Statistical Account of the World and its Various Divisions* (Glasgow, 1832), VI 501. All further references are incorporated in the text as *System.*
3. Robert Stuart, *Party and Politics, 1830–1852* (London: Macmillan, 1989), p. 61. Stuart is referring to 1841, but his characterisation certainly has weight a decade previously when Bell wrote.
4. Clark, *English Society, 1688–1832*, Ch. 6.
5. For which, see: Eric Evans, *Britain before the Reform Act: Politics and Society, 1815–1832* (London: Longman, 1989); Boyd Hilton, *Corn, Cash, Commerce: the Economic Policies of the Tory Governments 1815–1830* (Oxford: Clarendon Press, 1977); and Norman Gash, *Lord Liverpool: the Life and Political Career of Robert Banks Jenkinson, Second Earl of Liverpool 1770–1828* (London: Weidenfeld and Nicolson, 1984).
6. John McCaffrey, 'Political Issues and Developments' in W. Hamish Fraser and Irene Maver, eds, *Glasgow: Volume II: 1830 to 1912* (Manchester: Manchester University Press, 1996), pp. 186–226 at p. 187.
7. There is, of course, implied criticism of the British government of Ireland here, but Bell seems to emphasise that the Irish have rendered government impossible by their anti-Protestant hostility.
8. Sher and Hook, 'Introduction', p. 13.
9. Boyd Hilton, *The Age of Atonement: the Influence of Evangelicalism on Social and Economic Thought, 1785–1865* (Oxford: Clarendon Press, 1988); see also A.M.C. Waterman, *Revolution, Economics and Religion: Christian Political Economy, 1798–1833* (Cambridge: Cambridge University Press, 1991).
10. Bowen, *Empiricism and Geographical Thought*, p. 155.
11. See Robert A. Stafford, *Scientist of Empire: Sir Roderick Murchison, Scientific Exploration and Victorian Imperialism* (Cambridge: Cambridge University Press, 1989); and Morag Bell, Robin Butlin and Michael Heffernan, eds, *Geography and Imperialism, 1820–1940* (Manchester: Manchester University Press, 1995). Empire itself, of course, was not a single phenomenon: see Felix Driver, 'Geography's Empire: Histories of Geographical Knowledge', *Society and Space*, 10 (1992), pp. 23–40.

12. James Bell, *A New and Comprehensive Gazetteer of England and Wales* (Glasgow, 1834–5), I ii.
13. Ibid., I 113–14.
14. *Camden's Britannia*, Gibson, ed. (London, 1695), col. 270, note k.
15. Lawrence Goldman, 'The Origins of British "Social Science": Political Economy, Natural Science and Statistics, 1830–1835', *Historical Journal*, 26 (1983), pp. 587–616.
16. See D.P. O'Brien, *J.R. McCulloch: a Study in Classical Economics* (London: George Allen and Unwin, 1970), p. 16 for Scottish heritage and p. 91 for McCulloch as an early member of the Statistical Society.
17. For the doctrines of classical political economy generally, see D.P. O'Brien, *The Classical Economists* (Oxford: Clarendon Press, 1975).
18. Namely: J.R. McCulloch, *A Dictionary, Geographical, Statistical, and Historical, of the Various Countries, Places, and Principal Natural Objects in the World* (London, 1841), hereafter referred to in the text as *Dictionary*; J.R. McCulloch *et al.*, *A Statistical Account of the British Empire* (London, 1837); and J.R. McCulloch, *A Dictionary, Practical, Theoretical, and Historical, of Commerce and Commercial Navigation* (London, 1832).
19. See J.R. McCulloch, *A Catalogue of Books, the Property of a Political Economist* (London, 1862).
20. McCulloch, *Statistical Account*, I v.
21. McCulloch, *Catalogue*, p. 233.
22. Peter Mandler, *Aristocratic Government in the Age of Reform: Whigs and Liberals, 1830–1852* (Oxford: Clarendon Press, 1990), p. 119; see also Biancamaria Fontana, *Rethinking the Politics of Commercial Society: the* Edinburgh Review *1802–1832* (Cambridge: Cambridge University Press, 1985); and George Pottinger, *Heirs of the Enlightenment: Edinburgh Reviewers and Writers 1800–1830* (Edinburgh: Scottish Academic Press, 1992).
23. J.R. McCulloch, 'Census of the Population – Law of Mortality, etc', *Edinburgh Review*, 49 (1829), pp. 1–34 at pp. 33–4.
24. Cambridge fares no better: 'this [Cambridge] and the other university are institutions that belong to a different age'. *Dictionary*, I 513.
25. McCulloch, *Catalogue*, p. 55.
26. J.R. McCulloch, *A Discourse on the Rise, Progress, Peculiar Objects, and Importance, of Political Economy* (Edinburgh, 1824), p. 75. See also a similar passage on the use of historians and travellers to the political economist at pp. 17–18.
27. McCulloch, *Catalogue*, p. 319; see also pp. 11 and 117 for similar comments.
28. McCulloch, *Dictionary of Commerce and Commercial Navigation*, p. 945.
29. McCulloch, *Rise of Political Economy*, pp. 72–4.
30. J.R. McCulloch, *The Principles of Political Economy*, 4th edn. (Edinburgh, 1849), pp. 262–3.
31. O'Brien, *McCulloch*, pp. 88–9.
32. O'Brien, *Classical Economists*, p. 54.
33. McCulloch, *Statistical Account*, I vi.
34. See Gary Langer, *The Coming of Age of Political Economy, 1815–1825* (New York: Greenwood Press, 1987), p. 55; and Hilton, *Corn, Cash, Commerce*, p. 312.
35. McCulloch, *Catalogue*, pp. 347 and 205–6.

36. McCulloch, *Commercial Dictionary*, p. 493. Dundas's dealings with respect to the East India Company are discussed more favourably in Michael Fry, *The Dundas Despotism* (Edinburgh: Edinburgh University Press, 1992), pp. 125–8.
37. Fry, *Dundas Despotism*, p. 315.
38. O'Brien, *McCulloch*, pp. 33 and 57.
39. See Leslie Mitchell, *Holland House* (London: Duckworth, 1980); and Leslie Mitchell, *Charles James Fox* (Oxford: Oxford University Press, 1992).
40. McCulloch, *Catalogue*, p. 160. See also p. 6.
41. Mandler, *Aristocratic Government*, esp. Chapter 3; Richard Brent, *Liberal Anglican Politics: Whiggery, Religion, and Reform, 1830–41* (Oxford: Clarendon Press, 1987), p. 22.
42. McCulloch, *Catalogue*, p. 347.
43. This was the position of the *Edinburgh Review*: see Stefan Collini *et al.*, *That Noble Science of Politics: a Study in Nineteenth-Century Intellectual History* (Cambridge: Cambridge University Press, 1983), p. 98; and Fontana, *Rethinking the Politics of Commercial Society*, Ch. 5.
44. Stuart, *Party and Politics*, p. 61.
45. Fontana, *Rethinking the Politics of Commercial Society*, p. 7.
46. Member of the Royal Geographical Society, ed., *A Gazetteer of the World, or Dictionary of Geographical Knowledge, Compiled from the Most Recent Authorities, and Forming a Complete Body of Modern Geography, Physical, Political, Statistical, Historical, and Ethnographical* (Edinburgh, 1851–56) IV 282.
47. This term is used of McCulloch by Mandler, *Aristocratic Government*, p. 113.
48. Walter Blackie, *The Imperial Gazetteer* (Glasgow, 1855–68), I 493.
49. Ibid., I 505.
50. Ibid., I 507.
51. See E.R. Norman, *Anti-Catholicism in Victorian England* (London: George Allen and Unwin, 1968).
52. Blackie, *Imperial Gazetteer*, I 507. On the idiom of 'character', see: Stefan Collini, *Public Moralists: Political Thought and Intellectual Life in Britain, 1850–1930* (Oxford: Clarendon Press, 1991), Ch. 3; and John Burrow, *Whigs and Liberals: Continuity and Change in English Political Thought* (Oxford: Clarendon Press, 1988), Ch. 4.
53. *Gazetteer of the World*, II 63 and I 445.
54. For which, see: Felix Driver, *Geography Militant: Cultures of Exploration in the Age of Empire* (Oxford: Blackwell, 2000).

12 Coda: Halford Mackinder and the Empire of 'New' Political Geography, c.1887–1919

1. See especially H.C.G. Matthew, *The Liberal Imperialists: the Ideas and Politics of a Post-Gladstonian Élite* (Oxford: Oxford University Press, 1971), p. 224; and E.H.H. Green, *The Crisis of Conservatism: the Politics, Economics and Ideology of the British Conservative Party, 1880–1914* (London: Routledge, 1995), pp. 13–14.
2. On the Liberal League, see Matthew, *Liberal Imperialists*, pp. 86–91; on the Compatriots, see Green, *Crisis of Conservatism*, p. 161: both were collectivist think tanks.

3. Full biographical details on Mackinder are to be found in W.H. Parker, *Mackinder: Geography as an Aid to Statecraft* (Oxford: Clarendon Press, 1982); Brian Blouet, *Halford Mackinder: a Biography* (College Station, Texas: Texas A&M Press, 1987); and Gerry Kearns, 'Halford John Mackinder, 1861–1947', *Geographers: Biobibliographical Studies*, 9 (1985), pp. 71–86. On tariff reform, see Alan Sykes, *Tariff Reform in British Politics, 1903–1913* (Oxford: Clarendon Press, 1979).
4. David Brooks, *The Age of Upheaval: Edwardian Politics, 1899–1914* (Manchester: Manchester University Press, 1995), p. 6. See also G.R. Searle, *The Liberal Party: Triumph and Disintegration, 1886–1929* (London: Macmillan, 1992), p. 29.
5. Halford Mackinder, 'Man-Power as a Measure of National and Imperial Strength', *National Review*, 45 (1905), pp. 136–43 at p. 143. This was a lecture delivered to the conservative Compatriot club.
6. See Matthew, *Liberal Imperialists*, p. 200; and Frans Coetzee, *For Party or Country: Nationalism and the Dilemmas of Popular Conservatism in Edwardian England* (Oxford University Press: New York, 1990), p. 22.
7. Matthew, *Liberal Imperialists*, pp. 204–6 and 228; Coetzee, *For Party or Country*, p. 38; and Green, *Crisis of Conservatism*, p. 67.
8. G.R. Searle, *The Quest for National Efficiency: a Study in British Politics and Political Thought, 1899–1914* (Berkeley: University of California Press, 1971), pp. 1, 2 and 54.
9. See Searle, *Quest for National Efficiency*, pp. 13–14; Matthew, *Liberal Imperialists*, p. 228; and Geoffrey Sherington, *English Education, Social Change and War, 1911–20* (Manchester: Manchester University Press, 1981), p. 10.
10. See Matthew, *Liberal Imperialists*, p. vii; and Green, *Crisis of Conservatism*, Ch. 5, for national economics.
11. George Bernstein, *Liberalism and Liberal Politics in Edwardian England* (Boston: Allen and Unwin, 1986), p. 35, goes as far as to suggest that Home Rule was the only real distinguishing feature between Liberals and Liberal Imperialists.
12. See Matthew, *Liberal Imperialists*, p. 216; and Coetzee, *For Party or Country*, p. 7 and *passim* on the Navy League.
13. See Green, *Crisis of Conservatism*, p. 67; and Coetzee, *For Party or Country*, p. 38.
14. See Matthew, *Liberal Imperialists*, pp. 127, p. 140 and p. 227; Green, *Crisis of Conservatism*, pp. 69 and 179; Searle, *Quest for National Efficiency*, pp. 80–1; and Sykes, *Tariff Reform*, p. 117.
15. See especially Mackinder's paper on manpower (above, note 5), and its expanded version: Halford Mackinder, *Money-Power and Man-Power: the Underlying Principles rather than the Statistics of Tariff Reform* (London: Simpkin, Marshall, Hamilton, Kent and Co., 1906).
16. Halford Mackinder, 'On the Scope and Methods of Geography', *Proceedings of the Royal Geographical Society*, N.S. 9 (1887), pp. 141–74, at p. 141.
17. Halford Mackinder, 'The Geographical Pivot of History', *Geographical Journal*, 23 (1904), pp. 421–44, at p. 422. On the intellectual context of this theme, see Gerry Kearns, 'Closed Space and Political Practice: Frederick Jackson Turner and Halford Mackinder', *Society and Space*, 1 (1984), pp. 23–34.

18. Mackinder, 'Scope and Methods', p. 143.
19. Mackinder, *Money-Power and Man-Power*, esp. pp. 13–14. The subtitle of this work, 'the underlying principles rather than the statistics of tariff reform', neatly encapsulates Mackinder's hostility to statistics.
20. Halford Mackinder, 'The Physical Basis of Political Geography', *Scottish Geographical Magazine*, 6 (1890), pp. 78–84, at p. 83.
21. Halford Mackinder, 'Geography, an Art and a Philosophy', *Geography*, 27 (1942), pp. 122–30 at p. 126; see also Halford Mackinder, 'The Development of Geographical Teaching out of Nature Study', *Geographical Teacher*, 2 (1904), pp. 191–7 at p. 192; and Halford Mackinder, 'Geography in Education', *Geographical Teacher*, 2 (1903), pp. 95–101 at p. 96.
22. Halford Mackinder, *The Teaching of Geography and History: a Study in Method* (London: George Philip and Son, 1914), p. 44.
23. Ibid., p. 45; see also Mackinder, 'Art and Philosophy', p. 129.
24. Mackinder, 'Geography in Education', p. 96.
25. See Avril Maddrell, 'Discourses of Race and Gender and the Comparative Method in Geography School Texts, 1830–1918', *Society and Space*, 16 (1998), pp. 81–103, esp. pp. 97–8.
26. Mackinder, 'Scope and Methods', p. 143.
27. Mackinder, 'Development of Geographical Teaching', p. 192; see also Mackinder, *Teaching of Geography and History*, p. 3. On Mackinder and visualisation, see James Ryan, 'Visualizing Imperial Geography: Halford Mackinder and the Colonial Office Visual Instruction Committee, 1902–11,' *Ecumene*, 1 (1994), pp. 157–76; and Gearóid Ó Tuathail, *Critical Geopolitics: the Politics of Writing Global Space* (London: Routledge, 1996), Chapter 3.
28. Halford Mackinder, 'The Teaching of Geography from an Imperial Point of View, and the use which could and should be made of Visual Instruction', *Geographical Teacher*, 6 (1912), pp. 79–86 at p. 81; and Mackinder, 'Development of Geographical Teaching', p. 193. See also Mackinder, 'Scope and Methods', p. 159.
29. See for example, Green, *Crisis of Conservatism*, p. 195.
30. Coetzee, *For Party or Country*, p. 97.
31. See Mackinder, 'Teaching Geography from an Imperial Point of View', p. 83.
32. Halford Mackinder, 'Geography as a Pivotal Subject in Education', *Geographical Journal*, 57 (1921), pp. 376–84, at p. 377.
33. Mackinder, *Teaching of Geography and History*, p. 45; see also Halford Mackinder, 'The Teaching of Geography and History as a Combined Subject', *Geographical Teacher*, 7, (1913), pp. 4–9, at 6 and 8.
34. Mackinder, 'Development of Geographical Teaching', pp. 191–2.
35. Mackinder, 'Scope and Methods', p. 154.
36. Mackinder, 'Teaching Geography from an Imperial Point of View', p. 82; Mackinder, 'Geography as Pivotal Subject', p. 383; and Mackinder, 'Teaching Geography from an Imperial Point of View', p. 86.
37. Mackinder, 'Scope and Methods', p. 160. See also Mackinder, 'Geography, Art and Philosophy', p. 129 on the waning of 'greats' and geography's role as a 'modern alternative'.
38. Mackinder, 'Physical Basis of Political Geography', p. 83.
39. Halford Mackinder, *Britain and the British Seas* (London: William Heinemann, 1902); Mackinder, 'Geographical Pivot', Figure 5, p. 435; and

Halford Mackinder, *India. Eight Lectures prepared for the Visual Instruction Committee of the Colonial Office* (London: George Philip and Son, 1910).

40. Parker, *Mackinder*, p. 131.
41. Mackinder, 'Physical Basis of Political Geography', p. 78; Mackinder, 'Geographical Pivot', p. 422.
42. Ó Tuathail, *Critical Geopolitics, pp.* 99–108.
43. Mackinder, *Britain and the British Seas*, p. 20, accompanying Figures 12 and 13.
44. Ibid., p. 32, Figure 21.
45. See Parker, *Mackinder*, pp. 138–9.
46. Mackinder, 'Geographical Pivot', pp. 432–3.
47. Halford Mackinder, 'The Great Trade Routes', *Journal of the Institute of Bankers*, 21 (1900), pp. 1–6, 137–55 and 266–73, at p. 147. See also Mackinder, *Britain and the British Seas*, p. 12 on 'the simple physical fact' of the 'unity of the ocean'.
48. Mackinder, 'Great Trade Routes', pp. 153–4; Mackinder, 'Geographical Pivot', pp. 434–6.
49. Mackinder, *Britain and the British Seas*, p. 342; see also Halford Mackinder, *Democratic Ideals and Reality: a Study in the Politics of Reconstruction* (London: Constable and Company, 1919), p. 171 for an attack on free-market principles.
50. Mackinder, 'Great Trade Routes', p. 153; see also Halford Mackinder, *The Nations of the Modern World: an Elementary Study in Geography* (London: George Philip and Son, 1911), pp. 292–4; and Halford Mackinder, *The Rhine: Its Valley and History* (London: Chatto and Windus, 1908), esp. pp. 2–3.
51. Mackinder, 'Geographical Pivot', pp. 442–3.
52. Mackinder, *Britain and the British Seas*, p. 343; see also 'Geographical Pivot', p. 422. On Darwinism and Mackinder, see Kearns, 'Closed Space and Political Practice', pp. 25–8.
53. Mackinder, *Nations of the Modern World*, pp. 316–17; see also Mackinder, 'Geography from an Imperial Point of View', p. 79. Such views were central to tariff reform: see Green, *Crisis of Conservatism*, pp. 198–9.
54. Mackinder, *Democratic Ideals and Reality*, esp. p. 247 and pp. 263–7.
55. Mackinder, *Britain and the British Seas*, p. 358. On the navy as 'an integral part of our national organism', see Mackinder, *Nations of the Modern World*, pp. 288–90; and Halford Mackinder, *The Modern British State: an Introduction to the Study of Civics* (London: George Philip and Son, 1914), Ch. XVII.
56. Mackinder's 'Scope and Methods' paper was delivered to the Royal Geographical Society to explicate precisely what his usage of the epithet 'new' meant.
57. Parker, *Mackinder*, p. 108.
58. See Martin and James, *All Possible Worlds*, p. 205 and Mackinder as a paradigm setter, p. 208.
59. Livingstone, *Geographical Tradition*, p. 191.
60. See for example: Martin Ira Glassner and Harm de Blij, *Systematic Political Geography*, 3rd edn. (New York: John Wiley, 1980), p. 266; Geoffrey Parker, *Geopolitics: Past, Present and Future* (London: Pinter, 1998), p. 21; Joe Painter, *Politics, Geography and 'Political Geography'* (London: E.J. Arnold, 1995),

p. 22, which puts Mackinder's status as a 'great thinker' in scare quotes but does not alter the assessment; and Peter Taylor, *Political Geography: World-Economy, Nation-State and Locality*, 2nd ed. (London: Longman, 1989), p. 46.

61. Ó Tuathail, *Critical Geopolitics*, p. 22.

13 Enlightenments and Geography: Continuity and Change in the Politics of Early-Modern British Geography, c.1550–1850

1. Clark, *English Society, 1688–1832*, pp. 348 and 375.
2. See Waterman, 'Nexus between Theology and Political Doctrine'.
3. J.G.A. Pocock, 'Clergy and Commerce. The Conservative Enlightenment in England', in *L'Età dei Lumi: Studi Storici Sul Settecento Europeo in Onore di Franco Venturi* (Napoli: Jovene, 1985), pp. 525–62, at pp. 528–9 and p. 541; *idem*, 'Conservative Enlightenment and Democratic Revolutions: The American and French Cases in British Perspective', *Government and Opposition*, 24 (1989), pp. 81–105; and *idem*, *Barbarism and Religion* (Cambridge: Cambridge University Press, 1999). See also Young, *Religion and Enlightenment in Eighteenth Century England*; Roy Porter, 'The Enlightenment in England', in Roy Porter and Mikulaus Teich, eds, *The Enlightenment in National Context* (Cambridge: Cambridge University Press, 1981), pp. 1–18; and Gascoigne, *Joseph Banks and the English Enlightenment*, Ch. 2.
4. See Edward A. and Lillian D. Bloom, *Joseph Addison's Sociable Animal: In the Market Place, on the Hustings, in the Pulpit* (Providence: Brown University Press, 1971); and Lawrence Klein, *Shaftesbury and the Culture of Politeness: Moral Discourse and Cultural Politics in Early Eighteenth-Century England* (Cambridge: Cambridge University Press, 1994).
5. See Lawrence Klein, 'Politeness for Plebes. Consumption and Social Identity in Early Eighteenth-Century England', in Ann Bermingham and John Brewer, eds, *The Consumption of Culture 1600–1800: Image, Object, Text* (London: Routledge, 1995), pp. 362–82.
6. This argument should not be taken too far by itself, given that many radical figures certainly did rest their analyses on historical investigation: see Champion, *Pillars of Priestcraft Shaken*, Chapters 4–7.
7. J.G.A. Pocock, 'Empire, State and Confederation: the War of American Independence as a Crisis of Multiple Monarchy', in John Robertson, ed., *A Union for Empire: Political Thought and the British Union of 1707* (Cambridge: Cambridge University Press, 1995), pp. 318–48, at p. 320. For shifting imperial conceptions, see also C.A. Bayly, *Imperial Meridian: the British Empire and the World, 1780–1830* (London: Longman, 1989); and P.J. Marshall, 'Britain without America – A Second Empire?' in Marshall, ed., *Oxford History of the British Empire: Eighteenth Century*, pp. 576–95.
8. Ibid., p. 347.
9. David Allan, 'Protestantism, Presbyterianism and National Identity in Eighteenth-Century Scottish History' in Claydon and McBride, eds, *Protestantism and National Identity*, pp. 182–205, at p. 195; see also John Dwyer, *Virtuous Discourse: Sensibility and Community in Late Eighteenth-Century Scotland* (Edinburgh: Edinburgh University Press, 1987).

10. Kidd, 'North Britishness', p. 365. As will be apparent, I agree with Kidd about the fundamental differences created by the differing established churches of England and Scotland, but not with his Erastian characterisation of the Anglican Church.
11. As we are starting to see that Irish Protestant identities diverged from English ones (see the chapters by McBride and Barnard in Claydon and McBride, eds. *Protestantism and National Identity*), so the ways in which Irish geographical writings questioned the English politics of geography demand more attention.
12. Kidd, 'North Britishness', pp. 369 and 375.
13. John Robertson, 'Universal Monarchy and the Liberties of Europe: David Hume's Critique of an English Whig Doctrine', in Nicholas Phillipson and Quentin Skinner, eds, *Political Discourse in Early-Modern Britain* (Cambridge: Cambridge University Press, 1993), pp. 349–73, at p. 349; see also Kidd, 'North Britishness', pp. 374–7.
14. Ibid., p. 371.
15. Bayly, *Imperial Meridian*, p. 152.
16. Ibid., pp. 121–6.
17. See Cormack, *Charting an Empire*; and Canny, ed., *Oxford History of the British Empire: the Origins of Empire.*
18. See in particular, Pocock, 'Political Thought in the English-Speaking Atlantic'.
19. Clark, *English Society, 1688–1832*, p. 410.
20. See Richard Yeo, *Defining Science: William Whewell, Natural Knowledge and Public Debate in Early-Victorian Britain* (Cambridge: Cambridge University Press, 1993), Ch. 2; and Frank Turner, *Contesting Cultural Authority: Essays in Victorian Intellectual Life* (Cambridge: Cambridge University Press, 1993), Ch. 7.
21. Something which applied more generally: see Roy Porter, 'The Scientific Revolution and the Universities', in H. De Ridder-Symoens, ed., *A History of the University in Europe, Volume II: Universities in Early Modern Europe (1500–1800)* (Cambridge: Cambridge University Press, 1996), pp. 531–62, at pp. 542–4.

Bibliography

Manuscript sources

Cambridge University Library
Bohun Collection, Add.4521.237.
Minutes of the African Association, Add.7086–7.
Sel 3.237.357.

Primary sources

Abbot, George. *A Briefe Description of the Whole Worlde*. 3rd edn (London, 1605).
African Researches (London, 1799?).
Atterbury, Francis. *The Rights, Powers, and Privileges of an English Convocation*. 2nd edn (London, 1701).
Bailey, Nathan. *An Universal Etymological Dictionary* (London, 1721).
Baker, Richard [Continued. by Edward Phillips?]. *A Chronicle of the Kings of England* (London, 1674).
Barclay, James. *A Treatise on Education* (Edinburgh, 1743).
Barnard, John. *Theologo-Historicus, or the True Life of the Most Reverend Divine, and Excellent Historian Peter Heylyn* (London, 1683).
Barrow, William. *An Essay on Education* (London, 1802).
Bell, James. *A System of Geography, Popular and Scientific. Or, A Physical, Political, and Statistical Account of the World and its Various Divisions* (Glasgow, 1832).
Bell, James. *A New and Comprehensive Gazetteer of England and Wales* (Glasgow, 1834–35).
Bettesworth, J. and Fox, H. *Observations on Education: but Particularly on Naval Education* (London, 1782).
Blackie, Walter. *The Imperial Gazetteer* (Glasgow, 1855–68).
Blair, Hugh. *A Critical Dissertation on the Poems of Ossian* (London, 1763).
[Blount, Charles]. *Reasons Humbly Offered for the Liberty of Licens'd Printing. To Which is Subjoin'd, the Just and True Character of Edmund Bohun, the Licenser of the Press* (London, 1693).
Blount, Thomas. *Glossographia: Or, A Dictionary, Interpreting all such Hard Words of Whatsoever Language, now used in our Refined English Tongue* (London, 1661).
[Bohun, Edmund]. *A Defence of Sir Robert Filmer, against the Mistakes and Misrepresentations of Algernon Sidney, Esq.* (London, 1684).
B[ohun], E[dmund]. *An Apologie for the Church of England, Against the Clamours of the Men of No-Conscience* (London, 1685).
Bohun, Edmund. *A Geographical Dictionary* (London, 1688).
Bohun, Edmund. *A Geographical Dictionary*. 2nd edn (London, 1689 [1690]).
Bohun, Edmund. *A Geographical Dictionary*. 4th edn (London, 1710).

Bohun, Edmund. *The Doctrine of Non-Resistance or Passive Obedience No Way Concerned in the Controversies now Depending between the Williamites and the Jacobites* (London, 1689).
[Bohun, Edmund]. *The History of the Desertion, or An Account of all the Publick Affairs of England, from the Beginning of September 1688 to the Twelfth of February Following* (London, 1689).
Bohun, Edmund. *Three Charges Delivered at the General Quarter Sessions* (London, 1693).
Bohun, Edmund. *The Diary and Correspondence of Edmund Bohun.* S. Wilton Rix, ed. (Beccles: Privately Printed, 1853).
Boswell, James. *Life of Johnson.* G.B. Hill, ed., L.F. Powell, rev. (Oxford: Clarendon Press, 1934–50).
Bowen, Emanuel. *Britannia Depicta or Ogilby Improv'd* (London, 1720).
Büsching, Anton. *A New System of Geography.* Trans. (London, 1762).
Calamy, Edmund. *A Letter to Mr Archdeacon Echard, Upon Occasion of his History of England* (London, 1718).
Camden, William. *Britannia: sive Florentissimorum Regnorum, Angliæ, Scotiæ, Hiberniæ, et Insularum Adiacentum ex Intima Antiquitate Chorographica Descriptio* (London, 1586).
Camden, William. *Britannia: sive Florentissimorum Regnorum, Angliae, Scotiae, Hiberniae, et Insularum Adiacentum ex Intima Antiquitate Chorographica Descriptio* (London, 1607).
Camden, William. *Britain: Or, A Chorographical Description of the Most Flourishing Kingdomes, England, Scotland and Ireland.* Philemon Holland, trans. (London, 1610).
Cary, John. *Cary's New Itinerary; or, An Accurate Delineation of the Great Roads, both Direct and Cross, throughout England and Wales* (London, 1798).
Cawdrey, Robert. *A Table Alphabeticall* (London, 1604).
Chambers, Ephraim. *Cyclopædia: Or, an Universal Dictionary of Arts and Sciences.* 2nd edn (London, 1738).
Clare, Martin. *Youth's Introduction to Trade and Business.* 10th edn (London, 1769).
Critical Review, Or Annals of Literature, 49 (1779).
Cunningham, William. *The Cosmographical Glasse, Conteinyng the Pleasant Principles of Cosmographie, Geographie, Hydrographie, or Nauigation* (London, 1559).
Defoe, Daniel. *A Tour Thro' the Whole Island of Great Britain.* G.D.H. Cole, ed. (London: Frank Cass, 1968).
Dodsley, R. & J. [Printers]. *A New and Accurate Description of the Present Great Roads and the Principal Cross Roads of England and Wales* (London, 1756).
Drummond, William. *Polemo-Middinia.* Edward Gibson, ed. (Oxford, 1691).
Dyche, Thomas and Pardon, William. *A New General English Dictionary.* 2nd edn (London, 1737).
Echard, Laurence. *A Most Compleat Compendium of Geography.* 3rd edn (London, 1693).
Echard, Laurence. *A Most Compleat Compendium of Geography.* 4th edn (London, 1697).
Echard, Laurence. *The History of England* (London, 1707–18).

Echard, Laurence. *The History of the Revolution, and the Establishment of England in the Year 1688* (London, 1725).

Ferguson, Adam. *An Essay on the History of Civil Society*. Fania Oz-Salzberger, ed. (Cambridge: Cambridge University Press, 1995).

Fielding, Henry. *Joseph Andrews*. Douglas Brooks-Davies, ed. (Oxford: Oxford University Press, 1966).

Fielding, Henry. *The Jacobite's Journal and Related Writings*. W.B. Colley, ed. (Oxford: Clarendon Press, 1974).

Fielding, Henry. *The Covent Garden Journal*. Bertrand Goldgar, ed. (Oxford: Clarendon Press, 1988).

Filmer, Robert. *Patriarcha: Or the Natural Power of Kings*. E. Bohun, Preface and Postscript. 2nd edn (London, 1685).

Gentleman's Magazine, 45 (1775).

Gibson, Edmund, ed. *Chronicum Saxonicum* (Oxford, 1692).

Gibson, Edmund. *New Proposals for Printing by Subscription, Cambden's [sic] Britannia, English* (London, 1693).

Gibson, Edward, ed. *Camden's Britannia, Newly Translated into English: With Large Additions and Improvements* (London, 1695).

Gibson, Edmund, ed. *Britannia: Or, a Chorographical Description of Great Britain and Ireland*. 2nd edn (London, 1722).

[Gibson, Edmund]. *Synodus Anglicana: Or, the Constitution and Proceedings of an English Convocation* (London, 1702).

[Gibson, Edmund]. *A Short State of Some Present Questions in Convocation* (London, 1703).

Gibson, Edmund. *Codex Juris Ecclesiastici Anglicani* (London, 1713).

Gibson, Edmund. *Religion, the Best Security to Church and State* (London, 1715).

Gibson, Edmund. *The Deliverances and Murmurings, of the Israelites, and these Nations, Compar'd* (London, 1716).

[Gibson, Edmund]. *The Causes of the Discontents in Relation to the Plague* (London, 1721).

Gibson, Edmund. *Five Pastoral Letters* (London, 1760).

Gilpin, William. *Observations, Relative Chiefly to Picturesque Beauty, made in the Year 1776, on Several Parts of Great Britain; Particularly the High-Lands of Scotland*. 2nd edn (London, 1792).

Gordon, Patrick. *Geography Anatomiz'd: Or, The Geographical Grammar*. 7th edn (London, 1716).

Gough, Richard. *Anecdotes of British Topography* (London, 1768).

Gough, Richard. *Britannia: Or, A Chorographical Description of the Flourishing Kingdoms of England, Scotland, and Ireland* (London, 1789).

Graves, Richard. *The Spiritual Quixote: Or the Summer's Ramble of My Geoffry Wildgoose: A Comic Romance*. Clarence Tracy, ed. (London: Oxford University Press, 1967).

Guthrie, William. *A General History of Scotland* (London, 1767–68).

Guthrie, William. *A New Geographical, Historical, and Commercial Grammar*. 2nd edn (Dublin, 1771)

Guthrie, William. *A New Geographical, Historical, and Commercial Grammar*. 3rd edn (London, 1771).

Hawkins, John. *The Life of Samuel Johnson*. Bertram Davis, ed. (London: Jonathan Cape, 1962).

Heylyn, Peter. *Microcosmus, Or, A Little Description of the Great World. A Treatise Historicall, Geographicall, Politicall, Theologicall* (Oxford, 1621).

Heylyn, Peter. *A Full Relation of Two Journeys. The one into the Main-Land of France. The Other into some of the Adjacent Ilands* (London, 1656).

Heylyn, Peter. *Cosmographie in Four Books, Containing the Chorographie and Historie of the Whole World.* 2nd edn (London, 1657).

Heylyn, Peter. *Examen Historicum: Or a Discovery of the Mistakes, Falsities, and Defects in Some Modern Historians* (London, 1659).

Heylyn, Peter. *Ecclesia Restaurata: Or, The History of the Reformation of the Church of England* (London, 1661).

Heylyn, Peter. *Cyprianus Anglicus: Or, The History of the Life and Death, of the Most Reverend and Renowned Prelate William by Divine Providence, Lord Archbishop of Canterbury* (London, 1668).

Heylyn, Peter. *Keimhaia Ecclesiastica. The Historical and Miscellaneous Tracts of the Reverend and Learned Peter Heylyn* (London, 1681).

Heylyn, Peter. *Cosmographie in Four Books, Improv'd with an Historical Continuation to the Present Times by Edmund Bohun, Esq.* 7th edn (London, 1703).

Heylyn, Peter. 'Heylyn's Own Memoranda'. *Memorial of Bishop Waynflete.* John R. Bloxam, ed. (London: Caxton Society, 1851), pp. x–xxiv.

Hume, David. *Essays, Moral, Political and Literary.* Eugene Miller, ed., 2nd edn (Indianapolis: Liberty Press, 1987).

Hyde, Edward. *The Life of Edward Earl of Clarendon* (Oxford, 1759).

Johnson, Samuel. 'Review of Soame Jenyns's *Free Enquiry*'. *Literary Magazine, or Universal Review* (1756–57), XIII, pp. 171–5, XIV, pp. 251–3, XV, pp. 301–6.

Johnson, Samuel. *Sermons.* Jean Hagstrum and James Gray, eds (New Haven: Yale University Press, 1978).

Johnson, Samuel. *A Journey to the Western Islands of Scotland.* J.D. Fleeman, ed. (Oxford: Clarendon Press, 1985).

Johnson, Samuel. *A Dictionary of the English Language* [1755] (London: Longman facsimile, 1990).

Kennett, White. *Ecclesiastical Synods and Parliamentary Convocations in the Church of England Historically Stated and Justly Vindicated from the Misrepresentations of Mr Atterbury* (London, 1701).

Kennett, White. *The Happiness of this Church and Nation* (London, 1705).

Kennett, White. *A Complete History of England, Volume 3* (London, 1706).

Kennett, White. *A Thanksgiving Sermon for the Blessing of God, in Suppressing the Late Unnatural Rebellion* (London, 1716).

Knox, Vicesimus. *Liberal Education: Or, A Practical Treatise on the Methods of Acquiring Useful and Polite Learning* (London, 1781).

McCulloch, J.R. *A Discourse on the Rise, Progress, Peculiar Objects, and Importance, of Political Economy* (Edinburgh, 1824).

McCulloch, J.R. 'Census of Population – Law of Mortality, etc.' *Edinburgh Review,* 49 (1829), pp. 1–34.

McCulloch, J.R. *A Dictionary, Practical, Theoretical, and Historical, of Commerce and Commercial Navigation* (London, 1832).

McCulloch, J.R. *A Dictionary, Geographical, Statistical, and Historical, of the Various Countries, Places, and Principal Natural Objects in the World* (London, 1841).

McCulloch, J.R. *The Principles of Political Economy*. 4th edn (Edinburgh, 1849).
McCulloch, J.R. *A Catalogue of Books, the Property of a Political Economist* (London, 1862).
McCulloch, J.R. *et al. A Statistical Account of the British Empire* (London, 1837).
Mackinder, Halford. 'On the Scope and Methods of Geography'. *Proceedings of the Royal Geographical Society*, N.S. 9 (1887), pp. 141–74.
Mackinder, Halford. 'The Physical Basis of Political Geography'. *Scottish Geographical Magazine*, 6 (1890), pp. 78–84.
Mackinder, Halford. 'The Great Trade Routes'. *Journal of the Institute of Bankers*, 21 (1900), pp. 1–6, 137–55 and 266–73.
Mackinder, Halford. *Britain and the British Seas* (London: William Heinemann, 1902).
Mackinder, Halford. 'Geography in Education'. *Geographical Teacher*, 2 (1903), pp. 95–101.
Mackinder, Halford. 'The Geographical Pivot of History'. *Geographical Journal*, 23 (1904), pp. 421–44.
Mackinder, Halford. 'The Development of Geographical Teaching out of Nature Study'. *Geographical Teacher*, 2 (1904), pp. 191–7.
Mackinder, Halford. 'Man-Power as a Measure of National and Imperial Strength'. *National Review*, 45 (1905), pp. 136–43.
Mackinder, Halford. *Money-Power and Man-Power: the Underlying Principles rather than the Statistics of Tariff Reform* (London: Simpkin, Marshall, Hamilton, Kent and Co., 1906).
Mackinder, Halford. *The Rhine: Its Valley and History* (London: Chatto and Windus, 1908).
Mackinder, Halford. *India. Eight Lectures Prepared for the Visual Instruction Committee of the Colonial Office* (London: George Philip and Son, 1910).
Mackinder, Halford. *The Nations of the Modern World: an Elementary Study in Geography* (London: George Philip and Son, 1911).
Mackinder, Halford. 'The Teaching of Geography from an Imperial Point of View, and the use which could and should be made of Visual Instruction'. *Geographical Teacher*, 6 (1912), pp. 79–86.
Mackinder, Halford. 'The Teaching of Geography and History as a Combined Subject'. *Geographical Teacher*, 7 (1913), pp. 4–9.
Mackinder, Halford. *The Teaching of Geography and History: a Study in Method* (London: George Philip and Son, 1914).
Mackinder, Halford. *The Modern British State: an Introduction to the Study of Civics* (London: George Philip and Son, 1914).
Mackinder, Halford. *Democratic Ideals and Reality: a Study in the Politics of Reconstruction* (London: Constable and Company, 1919).
Mackinder, Halford. 'Geography as a Pivotal Subject in Education'. *Geographical Journal*, 57 (1921) pp. 376–84.
Mackinder, Halford. 'Geography, an Art and a Philosophy'. *Geography*, 27 (1942), pp. 122–30.
McNichol, Donald. *Remarks on Dr Samuel Johnson's Journey to the Hebrides* (London, 1779).
Martin, Benjamin. *A Course of Lectures in Natural and Experimental Philosophy, Geography and Astronomy* (Reading, 1743).

Member of the Royal Geographical Society, ed. *A Gazetteer of the World, or Dictionary of Geographical Knowledge, Compiled from the Most Recent Authorities, and Forming a Complete Body of Modern Geography, Physical, Political, Statistical, Historical, and Ethnographical* (Edinburgh, 1851–56).
Moll, Herman. *A System of Geography* (London, 1701).
Nichols, John, ed. *Letters on Various Subjects, Literary, Political, and Ecclesiastical, to and from William Nicolson* (London, 1809).
Nicolson, William. *The English Historical Library* (London, 1696–99).
Ogilby, John. *The Relation of His Majesties Entertainment Passing through the City of London, to his Coronation: With a Description of the Triumphal Arches, and Solemnity*. 2nd edn (London, 1661).
Ogilby, John. *The Entertainment of His Most Excellent Majestie Charles II, in his Passage through the City of London to his Coronation* (London, 1662).
Ogilby, John. *Africa* (London, 1670).
Ogilby, John. *America: Being the Latest, and Most Accurate Description of the New World* (London, 1671).
Ogilby, John. *The Fables of Æsop*. 3rd edn (London, 1673).
Ogilby, John. *Britannia, Volume the First: Or, An Illustration of the Kingdom of England and Dominion of Wales: By a Geographical and Historical Description of the Principal Roads Thereof* (London, 1675).
Ogilby, John and Morgan, William. *London Survey'd: Or, An Explanation of the Large Map of London* (London, 1677).
Oldmixon, John. *The Critical History of England, Ecclesiastical and Civil* (London, 1724).
Park, Mungo. *Travels in the Interior of Africa* (London, 1799).
Pennant, Thomas. *A Tour in Scotland; MDCCLXIX*. 3rd edn (Warrington, 1774).
Pennant, Thomas. *A Tour in Scotland, and Voyage to the Hebrides; MDCCLXXII* (Chester, 1774).
[Pennant, Thomas]. *American Annals; or, Hints and Queries for Parlement [sic] Men* (Darlington, 1778).
Pennant, Thomas. *A Tour in Wales* (London, 1781–84).
Pennant, Thomas. *The Journey from Chester to London* (Dublin, 1783).
Pennant, Thomas. *Some Account of London*. 2nd edn (London, 1791).
Pennant, Thomas. *The Literary Life of the Late Thomas Pennant* (London, 1793).
Pennant, Thomas. *The View of Hindoostan* (London, 1798–1800).
Pennant, Thomas. *A Journey from London to the Isle of Wight* (London, 1801).
Pennant, Thomas. *Tour on the Continent, 1765*. G.R. de Beer, ed. (London: Ray Society, 1948).
Pepys, Samuel. *Diary*. Robert Latham and William Matthews, transcribed. (London: G. Bell and Sons, 1970).
Phillips, Edward. *The New World of Words: Or, Universal English Dictionary*. John Kersey, rev., 6th edn (London, 1706).
[Pinkerton, John] 'Robert Heron'. *Letters of Literature* (London, 1785).
[Pinkerton, John] 'Philostor'. 'Letters to the People of Great Britain on the Cultivation of their National History'. *Gentleman's Magazine*, 58 (1788), February–December.
Pinkerton, John. *The History of Scotland from the Accession of the House of Stuart to that of Mary* (London, 1797).
Pinkerton, John. *Modern Geography* (London, 1802).

Pinkerton, John. *Recollections of Paris, in the Years 1802–3–4–5* (London, 1806).

Pinkerton, John, ed. *A General Collection of the Best and Most Interesting Voyages and Travels* (London, 1808–14).

Pinkerton, John. *Literary Correspondence* (London, 1830).

Piozzi, Hester. *Anecdotes of the Late Samuel Johnson*, in *Johnsonian Miscellanies*, G.B. Hill, ed. (Oxford: Clarendon Press, 1898).

Piozzi, Hester. *Thraliana: the Diary of Mrs Hester Lynch Thrale (later Mrs Piozzi), 1776–1809*. Katharine Balderston, ed. (Oxford: Clarendon Press, 1951).

Pitt, Moses, ed. *The English Atlas* (Oxford, 1680–83).

Potter, John. *The Traveller's Pocket-Book: Or, Ogilby and Morgan's Book of the Roads Improved and Amended* 20th edn (London, 1780).

Pope, Alexander. *Poems*. John Butt, ed. (London: Methuen, 1963).

Priestley, Joseph. *Miscellaneous Observations relating to Education* (Bath, 1778).

Priestley, Joseph. *Lectures on History and General Policy* (Birmingham, 1788).

Ptolemy. *Geography*. Edward Stevenson, ed. (New York: Dover Press, 1991).

Rennell, James. *Memoirs of a Map of Hindoostan* (London, 1783).

Rennell, James. *Memoirs of a Map of Hindoostan*. 2nd edn (London, 1788).

Rennell, James. *Memoirs of a Map of Hindoostan*. 3rd edn (London, 1793).

Rennell, James. *The Marches of the British Armies in the Peninsula of India, during the Campaigns of 1790 and 1791* (London, 1792).

[Rennell, James]. 'An Old Englishman'. *War with France the Only Security of Britain* (London, 1794).

Rennell, James. *The Geographical System of Herodotus Examined* (London, 1800).

Rennell, James. *Observations on the Topography of the Plain of Troy* (London, 1814).

Rennell, James. *Illustrations (Chiefly Geographical) of the History of the Expedition of Cyrus* (London, 1816).

Rennell, James. *A Treatise on the Comparative Geography of Western Asia* (London, 1831).

Rennell, James. *An Investigation of the Currents of the Atlantic Ocean* (London, 1832).

Rennell, James. 'On the Topography of Ancient Babylon' in Claudius Rich, *Narrative of a Journey to the Site of Babylon* (London, 1839).

Rennell, James. *Journals. Memoirs of the Asiatic Society of Bengal*, 3 (Calcutta, 1914). T.H.D. La Touche, ed.

Rollin, Charles. *New Thoughts Concerning Education* (London, 1735).

Salmon, Nathaniel. *A New Survey of England* (London, 1731).

Salmon, Thomas. *A Review of the History of England* (London, 1722–24).

Salmon, Thomas. *An Impartial Examination of Bishop Burnet's History of His Own Times* (London, 1724).

Salmon, Thomas. *The History of the Life and Reign of her Late Majesty Queen Anne* (London, 1740).

Salmon, Thomas. *Considerations on the Bill for a General Naturalization* (London, 1748).

Salmon, Thomas. *A New Geographical and Historical Grammar* (London, 1749).

Salmon, Thomas. *Salmon's Geographical and Astronomical Grammar*. 13th edn (London, 1785).

Smollett, Tobias. *Travels through France and Italy*. Frank Felsenstein, ed. (Oxford: Oxford University Press, 1981).

Strabo. *Geography*. H.L. Jones, trans. and ed. (Cambridge, Mass.: Harvard University Press, 1917–32).
Thoresby, Ralph. *Letters of Eminent Men, Addressed to Ralph Thoresby* (London, 1832).
Tucker, Josiah. *Instructions for Travellers* (Dublin, 1758).
Vernon, George. *The Life of the Learned and Reverend Dr Peter Heylyn* (London, 1682).
Walker, Edward. *A Circumstantial Account of the Preparations for the Coronation of His Majesty King Charles the Second* (London, 1820).
Walker, John. *Elements of Geography, with the Principles of Natural Philosophy, and Sketches of General History* (Dublin, 1788).
Waterland, Daniel. *Advice to a Young Student* (Oxford, 1755).
Watts, Thomas. *An Essay on the Proper Method of Forming the Man of Business* [1716]. Arthur H. Cole, ed. (Boston: Harvard Graduate School, 1946).
Webster, Noah. *An American Dictionary of the English Language* (New York, 1828).
Wells, Edward. *A Treatise of Antient and Present Geography* (Oxford, 1701).
Wells, Edward. *An Historical Geography of the New Testament* (London, 1708).
Wells, Edward. *An Historical Geography of the Old Testament* (London, 1711–12).
Wells, Edward. *Remarks on Dr Clarke's Introduction to his Scripture-Doctrin of the Trinity* (Oxford, 1713).
Wells, Edward. *Dr Wells's Controversial Tracts Against the Dissenters*. 6th edn (London, 1716).
Wesley, John. *Letters*. J. Telford, ed. (London: Epworth Press, 1931).
Wesley, John. *Journal*. Nehemiah Curnock, ed. (London: Epworth Press Reprint, 1938).
Wesley, John. *Sermons, Volume II*. Albert Outler, ed. (Nashville: Abingdon Press, 1985).
Whitefield, George. *Journals* (Edinburgh: Banner of Truth Trust, 1960).

Secondary sources

Abbattista, Guido. 'Establishing the "Order of Time and Place": "Rational Geography", French Erudition and the Emplacement of History in Gibbon's Mind'. *Edward Gibbon: Bicentenary Essays*. David Womersley, ed. Studies on Voltaire and the Eighteenth Century, 335 (Oxford: Voltaire Foundation, 1997), pp. 45–72.
Adams, Percy. *Travel Literature and the Evolution of the Novel* (Lexington: University of Kentucky Press, 1983).
Age of Johnson, Volumes 7 and 8 (1996–97).
Allan, David. *Virtue, Learning and the Scottish Enlightenment: Ideas of Scholarship in Early Modern History* (Edinburgh: Edinburgh University Press, 1993).
Allan, David. 'Protestantism, Presbyterianism and National Identity in Eighteenth Century Scottish History'. *Protestantism and National Identity: Britain and Ireland, c.1650–c.l850*. Tony Claydon and Ian McBride, eds (Cambridge: Cambridge University Press, 1998), pp. 182–205.
Alpers, Svetlana. *The Art of Describing: Dutch Art in the Seventeenth Century* (Harmondsworth: Penguin edn, 1989).

Armitage, David. 'Literature and Empire'. *The Oxford History of the British Empire: Volume 1: The Origins of Empire*. Nicholas Canny, ed. (Oxford: Oxford University Press, 1998), pp. 99–123.

Ayres, Philip. *Classical Culture and the Idea of Rome in Eighteenth-Century England* (Cambridge: Cambridge University Press, 1997).

Baker, J.N.L. *The History of Geography: Papers by J.N.L. Baker* (Oxford: Blackwell, 1963).

Batten, Charles. *Pleasurable Instruction: Form and Convention in Eighteenth-Century Travel Literature* (Berkeley: University of California Press, 1978).

Bayly, C.A. *Imperial Meridian: the British Empire and the World, 1780–1830* (London: Longman, 1989).

Beddard, Robert. 'The Unexpected Whig Revolution of 1688'. *The Revolutions of 1688: The Andrew Browning Lectures 1988*. Robert Beddard, ed. (Oxford: Oxford University Press, 1991), pp. 11–101.

Beddard, Robert. 'James II and the Catholic Challenge'. *The History of the University of Oxford: Volume IV: Seventeenth-Century Oxford*. Nicholas Tyacke, ed. (Oxford: Oxford University Press, 1997), pp. 917–54.

Bell, Morag, Butlin, Robin, and Heffernan, Michael, eds. *Geography and Imperialism, 1820–1940* (Manchester: Manchester University Press, 1995).

Bennett, G.V. *White Kennett 1660–1728. Bishop of Peterborough. A Study in the Political and Ecclesiastical History of the Early Eighteenth Century* (London: SPCK, 1957).

Bennett, G.V. *The Tory Crisis in Church and State 1688–1730: the Career of Francis Atterbury, Bishop of Rochester* (Oxford: Clarendon Press, 1975).

Bennett, G.V. 'The Era of Party Zeal, 1702–14'. *The History of the University of Oxford: Volume V: the Eighteenth Century*. L.S. Sutherland and L.G. Mitchell, eds (Oxford: Oxford University Press, 1988), pp. 61–98.

Berkowitz, David. *John Selden's Formative Years: Politics and Society in Early Seventeenth-Century England* (London: Associated University Press, 1988).

Bernstein, George. *Liberalism and Liberal Politics in Edwardian England* (Boston: Allen and Unwin, 1986).

Bevir, Mark. *The Logic of the History of Ideas* (Cambridge: Cambridge University Press, 1999).

Black, Jeremy. *The British Abroad: the Grand Tour in the Eighteenth Century* (Stroud: Alan Sutton, 1992).

Bloom Edward A. and Lillian D. *Joseph Addison's Sociable Animal: In the Market Place, on the Hustings, in the Pulpit* (Providence: Brown University Press, 1971).

Blouet, Brian. *Halford Mackinder: a Biography* (College Station, Texas: Texas A&M Press, 1987).

Bosher, Robert. *The Making of the Restoration Settlement: the Influence of the Laudians 1649–62* (London: Dacre Press, 1951).

Bowen, Margarita. *Empiricism and Geographical Thought: From Francis Bacon to Alexander von Humboldt* (Cambridge: Cambridge University Press, 1981).

Bradley, James. *Religion, Revolution and English Radicalism: Non-Conformity in Eighteenth-Century Politics and Society* (Cambridge: Cambridge University Press, 1990).

Bravo, Michael. 'Precision and Curiosity in Scientific Travel: James Rennell and the Orientalist Geography of the New Imperial Age (1760–1830)'. *Voyages and*

Visions: Towards a Cultural History of Travel. Jas Elsner and Joan-Pau Rubiés, eds (London: Reaktion, 1999), pp. 162–83.

Brent, Richard. *Liberal Anglican Politics: Whiggery, Religion and Reform, 1830–41* (Oxford: Clarendon Press, 1987).

Brooks, David. *The Age of Upheaval: Edwardian Politics, 1899–1914* (Manchester: Manchester University Press, 1995).

Brown, Ralph. 'The American Geographies of Jedidiah Morse'. *Annals of the Association of American Geographers*, 31 (1941), pp. 144–217.

Brown, Stewart, ed. *William Robertson and the Expansion of Empire* (Cambridge: Cambridge University Press, 1997).

Burgess, Glenn. *The Politics of the Ancient Constitution: an Introduction to English Political Thought, 1603–1642* (London: Macmillan, 1992).

Burrow, John. *Whigs and Liberals: Continuity and Change in English Political Thought* (Oxford: Clarendon Press, 1988).

Butlin, Robin. 'Ideological Contexts and the Reconstruction of Biblical Landscapes in the Seventeenth and Early Eighteenth Centuries: Dr Edward Wells and the Historical Geography of the Holy Land'. *Ideology and Landscape in Historical Perspective: Essays on the Meanings of Some Places in the Past*. Alan Baker and Gideon Biger, eds (Cambridge: Cambridge University Press, 1992), pp. 31–62.

Canny, Nicholas. ed. *The Oxford History of the British Empire: Volume I: The Origins of Empire* (Oxford: Oxford University Press, 1998).

Carnall, Geoffrey. 'Robertson and Contemporary Images of India'. *William Robertson and the Expansion of Empire*. Stewart Brown, ed. (Cambridge: Cambridge University Press, 1997), pp. 210–30.

Carter, Jennifer. 'The Revolution and the Constitution'. *Britain after the Glorious Revolution, 1689–1714*. Geoffrey Holmes, ed. (London: Macmillan, 1969), pp. 39–58.

Carter, Paul. *The Road to Botany Bay: an Essay in Spatial History* (London: Faber and Faber, 1987).

Champion, J.A.I. *The Pillars of Priestcraft Shaken: the Church of England and its Enemies, 1660–1730* (Cambridge: Cambridge University Press, 1992).

Chapin, Chester. *The Religious Thought of Samuel Johnson* (Ann Arbor: University of Michigan Press, 1968).

Charlesworth, Michael. 'The Ruined Abbey: Picturesque and Gothic Values'. *The Politics of the Picturesque: Literature, Landscape and Aesthetics since 1770*. Stephen Copley and Peter Garside, eds (Cambridge: Cambridge University Press, 1994), pp. 62–80.

Chartier, Roger. *The Order of Books: Readers, Authors and Libraries in Europe between the Fourteenth and Eighteenth Centuries* (Cambridge: Polity, 1994).

Clark, J.C.D. *English Society, 1688–1832: Ideology, Social Structure and Political Practice during the Ancien Régime* (Cambridge: Cambridge University Press, 1985).

Clark, J.C.D. *The Language of Liberty, 1660–1832: Political Discourse and Social Dynamics in the Anglo-American World* (Cambridge: Cambridge University Press, 1994).

Clark, J.C.D. *Samuel Johnson: Literature, Religion and English Cultural Politics from the Restoration to Romanticism* (Cambridge: Cambridge University Press, 1994).

Clark, J.C.D. 'Religious Affiliation and Dynastic Allegiance in Eighteenth-Century England: Edmund Burke, Thomas Paine and Samuel Johnson'. *English Literary History*, 64 (1997), pp. 1029–67.

Claydon, Tony. *William III and the Godly Revolution* (Cambridge: Cambridge University Press, 1996).

Claydon, Tony and McBride, Ian, eds. *Protestantism and National Identity: Britain and Ireland, c.1650–c.1850* (Cambridge: Cambridge University Press, 1998).

Coetzee, Frans. *For Party or Country: Nationalism and the Dilemmas of Popular Conservatism in Edwardian England* (New York: Oxford University Press, 1990).

Colley, Linda. *In Defiance of Oligarchy: The Tory Party, 1714–60* (Cambridge: Cambridge University Press, 1982).

Colley, Linda. *Britons: Forging the Nation, 1707–1837* (New Haven: Yale University Press, 1992).

Collini, Stefan, *et al. That Noble Science of Politics: a Study in Nineteenth-Century Intellectual History* (Cambridge: Cambridge University Press, 1983).

Collini, Stefan. *Public Moralists: Political Thought and Intellectual Life in Britain, 1850–1930* (Oxford: Clarendon Press, 1991).

Condren, Conal. *The Language of Politics in Seventeenth-Century England* (London: Macmillan, 1994).

Cormack, Lesley. '"Good Fences make Good Neighbours": Geography as Self-Definition in Early-Modern England'. *Isis*, 82 (1991), pp. 639–61.

Cormack, Lesley. *Charting an Empire: Geography at the English Universities, 1580–1620* (Chicago: Chicago University Press, 1997).

Cornwall, Robert. *Visible and Apostolic: the Constitution of the Church in High Church Anglican and Non-Juror Thought* (Newark: University of Delaware Press, 1993).

Cosgrove, Denis. *Social Formation and Symbolic Landscape* (Beckenham: Croom Helm, 1984).

Cosgrove, Denis, and Domosh, Mona. 'Writing and Authority: Writing the New Cultural Geography'. *Place/Culture/Representation*, James Duncan and David Ley, eds (London: Routledge, 1993), pp. 25–38.

Cunish, P. *et al. A History of Magdalene College Cambridge, 1428–1988* (Cambridge: Magdalene College Press, 1994).

Cunningham, Andrew and Williams, Perry. 'De-Centring the "Big Picture": *The Origins of Modern Science* and the Modern Origins of Science'. *British Journal for the History of Science*, 26 (1993), pp. 407–32.

Darnton, Robert. 'History of Reading'. *New Perspectives on Historical Writing*, Peter Burke, ed. (Cambridge: Polity, 1991), pp. 140–67.

Daston, Lorraine. *Classical Probability in the Enlightenment* (Princeton: Princeton University Press, 1988).

Domosh, Mona. 'Toward a Feminist Historiography of Geography'. *Transactions of the Institute of British Geographers*, N.S. 16 (1991), pp. 95–104.

Douglas, David. *English Scholars* (London: Jonathan Cape, 1939).

Downes, Alan. 'The Bibliographical Dinosaurs of Georgian Geography (1714–1830)'. *Geographical Journal*, 137 (1971), pp. 379–87.

Doyle, William. *The Oxford History of the French Revolution* (Oxford: Oxford University Press, 1989).

Drayton, Richard. 'Knowledge and Empire'. *The Oxford History of the British Empire: Volume 2: The Eighteenth Century*. P.J. Marshall, ed. (Oxford: Oxford University Press, 1998), pp. 231–52.

Driver, Felix. 'Geography's Empire: Histories of Geographical Knowledge'. *Society and Space*, 10 (1992), pp. 23–40.

Driver, Felix. *Geography Militant: Cultures of Exploration in the Age of Empire* (Oxford: Blackwell, 2000).

Dunn, John. 'From Applied Theology to Social Anaylsis: the Break between John Locke and the Scottish Enlightenment'. *Wealth and Virtue: the Shaping of Political Economy in the Scottish Enlightenment*. Istvan Hont and Michael Ignatieff, eds (Cambridge University Press: Cambridge, 1983), pp. 119–35.

Dwyer, John. *Virtuous Discourse: Sensibility and Community in late Eighteenth-Century Scotland* (Edinburgh: Edinburgh University Press, 1987).

East, W. Gordon. 'An Eighteenth-Century Geographer. William Guthrie of Brechin'. *Scottish Geographical Magazine*, 72 (1956), pp. 32–7.

Edney, Matthew. *Mapping an Empire: the Geographical Construction of British India, 1765–1843* (Chicago: Chicago University Press, 1997).

Eisenstein, Elizabeth. *The Printing Press as an Agent of Change: Communications and Cultural Transformations in Early-Modern Europe* (Cambridge: Cambridge University Press, 1979).

Eliot Hurst, Michael. 'Geography has Neither Existence nor Future'. *The Future of Geography*. R.J. Johnston, ed. (London: Methuen, 1985), pp. 59–91.

English Literary History, 64 (1997).

Erskine-Hill, Howard. *The Augustan Idea in English Literature* (London: E.J. Arnold, 1983).

Erskine-Hill, Howard. *Poetry and the Realm of Politics: Shakespeare to Dryden* (Oxford: Oxford University Press, 1996).

Erskine-Hill, Howard. *Poetry of Opposition and Revolution: Dryden to Wordsworth* (Oxford: Oxford University Press, 1996).

Evans, Eric. *Britain before the Reform Act: Politics and Society, 1815–1832* (London: Longman, 1989).

Everett, Nigel. *The Tory View of Landscape* (New Haven: Yale University Press, 1994).

Fincham, Kenneth. 'Oxford and the Early Stuart Polity'. *The History of the University of Oxford: Volume IV: Seventeenth-Century Oxford*. Nicholas Tyacke, ed. (Oxford: Oxford University Press, 1997), pp. 179–210.

Fontana, Biancamaria. *Rethinking the Politics of Commercial Society: the* Edinburgh Review *1802–1832* (Cambridge: Cambridge University Press, 1985).

Forbes, Duncan. *Hume's Philosophical Politics* (Cambridge: Cambridge University Press, 1975).

Fordham, Herbert. *Notes on British and Irish Itineraries and Road Books* (Hertford: Stephen Austin Press, 1912).

Foucault, Michel. 'What is an Author?' *The Foucault Reader*, Paul Rabinow, ed. (Harmondsworth: Penguin, 1984), pp. 101–20.

Fry, Michael. *The Dundas Despotism* (Edinburgh: Edinburgh University Press, 1992).

Gascoigne, John. *Cambridge in the Age of the Enlightenment: Science, Religion and Politics from the Restoration to the French Revolution* (Cambridge: Cambridge University Press, 1989).

Gascoigne, John. *Joseph Banks and the English Enlightenment: Useful Knowledge and Polite Culture* (Cambridge: Cambridge University Press, 1994).

Gash, Norman. *Lord Liverpool: the Life and Political Career of Robert Banks Jenkinson, Second Earl of Liverpool 1770–1828* (London: Weidenfeld and Nicolson, 1984).

Gerrard, Christine. *The Patriot Opposition to Walpole: Politics, Poetry and National Myth, 1725–42* (Oxford: Oxford University Press, 1994).

Gilbert, E.W. *British Pioneers in Geography* (Newton Abbot: David and Charles, 1972).

Glassner, Martin and De Blij, Harm. *Systematic Political Geography*. 3rd edn (New York: John Wiley, 1980).

Godlewska, Anne. 'Napoleon's Geographers (1797–1815): Imperialists and Soldiers of Modernity'. *Geography and Empire*. Neil Smith and Anne Godlewska, eds (Oxford: Blackwell, 1994), pp. 31–53.

Godlewska, Anne. 'Map, Text and Image. The Mentality of Enlightened Conquerors: A New Look at the *Description de l'Egypte*'. *Transactions of the Institute of British Geographers*, N.S. 20 (1995), pp. 5–28.

Goldgar, Anne. *Impolite Learning: Conduct and Community in the Republic of Letters, 1680–1750* (New Haven: Yale University Press, 1995).

Goldie, Mark. 'Edmund Bohun and *Jus Gentium* in the Revolution Debate, 1689–93'. *Historical Journal*, 20 (1977), pp. 569–86.

Goldie, Mark. 'The Revolution of 1689 and the Structure of Political Debate'. *Bulletin of Research in the Humanities*, 83 (1980), pp. 473–564.

Goldie, Mark. 'John Locke and Anglican Royalism'. *Political Studies*, 31 (1983), pp. 61–83.

Goldie, Mark. 'The Political Thought of the Anglican Revolution'. *The Revolutions of 1688: The Andrew Browning Lectures 1988*. Robert Beddard, ed. (Oxford: Oxford University Press, 1991), pp. 102–36.

Goldman, Lawrence. 'The Origins of British "Social Science": Political Economy, Natural Science and Statistics, 1830–1835'. *Historical Journal*, 26 (1983), pp. 587–616.

Grafton, Anthony. *Defenders of the Text: the Traditions of Scholarship in an Age of Science, 1450–1800* (Cambridge, Mass.: Harvard University Press, 1991).

Grafton, Anthony and Jardine, Lisa. *From Humanism to the Humanities: Education and the Liberal Arts in Fifteenth- and Sixteenth-Century Europe* (London: Duckworth, 1986).

Gray, James. *Johnson's Sermons* (Oxford: Clarendon Press, 1972).

Green, E.H.H. *The Crisis of Conservatism: the Politics, Economics and Ideology of the British Conservative Party, 1880–1914* (London: Routledge, 1995).

Green, I.M. *The Re-Establishment of the Church of England, 1660–1663* (Oxford: Oxford University Press, 1978).

Greene, Donald. *The Politics of Samuel Johnson*. 2nd edn (Athens, Ga.: University of Georgia Press, 1990).

Greene, Jack. *Peripheries and Center: Constitutional Development in the Extended Polities of the British Empire and the United States, 1607–1788* (Athens, Ga.: University of Georgia Press, 1986).

Gregory, Derek. *Geographical Imaginations* (Oxford: Blackwell, 1994).

Haakonssen, Knud. *The Science of the Legislator: the Natural Jurisprudence of David Hume and Adam Smith* (Cambridge: Cambridge University Press, 1981).

Haakonssen, Knud, ed. *Enlightenment and Religion: Rational Dissent in Eighteenth-Century Britain* (Cambridge: Cambridge University Press, 1996).

Harris, Robert. *A Patriot Press: National Politics and the London Press in the 1740s* (Oxford: Oxford University Press, 1993).
Hart, Vaughan. *Art and Magic in the Court of the Stuarts* (London: Routledge, 1994).
Hartshorne, Richard. *The Nature of Geography: a Critical Survey of Current Geographical Thought in the Light of the Past* (Lancaster, Pa.: AAAG, 1939).
Harvey, David. 'On the History and Present Condition of Geography: an Historical Materialist Manifesto'. *Professional Geographer*, 3 (1984), pp. 1–11.
Harvey, David. *The Condition of Postmodernity: an Enquiry into the Origins of Cultural Change* (Oxford: Blackwell, 1989).
Helgerson, Richard. *Forms of Nationhood: the Elizabethan Writing of England* (Chicago: Chicago University Press, 1992).
Hicks, Philip. *Neoclassical History and English Culture: From Clarendon to Hume* (London: Macmillan, 1996).
Hilton, Boyd. *Corn, Cash, Commerce: the Economic Policies of the Tory Governments 1815–1830* (Oxford: Clarendon Press, 1977).
Hilton, Boyd. *The Age of Atonement: the Influence of Evangelicalism on Social and Economic Thought, 1785–1865* (Oxford: Clarendon Press, 1988).
Hirst, Derek. 'The English Republic and the Meaning of Britain'. *The British Problem, c.1534–1707: State Formation in the Atlantic Archipelago*, Brendan Bradshaw and John Morrill, eds (London: Macmillan, 1996), pp. 192–219.
Holmes, Geoffrey. *Politics in the Age of Queen Anne* (London: Macmillan, 1967).
Holmes, Geoffrey. 'Religion and Party in Late Stuart England' in his *Politics, Religion and Society in England 1679–1742* (London: Hambledon Press, 1986), pp. 181–215.
Höpfl, H.M. 'From Savage to Scotsman: Conjectural History in the Scottish Enlightenment'. *Journal of British Studies*, 17 (1978), pp. 19–40.
Horwitz, Henry. 'The Structure of Parliamentary Politics'. *Britain After the Glorious Revolution, 1689–1714*. Geoffrey Holmes, ed. (London: Macmillan, 1969), pp. 96–114.
Horwitz, Henry. *Parliament, Policy and Politics in the Reign of William III* (Manchester: Manchester University Press, 1977).
Hudson, Brian. 'The New Geography and the New Imperialism: 1870–1918'. *Antipode*, 9.2 (1977), pp. 12–19.
Hutton, Ronald. *The Restoration: a Political and Religious History of England and Wales 1658–67* (Oxford: Clarendon Press, 1985).
Hutton, Ronald. *Charles the Second: King of England, Scotland and Ireland* (Oxford: Clarendon Press, 1989).
Hylson-Smith, Kenneth. *High-Churchmanship in the Church of England: From the Sixteenth Century to the Late Twentieth Century* (Edinburgh: T&T Clark, 1993).
Ignatieff, Michael. 'John Miller and Individualism'. *Wealth and Virtue: the Shaping of Political Economy in the Scottish Enlightenment*. Istvan Hont and Michael Ignatieff, eds (Cambridge University Press: Cambridge, 1983), pp. 317–43.
James, Francis. *North Country Bishop: a Biography of William Nicolson* (New Haven: Yale University Press, 1956).
Johns, Adrian. *The Nature of the Book: Print and Knowledge in the Making* (Chicago: Chicago University Press, 1998).
Jones, J.R. *Charles II: Royal Politician* (London: Allen and Unwin, 1987).

Kearns, Gerry. 'Closed Space and Political Practice: Frederick Jackson Turner and Halford Mackinder'. *Society and Space*, 1 (1984), pp. 23–34.

Kearns, Gerry. 'Halford John Mackinder, 1861–1947'. *Geographers: Biobibliographical Studies*, 9 (1985), pp. 71–86.

Keay, John. *The Honourable Company: a History of the English East India Company* (London: HarperCollins, 1991).

Kenyon, J.P. *Revolution Principles: the Politics of Party, 1689–1720* (Cambridge: Cambridge University Press, 1977).

Kernan, Alvin. *Samuel Johnson and the Age of Print*. New edn (Princeton: Princeton University Press, 1989).

Kidd, Colin. *Subverting Scotland's Past: Scottish Whig Historians and the Creation of an Anglo-British Identity, 1689–c.1830* (Cambridge: Cambridge University Press, 1993).

Kidd, Colin. 'North Britishness and the Nature of Eighteenth-Century British Patriotisms'. *Historical Journal*, 39 (1996), pp. 361–82.

Kidd, Colin. *British Identities before Nationalism: Ethnicity and Nationhood in the Atlantic World, 1600–1800* (Cambridge: Cambridge University Press, 1999).

Klein, Lawrence. *Shaftesbury and the Culture of Politeness: Moral Discourse and Cultural Politics in Early Eighteenth-Century England* (Cambridge: Cambridge University Press, 1994).

Klein, Lawrence. 'Politeness for Plebes. Consumption and Social Identity in Early Eighteenth-Century England'. *The Consumption of Culture 1600–1800: Image, Object, Text*. Ann Bermingham and John Brewer, eds (London: Routledge, 1995), pp. 362–82.

Kugler, Michael. 'Provincial Intellectuals: Identity, Patriotism, and Enlightened Peripheries'. *The Eighteenth Century: Theory and Interpretation*, 37 (1996), pp. 156–73.

Langer, Gary. *The Coming of Age of Political Economy, 1815–1825* (New York: Greenwood Press, 1987).

Laudan, Rachel. *From Mineralogy to Geology: the Foundations of a Science, 1650–1830* (Chicago: Chicago University Press, 1987).

Lefebvre, Henri. *The Production of Space* (Oxford: Blackwell, 1991).

Lee, Maurice. *The Cabal* (Urbana: University of Illinois Press, 1965).

Lenman, Bruce. '"From Savage to Scot" via the French and the Spaniards: Principal Robertson's Spanish Sources'. *William Robertson and the Expansion of Empire*. Stewart Brown, ed. (Cambridge: Cambridge University Press, 1997), pp. 196–209.

Levine, Joseph. *Humanism and History: Origins of Modern English Historiography* (Ithaca: Cornell University Press, 1987).

Levy, F.J. 'The Making of Camden's *Britannia*'. *Bibliothèque D'Humanisme et Renaissance*, 26 (1964), pp. 70–97.

Lewis, Jayne Elizabeth. *The English Fable: Aesop and Literary Culture, 1641–1750* (Cambridge: Cambridge University Press, 1996).

Livingstone, David. *The Geographical Tradition: Episodes in the History of a Contested Enterprise* (Oxford: Blackwell, 1992).

Loveridge, Mark. *A Historian of Augustan Fable* (Cambridge: Cambridge University Press, 1998).

Lyles, Albert. *Methodism Mocked: the Satiric Reaction to Methodism in the Eighteenth Century* (London: Epworth Press, 1960).

McCaffrey, John. 'Political Issues and Developments'. *Glasgow: Volume II: 1830 to 1912*. W. Hamish Fraser and Irene Maver, eds (Manchester: Manchester University Press, 1996), pp. 186–226.

McGann, Jerome. *A Critique of Modern Textual Criticism* (Chicago: Chicago University Press, 1983).

McKenzie, Donald. *Bibliography and the Sociology of Texts* (Cambridge: Cambridge University Press reprint, 1999).

Maddrell, Avril. 'Discourses of Race and Gender and the Comparative Method in Geography School Texts, 1830–1918'. *Society and Space*, 16 (1998), pp. 81–103.

Mandler, Peter. *Aristocratic Government in the Age of Reform: Whigs and Liberals, 1830–1852* (Oxford: Clarendon Press, 1990).

Manley, Lawrence. *Literature and Culture in Early Modern London* (Cambridge: Cambridge University Press, 1995).

Markham, Clements. *Major James Rennell and the Rise of Modern English Geography* (London: Cassell, 1895).

Marshall, John. 'The Ecclesiology of the Latitude-Men 1660–89: Stillingfleet, Tillotson and "Hobbism"'. *Journal of Ecclesiastical History*, 26 (1985), pp. 407–27.

Marshall, P.J. *East India Fortunes: the British in Bengal in the Eighteenth Century* (Oxford: Oxford University Press, 1976).

Marshall, P.J. 'Britain without America – A Second Empire?' *The Oxford History of the British Empire: Volume II: The Eighteenth Century*. P.J. Marshall, ed. (Oxford: Oxford University Press, 1998), pp. 576–95.

Marshall, P.J. and Williams, Glyndwr. *The Great Map of Mankind: British Perceptions of the World in the Age of Enlightenment* (London: Dent, 1982).

Martin, Geoffrey and James, Preston. *All Possible Worlds: a History of Geographical Ideas*. 3rd edn (New York: John Wiley, 1993).

Matthew, H.C.G. *The Liberal Imperialists: the Ideas and Politics of a Post-Gladstonian Élite* (Oxford: Oxford University Press, 1971).

Mayhew, Robert. 'Was William Shakespeare an Eighteenth-Century Geographer? Constructing Histories of Geographical Knowledge'. *Transactions of the Institute of British Geographers*, N.S. 23 (1998), pp. 21–37.

Mayhew, Robert. 'Geography in Eighteenth Century British Education'. *Paedagogica Historica*, N.S. 34 (1998), pp. 731–69.

Mayhew, Robert. 'Samuel Johnson's Intellectual Character as a Traveler: A Reassessment'. *Age of Johnson*, 10 (1999), pp. 35–67.

Mayhew, Robert. 'William Gilpin's Latitudinarian Picturesque'. *Eighteenth-Century Studies*, 33 (2000), pp. 349–66.

Mendyk, Stan A.E. *Speculum Britanniae: Regional Study, Antiquarianism, and Science in Britain to 1700* (Toronto: Toronto University Press, 1989).

Miller, David and Reill, Peter, eds. *Visions of Empire: Voyages, Botany, and Representations of Nature* (Cambridge: Cambridge University Press, 1996).

Miller, John. 'The Potential for "Absolutism" in Later Stuart England'. *History*, 69 (1984), pp. 187–207.

Miller, Peter. *Defining the Common Good: Empire, Religion and Philosophy in Eighteenth-Century Britain* (Cambridge: Cambridge University Press, 1994).

Milton, Anthony. *Catholic and Reformed: the Roman and Protestant Churches in English Protestant Thought, 1600–1640* (Cambridge: Cambridge University Press, 1995).

Mitchell. Leslie. *Holland House* (London: Duckworth, 1980).
Mitchell, Leslie. *Charles James Fox* (Oxford: Oxford University Press, 1992).
Monod, Paul. *Jacobitism and the English People, 1688–1788* (Cambridge: Cambridge University Press, 1989).
Nockles, Peter. *The Oxford Movement in Context: Anglican High Churchmanship, 1760–1857* (Cambridge: Cambridge University Press, 1994).
Norman, E.R. *Anti-Catholicism in Victorian England* (London: George Allen and Unwin, 1968).
O'Brien, D.P. *J.R. McCulloch: a Study in Classical Economics* (London: George Allen and Unwin, 1970).
O'Brien, D.P. *The Classical Economists* (Oxford: Clarendon Press, 1975).
O'Brien, Karen. 'Robertson's Place in the Development of Eighteenth-Century Narrative History'. *William Robertson and the Expansion of Empire*. Stewart Brown ed. (Cambridge: Cambridge University Press, 1997), pp. 74–91.
O'Gorman, Frank. *The Whig Party and the French Revolution* (London: Macmillan, 1967).
O'Gorman, Frank. *Voters, Patrons and Parties: the Unreformed Electorate in Hanoverian England, 1734–1832* (Oxford: Oxford University Press, 1989).
Ó Tuathail, Gearóid. *Critical Geopolitics: the Politics of Writing Global Space* (London: Routledge, 1996).
Oakeshott, Michael. *Experience and Its Modes* (Cambridge: Cambridge University Press, 1933).
Oakeshott, Michael. *On History and Other Essays* (Oxford: Blackwells, 1983).
Oakeshott, Michael. 'The Activity of Being an Historian'. *Rationalism in Politics and Other Essays*. Timothy Fuller, new edn (Indianapolis: Liberty Press, 1991), pp. 151–83.
Ogborn, Miles. *Spaces of Modernity: London's Geographies, 1680–1780* (New York: Guilford Press, 1998).
Okie, Laird. *Augustan Historical Writing: Histories of England in the English Enlightenment* (Lancham, Maryland: University Press of America, 1991).
Orgel, Stephen. *The Illusion of Power: Political Theatre in the English Renaissance* (Berkeley: University of California Press, 1975).
Ousby, Ian. *The Englishman's England: Taste, Travel and the Rise of Tourism* (Cambridge: Cambridge University Press, 1990).
Painter, Joe. *Politics, Geography and 'Political Geography'* (London: E.J. Arnold, 1995).
Parker, Geoffrey. *Geopolitics: Past, Present and Future* (London: Pinter, 1998).
Parker, W.H. *Mackinder: Geography as an Aid to Statecraft* (Oxford: Clarendon Press, 1982).
Parry, Graham. *The Trophies of Time: English Antiquarians of the Seventeenth Century* (Oxford, Clarendon Press, 1995).
Patterson, Annabel. *Fables of Power: Aesopian Writing and Political History* (Durham, NC: Duke University Press, 1991).
Phillipson, Nicholas. 'Providence and Progress: an Introduction to the Historical Thought of William Robertson'. *William Robertson and the Expansion of Empire*. Stewart Brown ed. (Cambridge: Cambridge University Press, 1997), pp. 55–73.
Pierce, Charles. *The Religious Life of Samuel Johnson* (London: Athlone Press, 1983).
Piggott, Stuart. 'William Camden and the *Britannia*'. *Proceedings of the British Academy*, 37 (1951), pp. 199–217.

Pincus, Steven. 'The English Debate over Universal Monarchy'. *A Union for Empire: Political Thought and the British Union of 1707*, John Robertson, ed. (Cambridge: Cambridge University Press, 1995), pp. 37–62.

Pincus, Steven. *Protestantism and Patriotism: Ideologies and the Making of English Foreign Policy, 1650–1668* (Cambridge: Cambridge University Press, 1996).

Plumb, J.H. *The Growth of Political Stability in England, 1657–1725* (London: Macmillan, 1967).

Pocock, J.G.A. *The Machiavellian Moment: Florentine Political Thought and the Atlantic Republican Tradition* (Princeton: Princeton University Press, 1975).

Pocock, J.G.A. *Virtue, Commerce and History: Essays on Political Thought and History, Chiefly in the Eighteenth Century* (Cambridge: Cambridge University Press, 1985).

Pocock, J.G.A. 'Clergy and Commerce. The Conservative Enlightenment in England'. *L'Età dei Lumi: Studi Storici Sul Settecento Europeo in Onore di Franco Venturi* (Napoli: Jovene, 1985), pp. 525–62.

Pocock, J.G.A. *The Ancient Constitution and the Feudal Law: a Study of English Historical Thought in the Seventeenth Century: A Reissue with a Retrospect* (Cambridge: Cambridge University Press, 1987).

Pocock, J.G.A. 'Conservative Enlightenment and Democratic Revolutions: The American and French Cases in British Perspective'. *Government and Opposition*, 24 (1989), pp. 81–105.

Pocock, J.G.A. 'Political Thought in the English-Speaking Atlantic, 1760–1790: (ii) Empire, Revolution and the End of Early Modernity'. *The Varieties of British Political Thought, 1500–1800*, J.G.A. Pocock *et al.*, eds (Cambridge: Cambridge University Press, 1993), pp. 283–317.

Pocock, J.G.A. 'Within the Margins: the Definitions of Orthodoxy'. *The Margins of Orthodoxy: Heterodox Writing and Cultural Response, 1660–1750*. R.D. Lund, ed. (Cambridge: Cambridge University Press, 1995), pp. 33–53.

Pocock, J.G.A. 'Empire, State and Confederation: the War of American Independence as a Crisis of Multiple Monarchy'. *A Union for Empire: Political Thought and the British Union of 1707*. John Robertson, ed. (Cambridge: Cambridge University Press, 1995), pp. 318–48.

Pocock, J.G.A. *Barbarism and Religion* (Cambridge, Cambridge University Press, 1999).

Porter, Roy. *The Making of Geology: Earth Science in Britain, 1660–1815* (Cambridge: Cambridge University Press, 1977).

Porter, Roy. 'The Enlightenment in England'. *The Enlightenment in National Context*, Roy Porter and Mikulaus Teich, eds (Cambridge: Cambridge University Press, 1981), pp. 1–18.

Porter, Roy. 'The Scientific Revolution and the Universities'. *A History of the University in Europe, Volume II: Universities in Early Modern Europe (1500–1800)*. H. De Ridder-Symoens, ed. (Cambridge: Cambridge University Press, 1996), pp. 531–62.

Pottinger, George. *Heirs of the Enlightenment: Edinburgh Reviewers and Writers 1800–1830* (Edinburgh: Scottish Academic Press, 1992).

Quinlan, Maurice. *Samuel Johnson: a Layman's Religion* (Madison: Wisconsin University Press, 1964).

Rack, Henry. *Reasonable Enthusiast: John Wesley and the Rise of Methodism.* 2nd edn (London: Epworth Press, 1992).
Ravenhill, William. 'The Honourable Robert Edward Clifford, 1767–1817: a Cartographer's Response to Napoleon'. *Geographical Journal,* 160 (1994), pp. 159–72.
Robertson, John. 'The Scottish Enlightenment and the Limits of the Civic Tradition'. *Wealth and Virtue: the Shaping of Political Economy in the Scottish Enlightenment.* Istvan Hont and Michael Ignatieff, eds (Cambridge University Press: Cambridge, 1983), pp. 137–78.
Robertson, John. *The Scottish Enlightenment and the Militia Issue* (Edinburgh: Donald, 1985).
Robertson, John. 'Universal Monarchy and the Liberties of Europe: David Hume's Critique of an English Whig Doctrine'. *Political Discourse in Early-Modern Britain.* Nicholas Phillipson and Quentin Skinner, eds (Cambridge: Cambridge University Press, 1993), pp. 349–73.
Robertson, John. 'Gibbon's Roman Empire as a Universal Monarchy: the *Decline and Fall* and the Imperial Idea in Early Modern Europe'. *Edward Gibbon and Empire.* Rosamond McKitterick and Roland Quinault, eds (Cambridge: Cambridge University Press, 1997), pp. 247–70.
Rockett, William. 'Historical Topography and British History in Camden's *Britannia*'. *Renaissance and Reformation,* 26 (1990), pp. 71–80.
Rogers, Pat. *Grubstreet: Studies in a Subculture* (London: Methuen, 1972).
Rose, Mark. *Owners and Authors: the Invention of Copyright* (Cambridge, Mass.: Harvard University Press, 1993).
Rudwick, Martin. 'Minerals, Strata and Fossils'. *Cultures of Natural History,* Nicholas Jardine *et al.*, eds (Cambridge: Cambridge University Press, 1996), pp. 266–86.
Ryan, James. 'Visualizing Imperial Geography: Halford Mackinder and the Colonial Office Visual Instruction Committee, 1902–11'. *Ecumene,* 1 (1994), pp. 157–76.
Sauer, Carl. 'The Morphology of Landscape'. *Land and Life: a Selection from the Writings of Carl Ortwin Sauer.* John Leighty, ed. (Berkeley: University of California Press, 1963), pp. 315–50.
Schaffer, Simon. 'The Consuming Flame: Electrical Showmen and Tory Mystics in the World of Goods'. *Consumption and the World of Goods,* John Brewer and Roy Porter, eds (London: Routledge, 1993), pp. 489–526.
Searle, G.R. *The Quest for National Efficiency: a Study in British Politics and Political Thought, 1899–1914* (Berkeley: University of California Press, 1971).
Searle, G.R. *The Liberal Party: Triumph and Disintegration, 1886–1929* (London: Macmillan, 1992).
Seaward, Paul. *The Cavalier Parliament and the Reconstruction of the Old Regime, 1661–1667* (Cambridge: Cambridge University Press, 1989).
Secord, Anne. 'Artisan Botany'. *Cultures of Natural History,* Nicholas Jardine *et al.*, eds (Cambridge: Cambridge University Press, 1996), pp. 378–93.
Shapin, Steven. *A Social History of Truth: Civility and Science in Seventeenth-Century England* (Chicago: Chicago University Press, 1994).
Sharpe, Kevin. *Sir Robert Cotton, 1586–1631: History and Politics in Early Modern England* (Oxford: Oxford University Press, 1979).

Sher, Richard. *Church and University in the Scottish Enlightenment: the Moderate Literati of Edinburgh* (Edinburgh: Edinburgh University Press, 1985).

Sher, Richard. '*Charles V* and the Book Trade: An Episode in Enlightenment Print Culture'. *William Robertson and the Expansion of Empire*. Stewart Brown ed. (Cambridge: Cambridge University Press, 1997), pp. 164–95.

Sher, Richard and Hook, Andrew. 'Introduction: Glasgow and the Enlightenment'. *The Glasgow Enlightenment*. Andrew Hook and Richard Sher, eds (East Linton: Tuckwell Press, 1995), pp. 1–17.

Sherington, Geoffrey. *English Education, Social Change and War, 1911–20* (Manchester: Manchester University Press, 1981).

Skinner, Quentin. 'Meaning and Understanding in the History of Ideas'. *Meaning and Context: Quentin Skinner and His Critics*. James Tully, ed. (Cambridge: Polity, 1988), pp. 29–67.

Skinner, Quentin. *Liberty before Liberalism* (Cambridge: Cambridge University Press, 1998).

Sitwell, O.F.G. *Four Centuries of Special Geography: an Annotated Guide to Books that Purport to Describe all the Countries of the World Published in English before 1888, with a Critical Introduction* (Vancouver: University of British Columbia Press, 1993).

Smith, Bernard. *European Vision and the South Pacific*. 2nd edn (New Haven: Yale University Press, 1985).

Smith, Jonathan. 'State Formation, Geography, and a Gentleman's Education'. *Geographical Review*, 86 (1996), pp. 91–100.

Smith, Neil and Godlewska, Anne, eds. *Geography and Empire* (Oxford: Blackwell, 1994).

Smith, R.J. *The Gothic Bequest: Medieval Institutions in British Thought, 1688–1863* (Cambridge: Cambridge University Press, 1987).

Sommerville, J.P. *Politics and Ideology in England 1603–1640* (London: Longman, 1986).

Sorrenson, Richard. 'Towards a History of the Royal Society in the Eighteenth Century'. *Notes and Records of the Royal Society of London*, 50 (1996), pp. 29–46.

Stafford, Barbara. *Voyage into Substance: Art, Science and the Illustrated Travel Account, 1760–1840* (Cambridge, Mass.: MIT Press, 1984).

Stafford, Robert. *Scientist of Empire: Sir Roderick Murchison, Scientific Exploration and Victorian Imperialism* (Cambridge: Cambridge University Press, 1989).

Stephan, Deborah. 'Laurence Echard – Whig Historian'. *Historical Journal*, 32 (1989), pp. 843–66.

Stewart, Larry. *The Rise of Public Science: Rhetoric, Technology and Natural Philosophy in Newtonian Britain, 1660–1750* (Cambridge: Cambridge University Press, 1992).

Stock, R.D. *The Holy and the Daemonic from Sir Thomas Browne to William Blake* (Princeton: Princeton University Press, 1982).

Stoddart, David. *On Geography and its History* (Oxford: Blackwells, 1986).

Strong, Roy. *Art and Power: Renaissance Festivals, 1450–1650* (Woodbridge: Boydell and Brewer, 1984).

Stuart, Robert. *Party and Politics, 1830–1852* (London: Macmillan, 1989).

Suarez, Michael. 'Johnson's Christian Thought'. *The Cambridge Companion to Samuel Johnson*, Greg Clingham, ed. (Cambridge: Cambridge University Press, 1997), pp. 192–208.

Sykes, Alan. *Tariff Reform in British Politics, 1903–1913* (Oxford: Clarendon Press, 1979).

Sykes, Norman. *Edmund Gibson, Bishop of London 1669–1748: A Study in Politics and Religion in the Eighteenth Century* (London: Oxford University Press, 1926).

Tapper, Alan. 'Priestley on Politics, Progress and Moral Theology'. *Enlightenment and Religion: Rational Dissent in Eighteenth-Century Britain*, Knud Haakonssen, ed. (Cambridge: Cambridge University Press, 1996), pp. 272–86.

Taylor, Peter. *Political Geography: World-Economy, Nation-State and Locality*. 2nd edn (London: Longman, 1989).

Thompson, E.P. *Whigs and Hunters: the Origin of the Black Acts* (Harmondsworth: Penguin, 1975).

Townend, G.M. 'Religious Radicalism and Conservatism in the Whig Party under George I: The Repeal of the Occasional Conformity and Schism Acts'. *Parliamentary History*, 7 (1988), pp. 24–44.

Tracy, Clarence. *A Portrait of Richard Graves* (Toronto: Toronto University Press, 1987).

Tuck, Richard. 'Grotius and Selden'. *The Cambridge History of Political Thought, 1450–1700*, J.H. Burns with Mark Goldie, eds (Cambridge: Cambridge University Press, 1991), pp. 499–529.

Turner, Frank. *Contesting Cultural Authority: Essays in Victorian Intellectual Life* (Cambridge: Cambridge University Press, 1993).

Tyacke, Nicholas. 'Science and Religion at Oxford before the Civil War'. *Puritans and Revolutionaries: Essays in Seventeenth Century History presented to Christopher Hill*, Donald Pennington and Keith Thomas, eds (Oxford: Clarendon Press, 1978), pp. 73–93.

Tyacke, Nicholas. *Anti-Calvinists: The Rise of English Arminianism, c.1590–1640* (Oxford: Clarendon Press, 1987).

Ullmann, Walter. 'This Realm of England is an Empire'. *Journal of Ecclesiastical History*, 20 (1979), pp. 175–203.

Van Eerde, Katherine. *John Ogilby and the Taste of His Times* (Folkestone: Dawson Press, 1976).

Wall, Cynthia. *The Literary and Cultural Spaces of Restoration London* (Cambridge: Cambridge University Press, 1998).

Walters, Gwyn and Emery, Frank. 'Edward Lhuyd, Edmund Gibson, and the Printing of Camden's *Britannia*, 1695'. *The Library*, 5th series 32 (1977), pp. 109–37.

Waterman, A.M.C. *Revolution, Economics and Religion: Christian Political Economy, 1798–1833* (Cambridge: Cambridge University Press, 1991).

Waterman, A.M.C. 'The Nexus between Theology and Political Doctrine in Church and Dissent'. *Enlightenment and Religion: Rational Dissent in Eighteenth-Century Britain*, Knud Haakonssen, ed. (Cambridge: Cambridge University Press, 1996), pp. 193–218.

Wilcock, A.A. '"The English Strabo": the Geographical Publications of John Pinkerton'. *Transactions of the Institute of British Geographers*, 61 (1974), pp. 35–45.

Wilson, Kathleen. *The Sense of the People: Politics, Culture and Imperialism in England, 1715–85* (Cambridge: Cambridge University Press, 1995).

Wilson, Kathleen. 'The Island Race: Captain Cook, Protestant Evangelism and the Construction of English National Identity, 1760–1800'. *Protestantism and*

National Identity: Britain and Ireland, c.1650–c.1850, Tony Claydon and Ian McBride, eds (Cambridge: Cambridge University Press, 1998), pp. 265–90.

Withers, Charles. 'How Scotland Came to Know Itself: Geography, National Identity and the Making of a Nation, 1680–1790'. *Journal of Historical Geography*, 21 (1995), pp. 371–97.

Withers, Charles. 'Encyclopaedism, Modernism and the Classification of Geographical Knowledge'. *Transactions of the Institute of British Geographers*, N.S. 21 (1996), pp. 275–98.

Withers, Charles. 'Geography, Science and National Identity in Early-Modern Britain: the Case of Scotland and the Work of Sir Robert Sibbald'. *Annals of Science*, 53 (1996), pp. 29–73.

Withers, Charles. 'Geography, Royalty and Empire: Scotland and the Making of Great Britain, 1603–61'. *Scottish Geographical Magazine*, 113 (1997), pp. 22–32.

Withers, Charles. 'Towards a History of Geography in the Public Sphere'. *History of Science*, 37 (1999), pp. 45–78.

Womersley, David. 'Gibbon's Unfinished History: the French Revolution and English Political Vocabularies.' *Historical Journal*, 35 (1992), pp. 63–89.

Woodward, David. *Maps as Prints in the Italian Renaissance: Makers, Distributors and Consumers* (London: British Library, 1996).

Wright, J.K. 'Some British "Grandfathers" of American Geography'. *Geographical Essays in Memory of Alan G. Ogilvie*. R. Miller and J. Wreford Watson, eds (London: Nelson, 1959), pp. 144–65.

Yeo, Richard. *Defining Science: William Whewell, Natural Knowledge and Public Debate in Early-Victorian Britain* (Cambridge: Cambridge University Press, 1993).

Young, B.W. *Religion and Enlightenment in Eighteenth-Century England; Theological Debate from Locke to Burke* (Oxford: Clarendon Press, 1998).

Young, B.W. 'A History of Variations: The Identity of the Eighteenth-Century Church of England'. *Protestantism and National Identity: Britain and Ireland, c.1650–c.1850*. Tony Claydon and Ian McBride, eds (Cambridge: Cambridge University Press, 1998), pp. 105–28.

Dissertations

Sherratt, Yvonne. *Adorno's Critique of Enlightenment: a Positive Interpretation* (Cambridge University, PhD., 1999).

Index